Digital System
Design with
FPGA

About the Authors

Cem Ünsalan, Ph.D., established the DSP Laboratory at Yeditepe University in Istanbul, Turkey, and is a microprocessor and digital signal processing professor there. He is the coauthor of *Programmable Micro-controllers with Applications: MSP430 LaunchPad with CCS and Grace.*

Bora Tar, Ph.D., is a postdoctoral researcher at The Ohio State University. His main research interests include analog and mixed-signal integrated-circuit design and energy harvesting and sensor networking applications.

Digital System Design with FPGA

Implementation Using Verilog and VHDL

Cem Ünsalan
Yeditepe University

Bora Tar
The Ohio State University

New York Chicago San Francisco
Athens London Madrid
Mexico City Milan New Delhi
Singapore Sydney Toronto

Digital System Design with FPGA:
Implementation Using Verilog and VHDL

1 2 3 4 5 6 7 8 9   QVS   21 20 19 18 17

ISBN 978-1-259-83790-6
MHID 1-259-83790-4

Sponsoring Editor Michael McCabe	**Copy Editor** Mohammad Taiyab Khan, MPS Limited
Editorial Supervisor Stephen M. Smith	**Proofreader** A. Nayyer Shamsi, MPS Limited
Production Supervisor Lynn M. Messina	**Indexer** Edwin Durbin
Acquisitions Coordinator Lauren Rogers	**Art Director, Cover** Jeff Weeks
Project Manager Poonam Bisht, MPS Limited	**Composition** MPS Limited

Contents

Preface

The world around us has become digital. Personal devices we use, houses we live in, and cars we drive contain digital systems to simplify life for us. Moreover, all these systems have started communicating with each other. Since digital systems have become one of the most important tools of our daily lives, besides engineers hobbyists have also started learning and using them.

There are four ways to realize a digital system. The first one is using discrete logic gates. This approach has become obsolete due to implementation issues. The second way is using a microcontroller, which has very desirable properties such as ease of programming and price. However, a microcontroller is static in terms of its configuration. The third one is using an application-specific integrated circuit (ASIC). For mass production, using ASICs is the solution. However, producing and testing an ASIC chip needs time, which limits its modification after it is designed. The fourth way is using a field-programmable gate array (FPGA). An FPGA can be configured easily such that it can be tailored for a specific application.

Managing an FPGA and getting the best out of it are slightly harder than for a microcontroller. However, if done appropriately the benefit will be enormous. Therefore, this book aims to guide the reader to mastering FPGAs through digital system design. While doing this, the main focus will be on implementation. Hence, the reader will grasp theoretical digital design concepts via implementing real-life applications. For this purpose, we pick two recent boards: Basys3 and Arty. Both boards have a Xilinx Artix-7 FPGA on them. Baysy3 has most of the required peripherals onboard. Hence, it is an excellent candidate for being used in digital design education. Arty has Arduino-compatible pins. Since Arduino is widely accepted as a microcontroller platform by hobbyists, it has a wide range of peripheral devices as shields. Arty allows us to benefit from these. Moreover, the hobbyist can switch from Arduino to Arty when a custom-made digital design is required. Throughout the book, we will provide practical application examples mostly on the Basys3 board due to its available resources onboard. However, these applications can be modified to work on the Arty board as well. Besides, we will use simulation for almost all applications. Hence, buying Basy3 or Arty is not a must to follow the book.

There are two popular hardware description languages (HDLs) used to implement a digital system on an FPGA. These are Verilog and VHDL. Each HDL has its advantages and disadvantages. Throughout the book, we will cover both HDLs in parallel. This will allow readers to choose the HDL he or she likes. Note that this is not a book on advanced Verilog or VHDL. We will focus only on important and necessary topics. This way, we expect the beginner or hobbyist to benefit from the book.

Before diving into the fascinating world of digital systems, we would like to remind the reader of one or two things. We did not intend to write a standard textbook for a digital design course. Therefore, we did not cover theoretical concepts in depth. Instead, we tried to explain all these concepts using real-life applications. This way, we hope the reader will grasp digital design concepts better. Moreover, we do not believe digital design is just a mandatory engineering course to be attended. It is a talent every engineering student should gain for the job market. Besides, it is fun to play with, as done by most hobbyists. So, let's enjoy digital design with the FPGA while mastering it.

Cem Ünsalan
Bora Tar

Acknowledgments

We would like to thank Cathal McCabe from Xilinx for his guidance and valuable comments. We would also like to thank Digilent Inc. for allowing us to use Basys3 and Arty board images and sample projects.

Artix is a trademark of Xilinx Inc. Vivado Design Suite is a trademark of Xilinx Inc. Basys3 is a trademark of Digilent Inc. Arty is a trademark of Avnet and Digilent Inc.

CHAPTER 1

Introduction

The world around us has become digital. Hence, digital systems have become the dominant part of our lives. Although most of us enjoy benefits offered by digital systems, it is the duty of a candidate engineer to learn how to design and analyze them. Besides, digital design concepts have become topics of interest to a hobbyist and the maker community due to their power in implementing systems. Therefore, we aim to introduce digital system design techniques throughout this book.

Although there are several ways to implement a digital system, we will focus only on implementation by field-programmable gate arrays (FPGAs) in this book. FPGA can be taken as a generic platform such that a digital system can be implemented on it. Recently, the price of a standard FPGA chip has become affordable. Moreover, evaluation boards using such chips became widespread. Hence, a hobbyist or an engineering student can implement his or her design on such a platform. The only requirement left is how to do it. This book aims to fill this gap. Therefore, we will guide the reader through the complex paths of FPGA usage for digital design. In doing this, we aim for an introductory approach to form a background that may open up ways to understand more advanced FPGA topics.

1.1 Hardware Description Languages

There are two popular hardware description languages (HDLs) to implement a digital system design on an FPGA. These are Verilog and VHDL. In literature, it is clearly emphasized that learning one HDL simplifies learning the other. Moreover, it is indicated that learning both HDLs is important to become an expert in this discipline. However, most books on digital design pick either Verilog or VHDL alone and explain the concepts using it. There is only a small group of books introducing both HDLs together. We prefer this strategy in this book. However, we suggest the reader to master one HDL first (possibly Verilog). Then, he or she can revisit the book to understand the second HDL (possibly VHDL). This way, the same digital system design concepts will be revisited twice. Hence, we expect repetition to make perfection.

We should warn the reader at this step. This is not a comprehensive book on Verilog or VHDL. Such a target is beyond our reach. However, we aim to introduce digital system design techniques using HDLs. Therefore, we cover HDL concepts falling in this area. In doing this, we provide practical applications. Afterward, the reader can consult comprehensive books to master his or her knowledge on advanced HDL topics.

1

1.2   FPGA Boards and Software Tools

Throughout the book, we will approach digital design concepts from a practical point of view. Hence, we need appropriate hardware and software platforms. Fortunately, there are several FPGA boards under different brands with various properties. In this book, we pick two such boards: Basys3 and Arty. Both boards have a Xilinx Artix-7 FPGA on them. Basys3 has most digital peripherals on it. Therefore, it is suitable for education purposes. On the other hand, Arty has Arduino compatible pins such that Arduino shields can be used with it. Therefore, it is suitable for hobbyists and the maker community. Throughout the book, we will provide practical application examples mostly on the Basys3 board due to its available resources onboard. However, these applications can be modified to work on the Arty board as well. Note that Basys3 and Arty boards have differences that are explored in detail in Chap. 3. In applications where such differences matter, it is advisable to use the suitable board.

We will use simulation tools while explaining digital system design concepts. Therefore, this book can also be of use without any FPGA board at hand. In the same line, most concepts to be explained throughout the book do not depend on a specific FPGA platform. Hence, a different FPGA platform can also be used to implement them. However, there are some concepts that require a specific FPGA platform. For these, minor modifications should be made by the reader for implementation. Bearing this in mind, we should also mention the software to be used throughout the book. We will use the Vivado design suite to implement the designed digital system on the Xilinx Artix-7 FPGA. This design suite is supported by Xilinx. As of the writing of this book, Vivado was available from Xilinx's website free of charge.

1.3   Topics to Be Covered in the Book

An FPGA is itself a digital electronic system. Therefore, first we have to introduce the basic digital electronics background. The second chapter of the book handles this. However, digital system concepts will be explained briefly in this chapter. They will be analyzed in detail in the following chapters. The third chapter of the book explores properties of Basys3 and Arty boards. Here, the aim is getting familiar with physical hardware to be used throughout the book. Related to this, the fourth chapter introduces the Vivado design suite. Hence, the reader gets familiar with digital design implementation issues. The first four chapters can be taken as preparatory steps for digital system implementation. Starting from the fifth chapter, HDL concepts will be the main focus of interest. Therefore, Chap. 5 introduces Verilog and VHDL. Then, the sixth chapter deals with data types and operators on these. The reader should remember these concepts since they will be extremely useful in the following chapters. Chapters 5 and 6 can also be taken as preparatory steps for digital system implementation on FPGA via HDL. Based on these, the seventh chapter focuses on combinational circuits. Here, HDL will be used to implement basic combinational circuits. The eighth chapter extends these concepts further such that more complex digital systems can be constructed via HDL. The ninth chapter is on data storage elements that are extensively used in constructing sequential circuits. As a follow-up, the tenth chapter introduces sequential circuits. Here, standard sequential digital systems such as counters and registers are evaluated. Therefore, Chaps. 7 to 10 can be taken as the building blocks of a generic digital system such as a microcontroller. The eleventh chapter introduces methods to embed a soft-core microcontroller on FPGA. Chapter 12 focuses on digital interfacing tools. Here, HDL implementation details of recent digital communication and interfacing methods are summarized. In all these chapters, we provide relevant real-life applications. However, some applications may cover more than one topic. Therefore, Chap. 13 provides such advanced applications using FPGA. Finally, Chap. 14 provides the path to be followed to learn more advanced topics on FPGA.

Sample Verilog and VHDL descriptions in this book and related testbench files are available for the reader on a companion website, www.mhprofessional.com/1259837904. For some real-life applications, we could not include VHDL descriptions in the book due to page limitations. However, these are available on the companion website, and we kindly ask the reader to download them. Course slides for the reader and instructor and the solution manual for the instructor are also available on this website.

CHAPTER 2

Field-Programmable Gate Arrays

The aim of this book is explaining field-programmable gate array (FPGA) usage for digital system implementation. Naturally, the first step in doing this is explaining what an FPGA is. An FPGA is itself a digital system composed of basic building blocks. Therefore, some digital logic background is necessary to understand the FPGA architecture. To do so, we adopt the following strategy in this book. We start with the basics of digital electronics in this chapter. Then, we explain the architecture of an FPGA using abstract building blocks. As we overview the FPGA architecture in this chapter, we focus on the digital system design and implementation philosophy using the FPGA next. Finally, we summarize the usage areas of the FPGA to motivate the reader.

2.1 A Brief Introduction to Digital Electronics

There are two main approaches in explaining digital systems. The first one starts with digital electronic representation and ends up with it. Here, all concepts are explained in transistor level. Although this approach is reasonable, it is not suitable for us since the reader does not need such a detailed explanation to use an FPGA. The second approach is not mentioning any hardware representation and explaining all concepts using binary representation and Boolean algebra. This approach is more refined and allows a more theoretical background. Unfortunately, it does not invoke physical device properties for implementation. Hence, all concepts will be in abstract level. We believe that a third approach, mixing digital device representation with abstract formalism, may be more helpful to the reader. Therefore, we briefly introduce digital electronics in this chapter. In the following chapters, we will not represent digital devices this way. However, we expect the reader to recall physical representations mentioned in this chapter.

2.1.1 Bit Values as Voltage Levels

All digital devices are based on binary representation. In other words, everything in a digital device is represented in terms of two logic levels as zero and one. At first, this may seem unreasonable. How is it possible to represent data processing in all complex digital devices (including computers, tablets, smart phones, etc.) in terms of zeros and ones? Well, this is the case. Throughout the book, we will try to convince the reader that all complex digital systems are composed of basic building blocks working on binary logic levels. Moreover, we will show that most parts of these devices can be implemented on an FPGA.

Next comes the second question. How is a binary digit (or a bit, in short) represented in a digital device? The answer to this question leads to understanding digital logic concepts in the physical level. In its basic sense, we have two voltage levels to represent a binary digit (either as zero or one). Let's call these ground (zero) and supply voltage (V_{CC}). These correspond to binary logic levels zero and one, respectively. Therefore, whenever we talk about a bit value as zero or one, we actually mean a voltage level as either ground or supply voltage.

2.1.2   Transistor as a Switch

A digital circuit can be constructed by transistors. A transistor is an active circuit element used either as an amplifier or a digital switch. The latter property is extremely important, since all binary logic operations can be performed this way. Instead of dealing with physical properties of a transistor, we can simplify its characteristics as follows.

Assume that there is a digital switch controlled by voltage V_{in}. When there is no voltage applied to the switch, it acts as an open circuit. In other words, the switch does not pass current on it as in Fig. 2.1a. Based on this setup, we can say that when $V_{in} = 0$, output voltage of the circuit will be $V_{out} = 0$. When the voltage V_{CC} is applied to the switch, it acts as a short circuit. Therefore, the switch passes current on it as in Fig. 2.1b. Based on this setup, we can say that when $V_{in} = V_{CC}$ output voltage of the circuit will be $V_{out} = V_{CC}$. These two characteristics will lead to logic gates. Note that R represents the resistor in Fig. 2.1 to limit current in the circuit.

2.1.3   Logic Gates from Switches

As mentioned in the previous section, by applying a suitable voltage level to the switch, the current (hence output voltage) can be controlled. This leads to the development of digital logic gates. Before exploring logic gates, let's start with the buffer.

2.1.3.1   The Buffer

The buffer can be taken as a logic gate which feeds its input to output without changing it. Therefore, it does not perform any logical operation. However, the buffer is extremely important in input/output ports of digital devices to minimize voltage loading effects between different elements. In other words, the buffer acts as a protective shield. We will see this usage extensively in the input/output ports of an FPGA implementation in the following chapters. We can represent the buffer in symbolic form as in Fig. 2.2. In this figure, in=out.

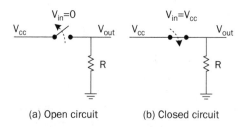

(a) Open circuit   (b) Closed circuit

FIGURE 2.1   Abstract representation of a transistor working as a switch.

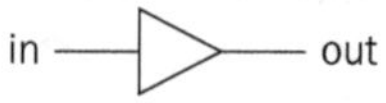

in ———▷——— out

FIGURE 2.2   The buffer symbol.

2.1.3.2 The NOT Gate

The NOT gate can be constructed by a switch with two input pins as in Fig. 2.3. In this setup, when input is equal to supply voltage ($V_{in} = V_{CC}$) the switch connects ground to output. Hence, output voltage will be zero ($V_{out} = 0$). When input voltage equals to ground ($V_{in} = 0$), the switch connects supply voltage to output. Hence, $V_{out} = V_{CC}$.

Now, let's represent V_{CC} as logic level one and ground as logic level zero. Furthermore, let's call V_{in} as in and V_{out} as out. Based on these simplifications, we can summarize working principle of the NOT gate as follows:

$$out = \begin{cases} 1 & \text{if } in = 0 \\ 0 & \text{if } in = 1 \end{cases} \tag{2.1}$$

As can be seen in Eq. (2.1), the NOT gate is a simple inverter in terms of binary logic. When a logic level zero is applied to its input, output will be logic level one. When a logic level one is applied to input of the NOT gate, output will be zero.

We can represent the NOT gate in symbolic form as in Fig. 2.4. In this figure, in and out values are the ones in Eq. (2.1). Hence, the relation between them is satisfied with this equation.

2.1.3.3 The OR Gate

The OR is the next logic gate to be considered. This gate can be constructed by two switches connected in parallel as in Fig. 2.5. In this setup, when either the first or the second input is equal to supply voltage ($V_{in1} = V_{CC}$ or $V_{in2} = V_{CC}$), output equals to supply voltage as well ($V_{out} = V_{CC}$). For all other cases, output voltage equals to ground ($V_{out} = 0$).

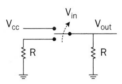

FIGURE 2.3 The NOT gate formed by a switch.

FIGURE 2.4 The NOT gate symbol.

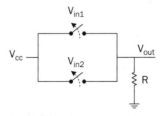

FIGURE 2.5 The OR gate formed by two parallel switches.

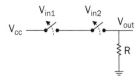

FIGURE 2.6 The OR gate symbol.

FIGURE 2.7 The AND gate formed by two series switches.

FIGURE 2.8 The AND gate symbol.

As in NOT gate, we can simplify working principle of the OR gate. Let's call V_{in1} as in1, V_{in2} as in2, and V_{out} as out. Based on these simplifications, we can summarize working principle of the OR gate as follows:

$$out = \begin{cases} 1 & \text{if } in1 = 1 \text{ or } in2 = 1 \\ 0 & \text{otherwise} \end{cases} \tag{2.2}$$

As can be seen in Eq. (2.2), the OR gate gives logic level one when any of the parallel switches has input logic level one. Otherwise, output of the gate will be logic level zero.

We can represent the OR gate in symbolic form as in Fig. 2.6. In this figure, in1, in2, and out values are the ones in Eq. (2.2). Hence, the relation between them is satisfied with this equation.

2.1.3.4 The AND Gate
The AND is the final logic gate to be considered in this chapter. This gate can be constructed by two switches connected in series as in Fig. 2.7. In this setup, when both inputs are equal to supply voltage ($V_{in1} = V_{CC}$ and $V_{in2} = V_{CC}$), then output equals to supply voltage as well ($V_{out} = V_{CC}$). For all other cases, output voltage will be equal to ground ($V_{out} = 0$).

As in OR gate, we can simplify working principle of the AND gate. Let's call V_{in1} as in1, V_{in2} as in2, and V_{out} as out. Based on these simplifications, we can summarize working principle of the AND gate as follows:

$$out = \begin{cases} 1 & \text{if } in1 = 1 \text{ and } in2 = 1 \\ 0 & \text{otherwise} \end{cases} \tag{2.3}$$

As can be seen in Eq. (2.3), the AND gate gives logic level one when both serial switches have input logic level one. Otherwise, the output of the gate will be logic level zero.

We can represent the AND gate in symbolic form as in Fig. 2.8. In this figure, in1, in2, and out values are the ones in Eq. (2.3). Hence, the relation between them is satisfied with this equation.

We have introduced only basics of digital logic gates in this section. The aim is to use these in explaining the FPGA architecture. We will analyze logic gates further in Chap. 7.

2.2 FPGA Building Blocks

The architecture of the FPGA should be known by the reader to appreciate its working principles. Although the reader will not directly interact with the architecture, this knowledge will lead to better usage of the FPGA. Besides, design principles to be applied in implementing a digital system on the FPGA will make sense. Therefore, we will introduce basic building blocks of the FPGA (Xilinx Artix-7 XC7A35T) available on the Basys3 and Arty boards in this section. These building blocks will be represented in abstract form. Since we do not want to go into detail of digital electronics, we believe this level is sufficient. We will start with layout of the Xilinx Artix-7 XC7A35T FPGA next.

2.2.1 Layout of the Xilinx Artix-7 XC7A35T FPGA

Basys3 and Arty boards have their FPGA from the Xilinx Artix-7 XC7A35T family. To be more specific, the FPGA on the Basys3 board is XC7A35TCPG236-1. Similarly, the FPGA on the Arty board is XC7A35TICSG324-1L. These two FPGAs share similar properties. Therefore, we will call them by their family name Xilinx Artix-7 XC7A35T from this point on. If there is a difference in the FPGA, then we specify it by the board name.

The Xilinx Artix-7 XC7A35T FPGA is basically composed of nine different components. These are input/output blocks, configurable logic blocks (CLBs), interconnect resources, block RAM, DSP slices, clock management block, XADC block, high-speed serial I/O transceivers, and PCIe interface. Layout of these blocks is as in Fig. 2.9. Most of these blocks can also be observed via Vivado design suite to be introduced in Chap. 4. Therefore, the reader will have chance to observe which of them are used in his or her digital system design. Mentioned blocks (or their variants) are almost standard in an FPGA. However, some of these may be missing or other extra blocks may be available in different FPGA families. The reader should keep this in mind while using another FPGA family.

2.2.2 Input/Output Blocks

A digital device interacts with the outside world through its input and output pins. This is also the case for the FPGA. Hence, data from the outside world is acquired through input pins. Output is fed to the outside world using output pins. Moreover, these input and output pins are located in input/output blocks within the FPGA.

The Artix-7 XC7A35T FPGA has input/output pins which can operate on standard voltage levels from 1.2 to 3.3 V. The FPGA on the Basys3 board has 106 such input/output pins. In a similar manner, the FPGA on the Arty board has 210 such pins. These input/output pins can be used as input, output, and both. In the first case, data will be taken from outside world through the pin. In the second case, voltage levels will be fed to outside world through the pin. In the third case, the same pin can be used for both input and output purposes.

Input/output pins are grouped into banks. Two pins in these banks are grouped as positive (P) and negative (N) pairs. These can be used in two modes as single-ended and differential. In the single-ended mode, input will be recognized as logic level zero when input voltage is near ground. It will be recognized as logic level one when input voltage is near V_{CC}. In the differential mode, input will be recognized as logic level zero when the voltage at pin P is lower than the voltage at pin N. When the voltage at pin P is higher than the voltage at pin N, then input will be taken as logic level one.

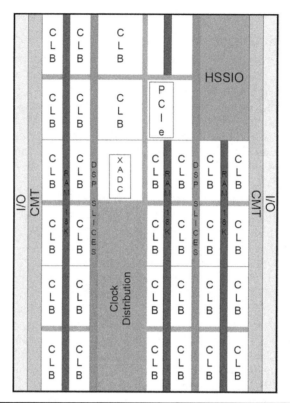

FIGURE 2.9 Basic building blocks of the Artix-7 XC7A35T FPGA.

Input/output pins can also be used in reference mode. Here, input will be taken as logic level zero when input voltage is below reference voltage. When input voltage is above reference voltage, it will be taken as logic level one.

Single-ended pins can also be used as output. When output is at logic level one, the corresponding voltage value at the pin will be V_{CC}. When output is at logic level zero, the corresponding voltage value at the pin will be ground.

Note that we are bound by input/output pins available on the Basys3 and Arty boards. Therefore, please see Chap. 3 for the actual pin layout on these boards. For more information on input/output blocks and their properties, please see [1].

2.2.3 Configurable Logic Blocks

Configurable logic blocks are the basic elements used to implement a digital system on an FPGA [2]. At the heart of CLBs lies look-up tables (LUTs), flip-flops, and multiplexers. We will try to explain working principles of these devices in generic form. Therefore, they may not correspond to actual implementation on an FPGA. Let's start with the multiplexer.

2.2.3.1 Multiplexer

A multiplexer is, in fact, a selector with N select bits (pins), 2^N input pins, and one output pin. One input pin at a time is connected to output. Hence, the value at that pin will be seen at output. Via select pins, we decide on which input pin will be connected to output.

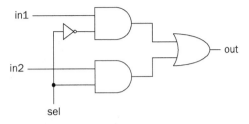

FIGURE 2.10 Circuit diagram of two-to-one multiplexer built from basic logic gates.

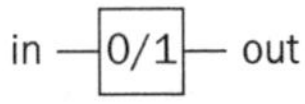

$$\text{in} \quad \boxed{0/1} \quad \text{out}$$

FIGURE 2.11 Abstract form of a flip-flop.

We can form a two input multiplexer by digital logic gates in Sec. 2.1.3. Here, the aim is to show basic layout of a multiplexer. We provide circuit diagram of the formed multiplexer in Fig. 2.10. Since there are two inputs in this device, it is called a two-to-one multiplexer.

We can summarize working principles of the two-to-one multiplexer as follows:

$$out = \begin{cases} in1 & \text{if } sel = 0 \\ in2 & \text{if } sel = 1 \end{cases} \tag{2.4}$$

The select pin (labeled as sel in Fig. 2.10) decides which input will be connected to output.

The two-to-one multiplexer is the simplest device of its kind. Let's consider a 32-to-1 multiplexer. This device has five select pins to map $2^5 = 32$ input pins. Assume that select pins have value 10001. Then, 17th input will be connected to output. Therefore, whatever the value of that pin is, it will be seen at output. We will explore working principles of multiplexers in detail in Chap. 8.

2.2.3.2 Flip-Flop

Flip-flop is the basic memory element in FPGA. It can store one bit of data. Although a flip-flop can be constructed by digital logic gates in Sec. 2.1.3, the layout will be slightly complex. Therefore, we postpone this operation till Chap. 9. As for now, it is important to remember that a flip-flop holds one bit of data which is fed to it. This data will be stored in the flip-flop till it is changed by the user. Let's represent the flip-flop in abstract form as in Fig. 2.11. In this figure, bit value to be stored in the flip-flop is set by in pin. The stored value in the flip-flop is obtained from out pin. Note that the flip-flop can only save one bit as either logic level zero or one.

2.2.3.3 Look-Up Table

There is no detail on the actual implementation of a LUT in the Artix-7 XC7A35T FPGA. Therefore, we will try to explain it using known digital devices. A LUT can be thought of as a collection of flip-flops connected to input pins of a multiplexer. Select pins of the multiplexer will be taken as address bits of the flip-flop to be reached. This architecture can be used to implement any combinational logic function which has total number of variables as select pins. We will see how this can be done in Chap. 7. The important

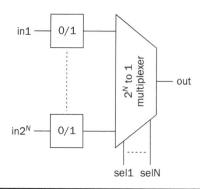

FIGURE 2.12   Abstract form of an *N* input LUT.

point here is that as the entry of flip-flops change, implemented logic function will also change. This will lead to reconfigurability of the FPGA.

A LUT will be called N input if it has 2^N entries. Therefore, it needs N select bits as explained previously. We provide such an abstract LUT composed of flip-flops and a multiplexer in Fig. 2.12. In the Artix-7 FPGA, two such five-input LUTs are decoupled. Each couple can be used either to implement two five-input combinational logic functions with the same input and different outputs or one six-input combinational logic function. Two such six-input LUTs can be combined by another multiplexer to form a seven-input LUT. Two such seven-input LUTs (hence four six-input LUTs) can be combined by another multiplexer to form an eight-input LUT. Hence, a combinational logic function with eight inputs can be formed by it.

2.2.3.4   *Slices*

LUTs, flip-flops, and multiplexers are grouped as slices in the CLB. Each slice has four six-input LUTs, eight flip-flops, multiplexers, and other support circuitry. There are two slice types as SLICEM and SLICEL. Both can be used to implement combinational logic functions. SLICEM can also be used as a distributed memory element. The Xilinx Artix-7 XC7A35T FPGA has a total of 5200 slices of which 3600 are SLICEL and 1600 are SLICEM. We will explore the usage of distributed memory in a digital system in detail in Chap. 9. Each SLICEM can also be used as a 32-bit shift register. We will explain working principles of this digital device in Chap. 10.

2.2.4   Interconnect Resources

What we mean by interconnect resources is a collection of wires and programmable switches. These are responsible for connecting CLBs and other building blocks within the FPGA. Interconnect is also called routing channels.

CLBs in the Artix-7 FPGA are placed in a grid structure which simplifies planning of interconnection usage. Note that it is not necessary to know interconnect features to use an FPGA at the beginner or intermediate level. The Vivado design suite to be introduced in Chap. 4 is responsible for efficient use of these resources.

2.2.5   Block RAM

Different from distributed memory elements composed of SLICEM blocks within CLBs, the Artix-7 FPGA also has block RAM modules. These can be used to store data. Moreover, they can form buffers, large LUTs, or shift registers. Usage of these block

RAMs will become mandatory when soft-core microcontrollers are considered in Chap. 11.

A block RAM in the Artix-7 XC7A35T FPGA can be used to store one block of 36-kbit or two blocks of 18-kbit data. There are 50 such blocks within the FPGA. Therefore, the total block RAM capacity for the FPGA is $50 \times 36 = 1800$ kbits. We will explore the usage of block RAM in a digital system in detail in Chap. 9.

Each 36-kbit block RAM can have 64-bit data width. Moreover, extra eight bits can be used for single-bit error correction or double-bit error detection during data read process. We will explain how error detection can be done in Chap. 8.

2.2.6 DSP Slices

There are dedicated blocks for arithmetic and logic operations in recent FPGAs. These are called digital signal processing (DSP) slices. In the Artix-7 FPGA, these slices are specifically called DSP48E1. There are a total of 90 such slices in the Artix-7 XC7A35T FPGA.

Each DSP slice can perform several arithmetic and logic operations. For our purposes, following operations are the most important ones: multiplying two binary numbers of length 25 and 18 bits; adding, subtracting, and accumulating two 48-bit numbers; applying logic operations on two 48-bit numbers. These operations would require complex algorithms for implementation unless a DSP slice was not available. Therefore, DSP slices will be very effective in implementation. Related to this, we will see how DSP slices can be used in arithmetic operations in Chap. 6. Vivado design suite will be responsible to add these slices to our design whenever needed. For more information on DSP48E1, please see [3].

2.2.7 Clock Management

Clock is basically a periodic square signal such that it stays at logic level zero and one for certain time intervals. Most digital systems need a clock signal to operate in synchronous manner. In such a setting, logic operations are done in the rising edge (from logic zero to one transition) or falling edge (from logic one to zero transition) of the clock signal. Hence, period of the clock signal indicates operation speed in the digital system. We will see clock-based operations in Chap. 10.

The Artix-7 FPGA does not have internal clock-generating circuitry. Therefore, the user should feed a clock signal to the FPGA. Some input/output pins are capable of receiving such clock signals. As the clock signal is fed to the FPGA, it can be processed by the clock management tile (CMT) and distributed through the FPGA. Basys3 and Arty boards have external clock sources to feed the FPGA. We will see their properties in Chap. 3.

The Artix-7 FPGA is divided into regions for clocking purposes. Each region includes most or all FPGA building blocks. There are six such clock regions in the Artix-7 XC7A35T FPGA. The user can observe these clock regions through the Vivado design suite. Moreover, Vivado is responsible to manage clock signals in the FPGA. For more information on clock management, please see [4].

2.2.8 The XADC Block

An analog signal can be processed by a digital system after being sampled and quantized. Module performing these operations is called the analog-to-digital converter (ADC). Since recent advances in digital systems require processing analog signals, the Artix-7 FPGA has a dedicated block called XADC.

The Artix-7 XC7A35T FPGA has one XADC block which consists of two ADC modules. Each module can acquire one million samples per second (MSPS). Each sample can be represented by 12 bits. Hence, a sample can be represented by a binary number in the range 0 to $2^{12} - 1$. The two ADC modules in the XADC block can process two different analog signals simultaneously.

Since we are using Basys3 and Arty boards, we are limited by analog input pins provided by them. Please see Chap. 3 related to this issue. Moreover, for more information on the XADC block and how it can be used in practical applications, please see [5–7].

2.2.9   High-Speed Serial I/O Transceivers

High-speed serial I/O transceivers (HSSIOs) are specialized circuitry to transfer and receive serial data. These transceivers are necessary to transfer data at speeds around gigabits per second (Gb/s). The FPGA on the Basys3 board has two such transceiver blocks which can transfer data up to speed of 3.75 Gb/s. Unfortunately, the FPGA on the Arty board does not have such a block. For more information on transceiver blocks, please see [8].

2.2.10   Peripheral Component Interconnect Express Interface

Peripheral component interconnect express (PCIe) is a high-speed serial connection bus standard. The Artix-7 XC7A35T FPGA has one integrated block for PCIe interfacing. For more information on PCIe interfacing, please see [9].

2.3   FPGA-Based Digital System Design Philosophy

A digital system may be implemented by using different design strategies and resources. This section deals with digital system design philosophy using FPGAs. In other words, the aim of this section is emphasizing the usage of FPGAs in an effective manner.

2.3.1   How to Think While Using FPGAs

The first important point to remember while using an FPGA for digital system design is that the user is free to choose the design methodology. In other words, the same digital system can be implemented in more than one way. Therefore, it is the designer's responsibility to pick the optimal or best design style for his or her needs.

The second important point to remember while using FPGAs is that in the beginning there is no predefined block to do the job. The designer has a powerful and unconstrained resource (within limits) to construct required design blocks. Therefore, a strong digital logic knowledge is required to design efficient and optimized FPGA designs. Vendors are also providing intellectual property (IP) blocks to simplify the FPGA usage. These are valuable sources used extensively in practical applications. We will introduce how to use them in Chap. 4.

The third important point to remember while using FPGAs is in terms of their programming. There are hardware description languages (HDLs) for this purpose. We will introduce two popular HDLs in Chap. 5. Although we can use the phrase "programming an FPGA" in some parts of the book, the user should always bear in mind that we are implementing a specific digital system. Therefore, a C like sequential code will not be prepared in HDL. On the contrary, design philosophy should be based on block-based digital system implementation. These blocks should be implemented in parallel whenever possible to get the best performance from an FPGA.

The fourth important point to remember while using an FPGA is its reconfigurability. Since an FPGA can be reconfigured after initial design has been done, this property can be used whenever needed. Therefore, the user can benefit from the reconfigurability property of the FPGA to improve and modify the design even after it has been finalized and embedded on the device.

2.3.2 Advantages and Disadvantages of FPGAs

We can categorize digital system design and implementation resources into four groups as discrete element, application-specific integrated circuit (ASIC), the FPGA, and microcontroller based. The standard question arises. When should we use an FPGA instead of other design options? Or, what are the advantages and disadvantages of using the FPGA over other design options? Let's try to answer this question by comparing the FPGA with other design options.

A digital system can be implemented using discrete elements. This has been the design strategy for a long time. The advantage here is that the designer only uses needed logic gates or discrete elements. Moreover, using these does not require any expertise besides basic logic knowledge. On the other hand, using discrete elements in logic design is not feasible in most cases. First, physical space needed to implement them may be limited. Second, wire connections between discrete elements may become prohibitive in implementation. Third, the design will be static once implemented. The FPGA provides a neat solution to these problems. Size of an FPGA chip is fixed independent of logic elements inside it. Moreover, interconnection of these elements is implicit in the FPGA. Therefore, wiring of logic elements is not an issue. The most important advantage of the FPGA comes when design needs to be reconfigured. Here, using the FPGA simplifies life for the designer. The design can be reconfigured by altering the corresponding HDL section. The only issue here is the need of expertise in HDL.

ASICs provide a good alternative to discrete implementation. They overcome the space and wiring problems. When mass-produced, an ASIC chip becomes cheaper. Moreover, the ASIC chip will be specific to the design. Therefore, it will only use the required number of digital logic elements. Note that an FPGA chip can also be taken as ASIC. In this section, we specifically call a digital circuit as ASIC which is designed for a specific purpose. Therefore, once designed the topology will be fixed. This is the drawback of ASIC design. The biggest problem in using ASIC is its fabrication time. FPGAs provide a clear advantage here. In fact, most ASIC designs are prototyped and verified on the FPGA before mass production for this purpose.

A microcontroller can be used instead of FPGA in most cases. They share similar characteristics such as reconfigurability, compactness, and cheapness. The first difference between them is that the microcontroller has a unique set of commands (instruction set) to perform an action. Therefore, the user should adjust his or her design accordingly. This is not an issue to an FPGA user. As we have mentioned previously, the FPGA can be taken as a free design environment within limits. Therefore, an FPGA is more flexible compared to the microcontroller. However, we should admit that programming a microcontroller is fairly easy compared to managing an FPGA. The second difference between the microcontroller and FPGA is power consumption in which the FPGA has a clear advantage. The third difference between the microcontroller and FPGA is in the inherent parallel implementation capacity of the FPGA. A microcontroller is a sequential device such that commands are performed step by step. However, the FPGA can be reconfigured as a parallel device. Hence, desired operations can be performed faster in orders of magnitude in the FPGA. Note that a microcontroller can be implemented using an FPGA. We will introduce this concept in Chap. 11.

2.4   Usage Areas of FPGAs

FPGAs can be used in almost all areas where digital systems are needed. To motivate the reader and show why learning digital design using the FPGA is important, we list possible usage areas as follows: aerospace, automotive, broadcast, consumer electronics, defense, high-performance computing, industrial applications, medical applications, and wireless and wired communications. These are not the only usage areas of FPGAs. New applications may emerge in time.

2.5   Summary

An FPGA is a good alternative to implement a digital system. However, the reader should understand what an FPGA is before using it in his or her design. This chapter introduced key FPGA concepts for this purpose. Therefore, we started with digital electronics and explored how basic digital logic gates can be constructed from these. Then, we evaluated basic building blocks of an FPGA. Here, we focused on CLB since it is the basic building block used in digital system implementation on an FPGA. Finally, we considered the design philosophy to be followed while using an FPGA. We believe these topics will be of great use in understanding concepts to be introduced in the following chapters. Therefore, we suggest the reader to grasp them fully before leaving this chapter.

2.6   Exercises

2.1 Besides the OR and AND logic gates, there are also NOR (NOT-OR) and NAND (NOT-AND) gates. Use basic logic gate structures in Figs. 2.3, 2.5, and 2.7 to construct them.

2.2 There is also an XOR gate used in some applications. Construct this gate using OR and AND logic gates.

2.3 The FPGA is not the only device for digital system implementation. Make research for similar devices developed in the past.

2.4 The Artix-7 FPGA is the family we consider in this book. However, Xilinx has other FPGA families as well. Pick two such families and compare their properties with the Artix-7 FPGA.

2.5 Xilinx is not the only FPGA producer in the market. Make research on other producers.
   a. Comment on market share of the FPGA developers.
   b. Compare general properties of developed FPGAs by different producers, if possible.

2.6 What is the main difference between a microcontroller and an FPGA?

CHAPTER 3

Basys3 and Arty FPGA Boards

Throughout the book, we will use two different field-programmable gate array (FPGA) boards: Basys3 and Arty. Both boards have the Xilinx Artix-7 FPGA on them. Although these boards have similar characteristics, Basys3 is more suitable for education purposes since it has several input/output connections. On the other hand, Arty is primarily developed for soft-core microcontrollers to be introduced in Chap. 11. Moreover, it has Arduino compatible pins. Hence, shields available for Arduino can be used with Arty.

In this chapter, we will briefly explore the properties of Basys3 and Arty boards. We will also analyze peripheral devices and connectors on each board besides the FPGA. While doing this, we will not go into the details of the connection diagrams and pin correspondence between a device (or connector) and an FPGA. Since this correspondence will be done by the Vivado design suite (to be introduced in Chap. 4), it is not necessary to add extra complexity at this level. Note that we have introduced general properties of FPGAs on the Basys3 and Arty boards in Chap. 2. Explanations in this chapter will be closely related to information given there.

3.1 The Basys3 Board

The first board to be considered in this chapter is Basys3 developed by Digilent Inc. [10]. As mentioned previously, this board is suitable for education purposes since it has several input/output connections. Let's start with the board layout in Fig. 3.1. In this figure, each important block is labeled by a number. Explanation of each label is given in Table 3.1. Since the SPI flash, power supply regulator, and the oscillator/clock circuitry are not visible in Fig. 3.1, they are labeled B1, B2, and B3 in the table.

Besides the Artix-7 FPGA (label 17), blocks in Table 3.1 can be categorized into six groups. These are powering the board, input/output, configuring the FPGA, advanced connectors, external memory, and oscillator/clock. Next, we explain each category in detail.

3.1.1 Powering the Board

The Basys3 board can be powered either from the USB port (label 13) or from an external power supply which should be connected to the external power connector (label 14). If an external power supply is used, it should be able to deliver a DC voltage between 4.5 and 5.5 V with at least 1-A current. The power source select jumper (label 16) can be used to select the power source to be fed to the board. Input voltage (either from the USB or

FIGURE 3.1 The Basys3 board layout.

Label	Explanation
1	Power LED
2	Three Pmod connectors
3	Analog signal Pmod connector
4	Four-digit seven-segment display
5	Sixteen slide switches
6	Sixteen LEDs
7	Five push buttons
8	FPGA programming done LED
9	FPGA configuration reset button
10	Programming mode jumper
11	USB host connector
12	VGA connector
13	Shared UART/JTAG USB port
14	External power connector
15	Power switch
16	Power source select jumper
17	Artix-7 FPGA
18	USB-UART bridge
19	Auxiliary function microcontroller
B1	SPI flash
B2	Power supply regulator
B3	Oscillator/clock

TABLE 3.1 Explanation of Labels in Fig. 3.1

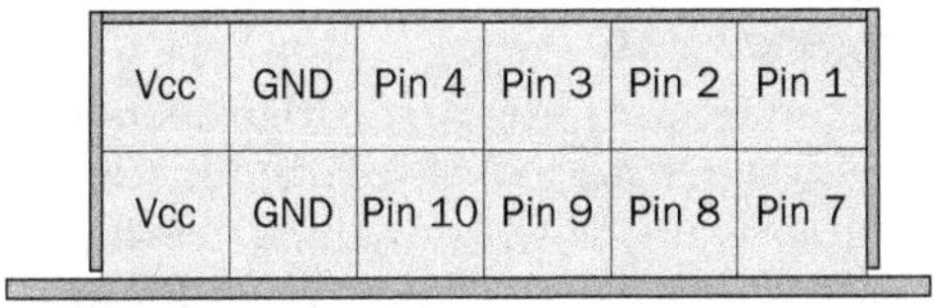

Vcc	GND	Pin 4	Pin 3	Pin 2	Pin 1
Vcc	GND	Pin 10	Pin 9	Pin 8	Pin 7

FIGURE 3.2   The Basys3 board Pmod connector pin layout.

external source) is regulated by power supply regulators (label B2). The power switch (label 15) turns on and off the board. The power LED (label 1) indicates that the board is turned on and operating normally. Connection diagram between all these elements can be found in [10].

3.1.2   Input/Output

There are several digital input/output connections on the Basys3 board. These can be summarized as peripheral module (Pmod) connectors, four-digit seven-segment display, 16 slide switches, 16 LEDs, and five push buttons. Let's explain these in detail.

There are three Pmod connectors (label 2) for digital input/output. Pin layout of a Pmod connector is as in Fig. 3.2. As can be seen in this figure, there are 2×10 female pins in the connector. Among these, 2×1 pins are for ground and 2×1 pins are for V_{CC} supply voltage. The FPGA receives and transmits digital data through the remaining Pmod pins. There is also an analog signal Pmod connector (label 3) on the Basys3 board. Pins in this connector are connected to analog input pins of the FPGA. The XADC block in the FPGA (introduced in Sec. 2.2.8) receives analog signals through these pins. Pin assignments between four Pmod connector pins and the Artix-7 FPGA on the Basys3 board can be found in [10].

There is a four-digit seven-segment display (label 4) on the Basys3 board. Each digit in this display is composed of seven segments arranged in a squarish 8 form. These segments are connected in common anode form [10]. Hence, when a logic level zero is applied to a segment, it turns on. Pin connection between seven-segment display and the FPGA can be found in [10].

There are 16 slide switches (label 5) on the Basys3 board. These are connected to the FPGA through series resistors. These switches can be used as input to the FPGA. Depending on the state of a switch, it can either generate constant input of logic level zero or one to the FPGA. Pin connection between these switches and the FPGA can be found in [10].

There are 16 LEDs (label 6) on the Basys3 board. These are connected to the FPGA through resistors. These LEDs can be used as output from the FPGA. When a logic level one is applied to an LED, it turns on. When a logic level zero is applied to an LED, it turns off. Pin connection between these 16 LEDs and the FPGA can be found in [10].

There are five push buttons (label 7) on the Basys3 board. These can be used as input to the FPGA. Push buttons are arranged in active high setup such that when pressed they provide logic level one. At rest, they provide logic level zero. Pin connection between these push buttons and the FPGA can be found in [10].

To remind again, we will use the Vivado design suite to interact with all these input/output connections in the following chapters. Therefore, it is not mandatory to learn which FPGA pin is connected to which Basys3 block. We expect this abstraction to simplify design steps.

3.1.3 Configuring the FPGA

The FPGA should be configured (programmed) to operate. The configuration file will be generated by Vivado design suite to be explained in Chap. 4. The generated file can be fed to the FPGA in three ways. The first method is using the shared UART/JTAG USB port (label 13). We will use this method while configuring the FPGA through Vivado. The second method is using the SPI flash (label 18). To do so, the configuration file should have been stored in flash beforehand. The third method is storing the configuration file in a USB stick and using it through the USB host connector (label 11). By the help of an auxiliary function microcontroller, programming can be done. On Basys3, there is a PIC24FJ128 microcontroller (label 19) for this purpose [11]. The programming mode jumper (label 10) can be used to set the FPGA programming method. More information on the second and third methods can be found in [10]. When the FPGA is successfully configured by any of the mentioned three methods, the programming done LED (label 8) turns on. Note that the "programming done" signal is fed by the FPGA. The FPGA configuration reset button (label 9) can be used to reset the FPGA configuration.

3.1.4 Advanced Connectors

There are advanced connectors on the Basys3 board. These are the USB host connector, VGA connector, and shared UART/JTAG USB port. Let's briefly explain them.

The USB host connector (label 11) can be used to transfer the configuration file to the FPGA. The connector also has USB human interface device (HID) capability. These two properties can be performed through the PIC24FJ128 microcontroller (label 19) connected to the connector. We will use the USB HID capability to connect keyboard and mouse to the Basys3 board in Chap. 12. More information on the usage of the USB host connector and PIC microcontroller can be found in [10].

There is a VGA connector (label 12) on the Basys3 board. This connector allows 12-bit data transfer (four bits for red, four bits for blue, four bits for green pins) to a VGA display device. More information on VGA can be found in [10]. We will use the VGA connector to display an image on a monitor in Chap. 12.

The shared UART/JTAG USB port (label 13) is mainly used to configure (program) the FPGA via Vivado. We will explore how to do this in Chap. 4. The shared UART/JTAG USB port also has a USB-UART bridge (label 18) connected to it [12]. Therefore, it can also be used as a UART medium to communicate the FPGA with PC or another device. We will explore how to use this property in Chap. 12. More information on the usage of shared UART/JTAG USB port and USB-UART bridge can be found in [10].

3.1.5 External Memory

The Basys3 board has a 32-Mbit non-volatile serial flash (label B1) as external memory developed by Spansion [13]. This device is connected to Artix-7 FPGA through a dedicated SPI bus. Pin connections between the FPGA and SPI flash can be found in [10]. The FPGA configuration files can be saved in this flash memory. Moreover, the FPGA can be set to read these files automatically at start up. The Artix-7 FPGA configuration file needs over 16 Mbits of memory space. Therefore, the remaining memory space (approximately 16 Mbits) will be available to the user.

3.1.6 Oscillator/Clock

The Basys3 board has an onboard oscillator/clock circuitry (label B3) working at 100 MHz. The clock signal generated by the oscillator is fed to the Artix-7 FPGA through its pins. Therefore, this onboard oscillator allows user to generate a required clock (within limits) in the design.

3.2   The Arty Board

The second board to be considered in this chapter is Arty. This evaluation kit is jointly developed by Digilent Inc and Avnet [14]. As mentioned previously, this board is more suitable for soft-core microcontrollers to be introduced in Chap. 11. Let's start with the board layout given in Fig. 3.3. In this figure, each important block is labeled by a number. Explanation of each label is given in Table 3.2. Since the oscillator/clock circuitry is not visible in Fig. 3.3, it is labeled as B1 in the table.

Besides the Artix-7 FPGA (label 18), blocks in Table 3.2 can be categorized into six groups. These are powering the board, input/output, configuring the FPGA, advanced connectors, external memory, and oscillator/clock. Next, we explain each category in detail.

3.2.1   Powering the Board

The Arty board can be powered in three ways as using the shared UART/JTAG USB port (label 2), external power jack, and Arduino/chipKIT connectors (label 10). Throughout the book, we will assume that the shared UART/JTAG USB port is used for powering the board. For external power usage, please see [14]. The power source select jumper (label 4) can be used to select the power source to be fed to the board. Input voltage (either from the USB or external source) is regulated by the power supply regulator

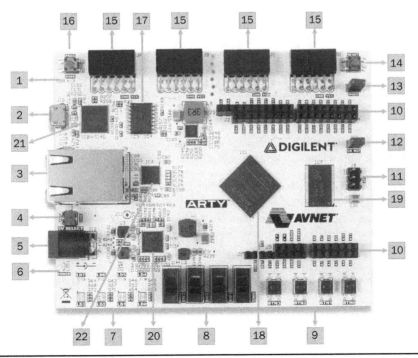

FIGURE 3.3 The Arty board layout.

Label	Explanation
1	FPGA programming done LED
2	Shared UART/JTAG USB port
3	Ethernet connector
4	Power source select jumper
5	Power jack
6	Power LED
7	Eight LEDs
8	Four slide switches
9	Four push buttons
10	Arduino/chipKIT shield connectors
11	SPI header (Arduino/chipKIT compatible)
12	chipKIT processor reset jumper
13	Programming mode jumper
14	chipKIT processor reset button
15	Four Pmod connectors
16	FPGA configuration reset button
17	SPI flash
18	Artix-7 FPGA
19	DDR3 memory
20	Power supply regulator
21	USB-UART bridge
22	Ethernet transceiver
B1	Oscillator/clock

TABLE 3.2 Explanation of Labels in Fig. 3.3

(label 20). The power LED (label 6) indicates that the board is turned on and operating normally. Connection diagram between all these elements can be found in [14].

3.2.2 Input/Output

There are several digital input/output connections on the Arty board. These can be summarized as four Pmod connectors, Arduino/chipKIT shield connector, four tricolor LEDs, four LEDs, four slide switches, four push buttons, and chipKIT processor reset button and jumper. Let's explain these in detail.

There are four Pmod connectors (label 15) for digital input/output. Pin layout of a Pmod connector is the same as in Fig. 3.2. In other words, Pmod connectors used in Arty are the same as in Basys3. However, Pmod connectors in the Arty board are grouped into two categories as standard (labeled as JA and JD on the board) and high speed (labeled as JB and JC on the board). Standard Pmod connectors are connected to the FPGA via series resistors which prevent accidental short circuit. High-speed connectors do not have such resistors. Hence, they should be used with care. More information on Pmod connectors can be found in [14].

Different from Basys3, Arty has Arduino/chipKIT shield connectors (label 10). These allow user to connect available Arduino and chipKIT shields. More information on Arduino/chipKIT shield connectors can be found in [14].

There are four tricolor LEDs (labeled as LD0–LD3 on the board) and four standard LEDs (labeled as LD4–LD7 on the board). All of these LEDs are indicated by label 4 in Fig. 3.3. Four standard LEDs operate as the ones on Basys3 board. Each tricolor LED is composed of three LEDs with red, green, and blue colors. Each internal LED can be turned on as if using the standard LED. However, Digilent suggests using pulse width modulation (PWM) signals to use tricolor LEDs. More information on standard and tricolor LEDs on the Arty board can be found in [14].

There are four slide switches (label 8) and four push buttons (label 9) on the Arty board. They have the same characteristics as in the Basys3 board. Therefore, we direct the reader to previous section. More information on slide switches and push buttons on the Arty board can be found in [14].

The chipKIT processor reset jumper (label 12) and button (label 14) are available to be used in soft-core microcontroller designs. Specifically, they can be used to reset the designed microcontroller. Hence, these can be of use while designing a soft-core microcontroller in Chap. 11.

To remind again, we will use the Vivado design suite to interact with all these input/output connections in the following chapters. Therefore, it is not mandatory to learn which FPGA pin is connected to which Arty block. We expect this abstraction to simplify design steps.

3.2.3 Configuring the FPGA

Configuring the FPGA on the Arty board is similar to Basys3. Therefore, we only explain the labels in Fig. 3.3 related to the FPGA configuration. The programming mode jumper (label 13) can be used to set the FPGA programming method. When the FPGA is successfully configured, the "programming done" LED (label 1) turns on. Note that the "programming done" signal is fed by the FPGA. The FPGA configuration reset button (label 16) can be used to reset the FPGA configuration. More information on these can be found in [14].

3.2.4 Advanced Connectors

There are advanced connectors on the Arty board. These are the shared UART/JTAG USB port, ethernet connector, and Arduino/chipKIT compatible SPI header. Let's briefly explain them.

The shared UART/JTAG USB port (label 2) is mainly used to configure (program) the FPGA. We will explore how to do this in Chap. 4. The shared UART/JTAG USB port also has a USB-UART bridge (label 21) connected to it [12]. Therefore, it can also be used as a UART medium to communicate the FPGA with PC or another device. We will explore how to use this property in Chap. 12. More information on the usage of shared UART/JTAG USB port and USB-UART bridge can be found in [14].

The Arty board has an ethernet connector (label 3) and transceiver chip (label 22) by Texas Instruments [15]. The transceiver chip is also called physical layer (PHY). Through the connector and transceiver, ethernet communication can be done. We will explore how to do this in Chap. 12. More information on the ethernet connector and transceiver chip can be found in [14].

Arty also has an Arduino/chipKIT compatible SPI header (label 11). This header can be used in connection with Arduino/chipKIT compatible shields. More information on the SPI header can be found in [14].

3.2.5   External Memory

Arty has two different external memory blocks. The first one is a 128-Mbit non-volatile serial flash memory (label 17) developed by Micron [16]. This device is connected to the Artix-7 FPGA through a dedicated SPI bus. Pin connections between the FPGA and SPI flash can be found in [14]. The FPGA configuration files can be saved in this flash memory. Moreover, the FPGA can be set to read these files automatically at start-up. The Artix-7 FPGA configuration file needs over 16 Mbits of memory space. Therefore, remaining memory space (approximately 14 MB) will be available to the user.

The second memory block on the Arty board is a 256-MB DDR3L SDRAM (label 19) developed by Micron [17]. More information on the DDR3 SDRAM and its connection to the Artix-7 FPGA can be found in [14].

3.2.6   Oscillator/Clock

The Arty board has an onboard oscillator/clock circuitry (label B1) working at 100 MHz. Clock signal generated by the oscillator is fed to the Artix-7 FPGA through its pins. Therefore, this onboard oscillator allows user to generate a required clock (within limits) in the design.

3.3   Summary

Topics introduced in this chapter are specific to the FPGA boards to be used throughout the book. Therefore, they will be needed when a real-life application is developed. We did not provide detailed connection diagrams in this chapter. Instead, we directed the reader to related references. However, the reader should bear in mind that connection between the FPGA and peripherals on the Basys3 and Arty boards will be done via the Vivado design suite. Therefore, it is not mandatory to memorize them. Finally, the reader can consult information in this chapter while exploring the following chapters of the book.

3.4   Exercises

3.1 We have two boards Basys3 and Arty. Compare properties of peripherals on these boards.

3.2 When should we choose the Basys3 board? Why?

3.3 When should we choose the Arty board? Why?

CHAPTER **4**

The Vivado Design Suite

Vivado design suite is the software environment we will be using throughout the book. Therefore, we will explain its properties starting from installation step. Then, we will explain how to create a new project containing either the Verilog or VHDL description of a simple digital system. Afterward, we will introduce tools necessary to synthesize and implement the HDL description. While doing this, we will emphasize how the FPGA building blocks introduced in Sec. 2.2 can be observed in Vivado. This way, we aim to show the reader to analyze his or her HDL design in detail. Then, we will explain how to program the FPGA on the Basys3 and Arty boards through Vivado. Finally, we will introduce IP management methods in Vivado.

4.1 Installation and the Welcome Screen

The Vivado design suite has several editions with different properties. For our purposes, the free HL WebPACK edition is sufficient. Installing this edition is straightforward. However, the reader should first create a Xilinx account for this purpose. Then, Vivado can be installed following the commands on the screen. Here, we assume that the reader uses Vivado on a PC with the Microsoft Windows operating system. Please consult the user guide for using Vivado design suite on other operating systems.

As of the writing of this book, the Vivado design suite available at Xilinx's Web page was version 2016.3. Therefore, we will use it throughout the book. After installation, Vivado starts as in Fig. 4.1. This screen tells us that we are ready to go.

4.2 Creating a New Project

Let's create our first HDL project in Vivado. We can start by clicking on 'Create New Project' on the start page of Vivado as in Fig. 4.1. Skip the first welcome popup window by clicking Next. Now, you should see a page where you can set the name and location of your new project as in Fig. 4.2.

Let's call our project as *first_project*. This project will be created under directory `.../Xilinx_Projects`. Click Next and select "RTL Project" in the upcoming window. Afterward, "Add Sources" window will pop-up as in Fig. 4.3. At this point, we will not add any sources to the project. However, we should select the "Target language" as either Verilog or VHDL at the bottom of this window. In a similar manner, we should also set the "Simulation language" as Verilog, VHDL, or Mixed here.

FIGURE 4.1   Vivado welcome screen.

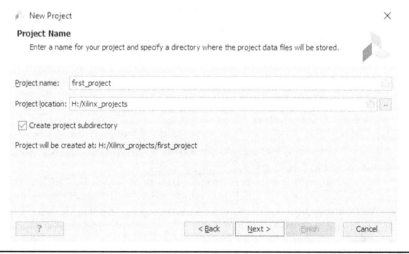

FIGURE 4.2   Create a new project window.

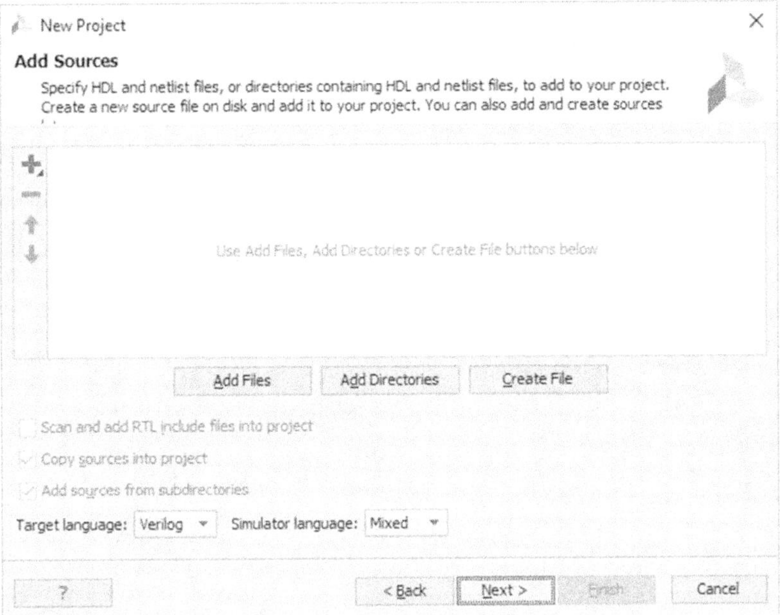

FIGURE 4.3 Add sources window.

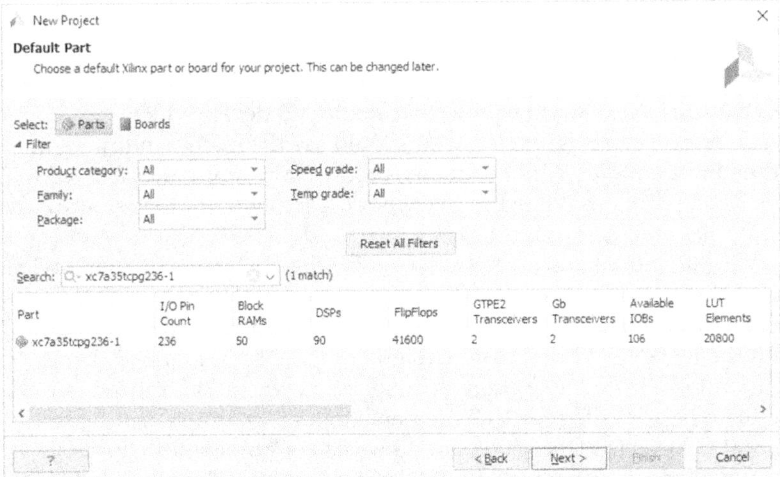

FIGURE 4.4 FPGA selection window.

We can skip the following two optional selection windows (Add Existing IP and Add Constraints) as for now. The next window will be on selecting the FPGA (called the default part) as in Fig. 4.4. The Artix-7 FPGA on the Basys3 board has full name "XC7A35TCPG236-1". The Artix-7 FPGA on the Arty board has full name "XC7A35TICS G324-1L". Depending on the application, one of these FPGAs can be picked. Note that

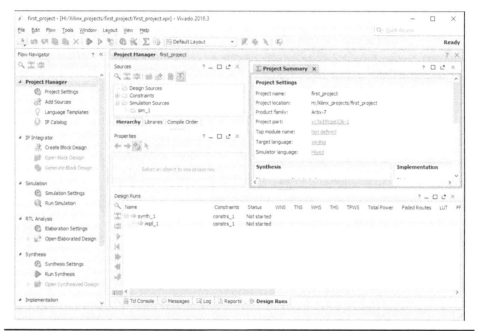

FIGURE 4.5 Vivado project main window.

the selection window in Fig. 4.4 also summarizes the FPGA properties introduced in Sec. 2.2. Click Next and the project for dedicated device will be created.

Once the project is created, you should see the main window as in Fig. 4.5. On the left-hand side of this window, there is Flow Navigator panel. Through it, the user can control all processes related to the project. On the top of the window, the user can see source files and their properties.

We can add a source file to the project by clicking on Flow Navigator → Project Manager → Add Sources. Then, we should select "Add or create design sources" from the popup menu as in Fig. 4.6a. As we click Next, a new popup window should appear with the name "Add or Create Design Sources" as in Fig. 4.6b. Here, the user should click on the Create File button. A small window should appear as in Fig. 4.6c. Select the file type as Verilog (or VHDL), name the file as *first_system*. Choose the location as <Local to Project>. Upcoming window asks for ports within the project. Do not define any ports for now. Simply click OK to create your file. The generated file should be available under Sources → Design Sources directory which can be found in Vivado's Project Manager window.

4.2.1 Adding a Verilog File

Let's pick Verilog as the working HDL at this point. Following the above steps, the source file *first_system.v* should be visible in the Sources window. Open this file by double-clicking on it. Copy the Verilog description in Listing 4.1 to the opened file. We will explain Verilog commands in this description in the following chapters. Here, we will only use it to explain working principles of Vivado.

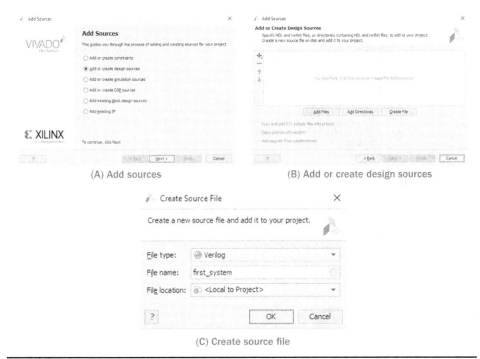

(A) Add sources

(B) Add or create design sources

(C) Create source file

FIGURE 4.6 Adding a source file to the project.

Listing 4.1 Verilog Description to be Used in Explaining Vivado

```verilog
module first_system(out1,out2,in1,in2);

// Port definitions
input in1,in2;
output out1,out2;

// Description of the digital system
// Dataflow modeling

wire and_out,or_out;

assign and_out = in1 & in2;
assign or_out  = in1 | in2;
assign out1    = and_out ^ or_out;
assign out2    = ~ in2;

endmodule
```

4.2.2   Adding a VHDL File

We can also pick VHDL as the working HDL. For this case, the source file *first_system.vhd* should have been created. As in the previous section, open this file by double-clicking on it. Copy the VHDL description in Listing 4.2 to the opened file. Again, we will explain VHDL commands in this description in the following chapters.

Listing 4.2   VHDL Description to be Used in Explaining Vivado

```vhdl
library ieee;
use ieee.std_logic_1164.all;

entity first_system is
port(in1 : in std_logic;
     in2 : in std_logic;
     out1 : out std_logic;
     out2 : out std_logic);
end first_system;

architecture dataflow_model of first_system is
begin
out1 <= (in1 and in2) xor (in1 or in2);
out2 <= not in1;
end dataflow_model;
```

4.3   Synthesizing the Project

The first step in realizing a digital system on the FPGA is synthesizing it. This means representing digital system's HDL description via the FPGA elements introduced in Sec. 2.2. In other words, this step transforms the system description from code to physical device. Note that Vivado is responsible for this operation. Therefore, synthesis steps are hidden to the user.

We can synthesize the HDL description added to the project by clicking on Flow Navigator → Synthesis → Run Synthesis. During this process, we can monitor events from the Log tab. Let's pick the description in Listing 4.1 in this section. Once the synthesis is finalized, a popup window will appear as in Fig. 4.7. Here, the user will have two choices. The first option is "Run Implementation." We will postpone it till the next section. Instead, we will select the "Open Synthesized Design" option.

As the "Open Synthesized Design" option is selected, Vivado subwindows will be as in Fig. 4.8. In these, the reader can observe almost all design specifications in terms of reports under Flow Navigator → Synthesis → Synthesized Design. The designed device

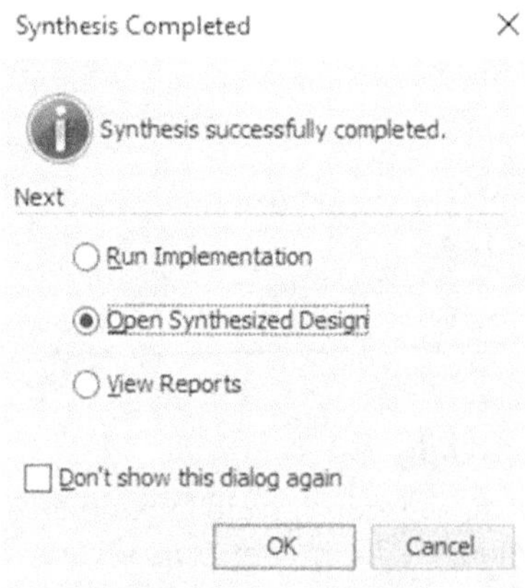

FIGURE 4.7   Synthesis completion window.

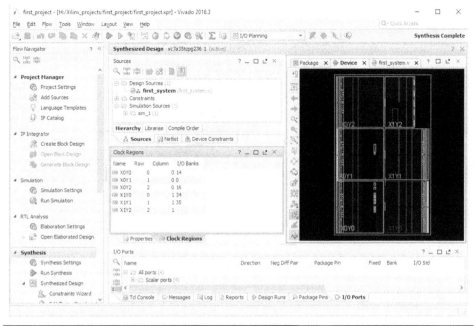

FIGURE 4.8 Vivado after synthesizing the project.

FIGURE 4.9 Utilization report after synthesizing the project.

can also be seen in the Design window. Here, placement of the synthesized design on the FPGA is provided. Unfortunately, it is not easy to see the layout of the used FPGA blocks in this window.

Although all generated project reports are important after synthesization, we will focus on the utilization report. This report will be as in Fig. 4.9 for the synthesized design. As can be seen in this figure, one slice and four input/output pins are used during synthesizing the HDL. The report also indicates that the LUT in the slice is used as a logic element.

The reader can observe the synthesized design by selecting Flow Navigator → Synthesis → Schematic. The result will be as in Fig. 4.10. As can be seen in this figure, the Verilog description in Listing 4.1 is realized by two LUTs (in the same slice) after synthesis.

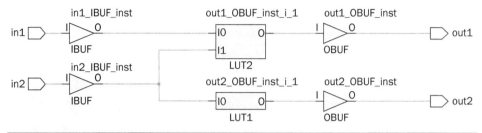

FIGURE 4.10   Schematic view of the design.

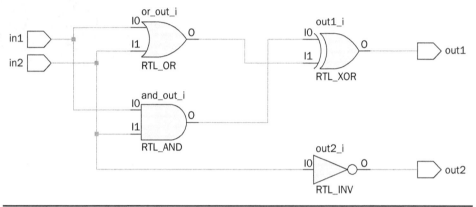

FIGURE 4.11   RTL schematic view of the design.

Schematic view of the design can be observed via selecting Flow Navigator → RTL Analysis → Elaborated Design → Schematic. The result will be as in Fig. 4.11. As can be seen in this figure, the schematic view is given in terms of basic logic gates. To be more specific, Verilog description of the first system in Listing 4.1 has two input ports as in1 and in2. The system has two output ports as out1 and out2. Basic logic gates AND, OR, NOT, and XOR are used to construct the system. Hence, schematic view under the RTL analysis option summarizes the overall system fairly well. This may be of great use in analyzing combinational and sequential digital systems to be introduced in the following chapters.

4.4   Simulating the Project

Synthesizing the project results in the generated digital system from its HDL description. To decide whether this system works as desired, we should test it. This can be done by feeding input to the system and observing the corresponding output. This is called simulating the system. The second step in realizing the project on the FPGA is simulating it.

We have to create a testbench file to simulate the designed digital system. Therefore, we should create and add a new file to the project. To do so, click on Flow Navigator → Project Manager → Add Sources. Then, select "Add or create simulation sources" from the popup menu as in Fig. 4.6a. As we click Next, a new popup window should appear with the name "Add or Create Simulation Sources." Here, the user should click on the

Create File button. A small window should appear as in Fig. 4.6c. Select the file type as Verilog (or VHDL), name the file as *first_system_tb*. Choose the location as <Local to Project>. The upcoming window asks for ports within the project. Do not define any ports for now. Simply click OK to create your file. The generated file should be available under Sources → Simulation Sources → Sim_1 directory which can be observed in Vivado's Project Manager window.

Vivado only creates an empty testbench file. The user should add all input, output, and call function declarations to test the digital system under consideration. Unfortunately, the testbench file is composed of HDL commands to be introduced in Chap. 5. Therefore, we will provide sample testbench files for Verilog and VHDL descriptions next.

4.4.1 Adding a Verilog Testbench File

We will first generate the Verilog testbench file for the description in Listing 4.1. To do so, we will benefit from the file in Listing 4.3. As a brief explanation, this testbench file generates input patterns changing at every 100 ns. These are fed to the digital system

Listing 4.3 Testbench File for the Given Verilog Description

```verilog
'timescale 1ns / 1ps

module first_system_tb;

// Inputs
reg in1t, in2t;

// Outputs
wire out1t, out2t;

// Instantiate the Unit Under Test (UUT)
first_system UUT (.out1(out1t),.out2(out2t),.in1(in1t),.in2(in2t));

//Providing input to the UUT
initial begin
// Initialize Inputs
in1t = 0;
in2t = 0;

// Wait 100 ns for global reset to finish
#100;

// Add stimulus here
repeat (4)
#100 {in1t,in2t} = {in1t,in2t} + 1'b1;
end

//Display the result on the Tcl console (Optional)
initial  begin
$display("   in1 in2 out1 out2");
$monitor("\t%b \t%b \t%b \t%b",in1t,in2t,out1t,out2t);
end

endmodule
```

FIGURE 4.12 Setting simulation properties.

under test and corresponding output is obtained. We will analyze the structure of this testbench file in detail in Sec. 5.2.

Just copy and paste all the lines in Listing 4.3 to the testbench file generated under Vivado. Make sure that the third line in the description reads as module first_system_tb;. In a similar manner, the module name under Unit Under Test (UUT) section should be read as first_system in this file. Now, the designed digital system is ready for simulation. Before that, we should set the runtime for simulation. To do so, click on Flow Navigator → Simulation → Simulation Settings → Simulation and change the xsim.simulate.runtime* to 490 ns as in Fig. 4.12. This runtime is suitable to view all input and output values for this simulation. For other simulations, the runtime should be set accordingly.

To start the simulation, click on Flow Navigator → Simulation → Run Simulation → Run Behavioral Simulation. When the simulation ends, Vivado opens a waveform window in the workspace named "Untitled1." The reader can use zoom tools on the left-hand side and fit waveforms in the window to check all input and output combinations in time. The simulation result should appear as in Fig. 4.13 once it fits into the window. Note that the default background color was set as black for this window. We had to change it to white for ease of observation. The user can check the simulation results to observe whether the designed system acts as desired.

Behavioral simulation is not the only option in observing results. Vivado also offers post-synthesis functional, post-synthesis timing, post-implementation functional, and post-implementation timing simulations. The reader can pick the most suitable one for his or her needs. We will only use behavioral simulation throughout the book.

first_system.v
first_system_tb.v
Untitled 1
Name
Value
0 ns
200 ns
400 ns
in1t 1
in2t 1
out1t 0
out2t 0

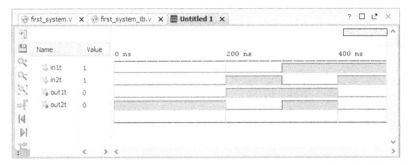

FIGURE 4.13 Simulation results in terms of input and output waveforms in time.

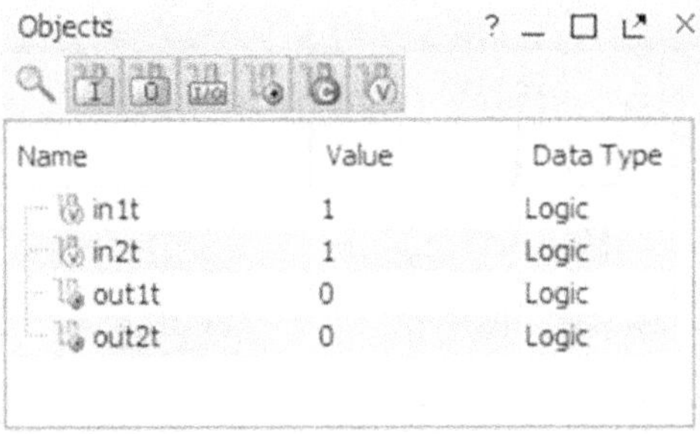

Objects
Name
Value
Data Type
in1t 1 Logic
in2t 1 Logic
out1t 0 Logic
out2t 0 Logic

FIGURE 4.14 Final simulation results in the Objects window.

The final simulation result (in the latest simulation time) can also be observed in Vivado's Objects window. We provide the final simulation result for the present example in Fig. 4.14. This window will be extremely helpful in Chap. 6.

4.4.2 Adding a VHDL Testbench File

We will next generate the testbench file for the VHDL description in Listing 4.2. As in previous section, we will benefit from the previously prepared file in Listing 4.4. We will analyze the structure of this testbench file in detail in Sec. 5.4. Just copy and paste all the lines in Listing 4.4 to the testbench file generated under Vivado as for now. Afterward, follow the steps given in previous section for simulation. After simulation ends, the same waveforms should be obtained as in Fig. 4.13.

4.5 Implementing the Synthesized Project

The third step in realizing the digital system on the FPGA is implementing it. Here, the synthesized HDL design is prepared to be implemented to target the FPGA platform. Besides, optimization and minimization tools are used on the synthesized design to decrease the FPGA resource usage. Physical properties of the FPGA (such as temperature in the device) are also taken into account at this step. We will talk about minimization tools in Sec. 7.3.3. However, the actual optimization and minimization tools working under Vivado are hidden to the user. Therefore, we are bound by Vivado in these operations.

Listing 4.4 Testbench File for the Given VHDL Description

```vhdl
library ieee;
use ieee.std_logic_1164.all;

entity first_system_tb is
end first_system_tb;

architecture dataflow of first_system_tb is

signal  in1t : std_logic := '0';
signal  in2t : std_logic := '0';
signal out1t : std_logic := '0';
signal out2t : std_logic := '0';

component first_system
port(in1 : in std_logic;
     in2 : in std_logic;
    out1 : out std_logic;
    out2 : out std_logic);
end component;

begin
UUT: first_system port map (in1 => in1t,in2 => in2t, out1 =>out1t, out2
     => out2t);

process
begin
wait for 100 ns;
in1t <= '0'; in2t <= '0';

wait for 100 ns;
in1t <= '0'; in2t <= '1';

wait for 100 ns;
in1t <= '1'; in2t <= '0';

wait for 100 ns;
in1t <= '1'; in2t <= '1';
wait;
end process;

end dataflow;
```

To implement the design, click on Flow Navigator → Implementation → Run Implementation. When the implementation ends, Vivado opens a window as in Fig. 4.15. As in the synthesis step, the reader can check all related reports from the Flow Navigator → Implementation → Implemented Design section.

Although all generated project reports are important in the Flow Navigator → Implementation → Implemented Design section, we will focus on the utilization report as in the synthesis step. This report will be as in Fig. 4.16. As can be seen in this figure, the utilization report after implementation is more detailed compared to the one obtained after synthesis step. Here, the reader can observe that one SLICEL is used in implementation.

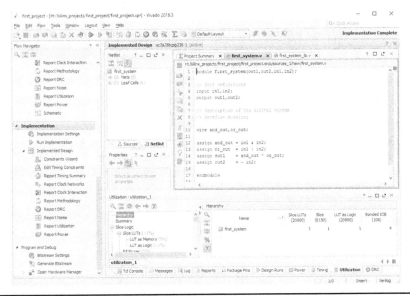

FIGURE **4.15** Vivado after implementing the project.

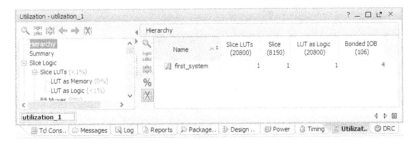

FIGURE **4.16** Utilization report after implementing the project.

4.6   Programming the FPGA

The fourth and final step in realizing the digital system on the FPGA is programming it to the target device. This can be done by clicking Flow Navigator → Program and Debug → Generate Bitstream. This way, Vivado translates the implemented design to the format (in terms of a bitstream) such that it can be fed to the FPGA. The FPGA on the Basys3 or Arty board can be programmed this way as explained in Chap. 3. However, the project should be altered beforehand such that input and output ports can be assigned to appropriate devices on the target board. Therefore, let's focus on this issue first.

4.6.1   Adding the Basys3 Board Constraint File to the Project

If we want to realize the implemented digital system on the Basys3 board, we should assign its peripheral devices as input and output ports first. As briefly explained in Sec. 3.1, the hardware–software interface between the Basys3 board and the implemented design can be set up by a constraint file.

The constraint file `Basys3_Master.xdc` for the Basys3 board can be obtained from [18]. There, the user should download the "Master Xilinx Design Constraint (XDC)" file under "Docs & Designs" tab. As the downloaded zip file is extracted, the `Basys3_Master.xdc` should be recovered. This file has pin information about clock, switches, LEDs, seven-segment display, buttons, Pmod headers, VGA connector, USB-RS232 interface, USB HID, and quad SPI flash on the Basys3 board.

To use the constraint file `Basys3_Master.xdc`, move it to your project directory. Click on Add Sources under Project Manager and select "add or create constraint" from the menu. Click Next. Then, click on Add Files in the opened window. Browse and locate the constraint file added to the project folder. As this file is added to the project, it can be seen in the Sources window under the Constraints → constrs_1 folder. Double-click on the `Basys3_Master.xdc` file to edit it. As can be seen, all the lines are commented out by the # sign in the beginning. We will use switches `sw[0]` and `sw[1]` as inputs in1 and in2 in Listing 4.1. In the same description, we will use LEDs `led[0]` and `led[1]` as outputs out1 and out2. Therefore, uncomment these parts in the constraint file and save it.

Since input and output ports are assigned to the Basys3 switches and LEDs, we should also apply these changes to the description in Listing 4.1. The new description file can be obtained by replacing in1 and in2 with `sw[0]` and `sw[1]`, respectively. Also, out1 and out2 should be replaced by `led[0]` and `led[1]`. The modified description file will be as in Listing 4.5. Apply these changes to the source file `first_system.v` in the project.

4.6.2 Programming the FPGA on the Basys3 Board

Now, we have all the necessary files to realize the Verilog description in Listing 4.5 on the FPGA of Basys3 board. To do so, synthesize and implement the HDL description as explained in previous sections. As implementation is complete, click on Flow Navigator → Program and Debug → Generate Bitstream. Select Open Hardware Manager from the popup window as in Fig. 4.17 when the bitstream is generated.

Listing 4.5 Verilog Description of the First System with Switches and LEDs as Input and Output

```verilog
module first_system(led,sw);

// Port definitions
input [1:0]sw;
output [1:0]led;

// Description of the digital system
// Dataflow modeling

wire and_out,or_out;

assign and_out = sw[0] & sw[1];
assign or_out  = sw[0] | sw[1];
assign led[0]  = and_out ^ or_out;
assign led[1]  = ~ sw[1];

endmodule
```

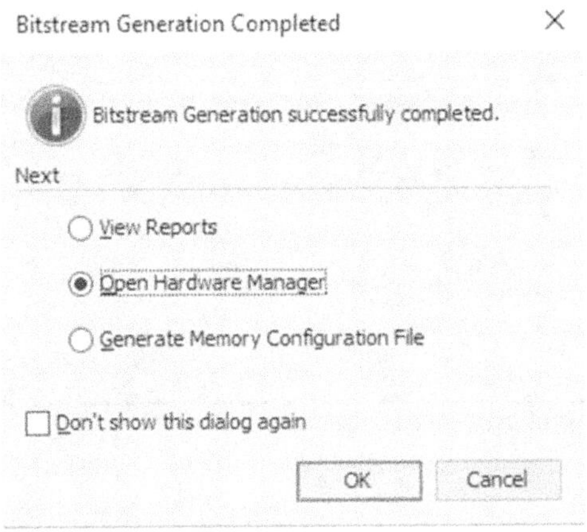

FIGURE 4.17 Generate bitstream completion window.

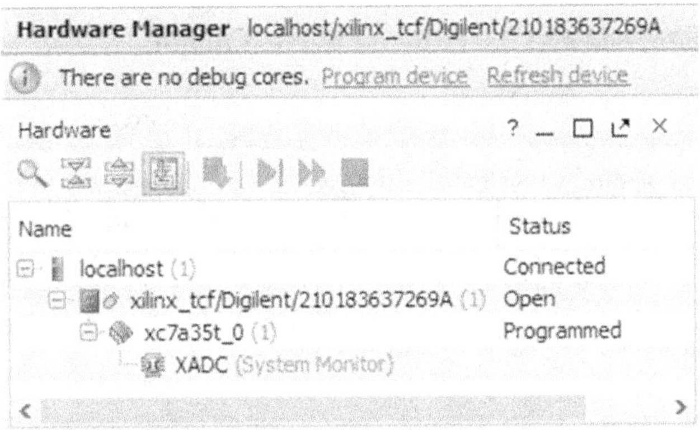

FIGURE 4.18 The Hardware Manager window after Basys3 board is automatically detected.

Hardware Manager window launches in the middle of the screen. By the way, this window can also be opened by clicking on Flow Navigator → Program and Debug → Hardware Manager. The title of the window appears as *Hardware Manager - unconnected*. Beneath the title you will see a warning as *No hardware target is open. Open target*. Click on Open target → Auto Connect after you connect the Basys3 board via USB port to the computer. Now, you should see localhost/xilinx_tcf/Digilent/21083637269A near the Hardware Manager title. If the Basys3 board is automatically detected, the Hardware Manager window will be as in Fig. 4.18.

Click on the program device link beneath the title and select `xc7a35t_0`. The popup window in Fig. 4.19 should appear. Click Program to program the board. As this operation finalizes successfully, implemented HDL description should be running on the Basys3 board.

4.6.3 Adding the Arty Board Constraint File to the Project

The project in Sec. 4.6.2 can also be realized on the Arty board. To do so, we should first add the constraint file for this board to the project instead of Basys3 board's constraint file. Besides, the same Verilog description in Listing 4.5 will be used here.

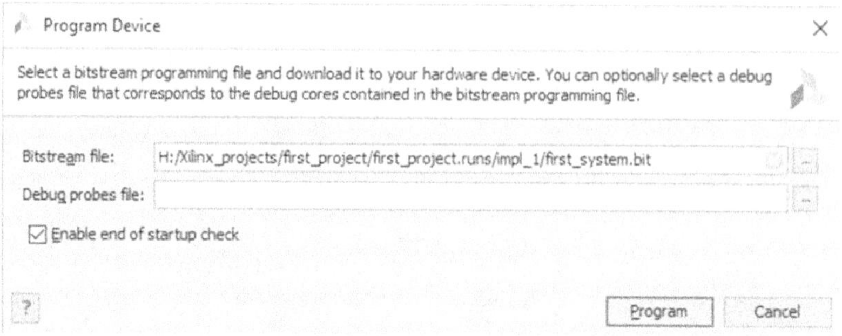

Figure 4.19 Hardware programming window.

The constraint file for the Arty board can be downloaded from [19]. After extracting this zip file, rename the file `Arty_sw_btn_Demo.xdc` as `Arty_Master.xdc` for consistency.

4.6.4   Programming the FPGA on the Arty Board

We will follow the same steps in Sec. 4.6.2 to program the FPGA on the Arty board. If everything goes as expected while generating the bitstream, then the FPGA should be programmed correctly.

There is one minor issue due to Vivado. Sometimes, programming the FPGA cannot be done automatically. Then, the bitstream file location in Fig. 4.19 will be empty. The reader should manually enter this location. For the present design, the location to be entered will be as `H:/Xilinx_Projects/first_project/first_project.runs/impl_1/first_system.bit`. Project root folder is `H:/` for our case. Then, programming can be done as expected.

4.7   Vivado Design Suite IP Management

We can benefit from existing intellectual property (IP) blocks available in Vivado for our design. We can also convert a Verilog or VHDL description to an IP block as well. In this section, we will make a brief introduction to these topics. Then, we will extensively use these options in the following chapters. For further information on IP management in Vivado, please see [20–23].

4.7.1   Existing IP Blocks in Vivado

Vivado has extensive IP blocks available to be used in a project. These can be reached from IP Catalog under Project Manager window. As we press the corresponding button, a new window appears as in Fig. 4.20. The reader can select the desired IP block from this list. In the following chapters, we will use these IP blocks in our projects.

4.7.2   Generating a Custom IP

A Verilog or VHDL description can be converted to a custom IP block in Vivado. This increases reusability of the description. Let's take the `first_system` in Listing 4.1 in our first project. We can create a custom IP from this description. To do so, we should first select "Create and Package IP ..." option under the Tools section in Vivado. A new window appears titled as "Create and Package New IP." As we click Next, a new window

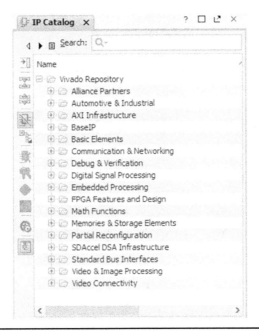

FIGURE 4.20   IP Catalog window.

FIGURE 4.21   Create and Package New IP window.

appears as in Fig. 4.21. Here, select the "Package your current project" under the "Packaging Options."

The next window summarizes location of the generated IP and include options. Here, select "Include .xci files" option. As we press next, a new window appears summarizing the IP block generation process. Pressing Finish in this window generates a

FIGURE 4.22 Package IP - first system.

segment as in Fig. 4.22. Within this section, the reader can make necessary adjustments related to the generated IP. To finalize IP generation, we should select the Review and Package option. In default settings, the generated custom IP will not be archived for future use. Only the current project can use it. To change this option, we should select "edit packaging settings." In the opened project settings window, we should select the IP tab. Then, "create archive of IP" should be checked under the "After Packaging" part. Within the window, we can also set the archive name and location. This information will be important while using the generated IP in another project. As we press OK, Package IP button appears. Pressing this button generates the IP block for the first system.

Generated IP block for the first system can be seen in IP catalog under the UserIP section as in Fig. 4.23. We will show how to use this IP block in a description in Chap. 5.

4.8 Application on the Vivado Design Suite

We will introduce an application to get familiar with Basys3 and Arty boards in this section. Moreover, topics introduced in this application will be of use in the following chapters. Let's start with the Basys3 board.

In Listing 4.6, we provide the Verilog description in which LEDs and switches on the Basys3 board are connected. Therefore, the reader can turn on/off a LED by the corresponding switch. To run this application, generate a new project as explained in this chapter. Include the Verilog description in Listing 4.6 to the project. Do not forget to include the Basys3 board XDC file to the project. Within this file, enable all LED and switch-based lines.

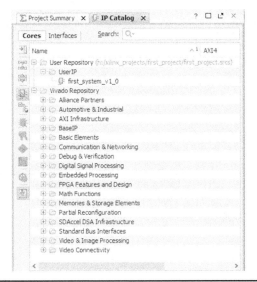

FIGURE 4.23 Modified IP Catalog.

Listing 4.6 Switches to LEDs Application on the Basys3 Board in Verilog

```verilog
module application(led,sw);

output [15:0] led;
input [15:0] sw;

assign led=sw;

endmodule
```

Listing 4.7 Switches to LEDs Application on the Basys3 Board in VHDL

```vhdl
library ieee;
use ieee.std_logic_1164.all;

entity application is
port(sw : in std_logic_vector (15 downto 0);
    led : out std_logic_vector (15 downto 0));
end application;

architecture dataflow_model of application is
begin
led <= sw;
end dataflow_model;
```

We can also generate the same project using the VHDL description in Listing 4.7. Again, all steps for the Verilog description should be applied to this project as well.

The same project can be implemented on the Arty board as well. To do so, modified Verilog and VHDL descriptions are as in Listings 4.8 and 4.9. As in the Basys3 board–based application, do not forget to add the Arty XDC file to the project.

Listing 4.8 Switches to LEDs Application on the Arty Board in Verilog

```verilog
module application(led,sw);

output [3:0] led;
input [3:0] sw;

assign led=sw;

endmodule
```

Listing 4.9 Switches to LEDs Application on the Arty Board in VHDL

```vhdl
library ieee;
use ieee.std_logic_1164.all;

entity application is
port(sw : in std_logic_vector (3 downto 0);
    led : out std_logic_vector (3 downto 0));
end application;

architecture dataflow_model of application is
begin
led <= sw;
end dataflow_model;
```

4.9 Summary

Vivado is a design platform to synthesize, simulate, and implement HDL descriptions. It can also be used to program a target FPGA. This chapter introduced Vivado such that it can be used in realizing digital systems in the following chapters. To do so, we started from scratch and developed a project using provided Verilog and VHDL descriptions. At this stage, the reader may not know the structure of the description provided. Such a strategy was necessary to coherently explain the working principles of Vivado. We will explain how these descriptions are constructed in detail in Chap. 5. Therefore, we kindly ask the reader to focus on Vivado usage in this chapter. The final stage here was realizing the given description on the Basys3 or Arty board. Afterward, we also introduced methods on IP management in Vivado. We will also analyze these in detail in the following chapters.

4.10 Exercises

4.1 Download the latest version of Vivado HL WebPACK edition to your computer and install it.

4.2 Create an empty project;
   a. add the Verilog description in Listing 4.1 to the project.
   b. synthesize and simulate the project.
   c. observe simulation results.

4.3 Create an empty project;
   a. add the VHDL description in Listing 4.2 to the project.
   b. synthesize and simulate the project.
   c. observe simulation results.

4.4 Create an empty project. Use Basys3 as the target board;
   a. add the Verilog description in Listing 4.5 to the project.
   b. add the constraint file for the Basys3 board to the project.
   c. implement the project and generate bitstream to program the FPGA.
   d. run the project on the FPGA.

4.5 Repeat Exercise 4.4 using the Arty board.

4.6 Repeat Exercise 4.4 using the VHDL description in Listing 4.10.

4.7 Repeat Exercise 4.6 using the Arty board.

Listing 4.10   VHDL Description of the First System with Switches and LEDs as Input and Output

```vhdl
library ieee;
use ieee.std_logic_1164.all;

entity first_system is
port(sw : in std_logic_vector (1 downto 0);
     led : out std_logic_vector (1 downto 0));
end first_system;

architecture dataflow_model of first_system is
begin
led(0) <= (sw(0) and sw(1)) xor (sw(0) or sw(1));
led(1) <= not sw(0);
end dataflow_model;
```

CHAPTER 5

Introduction to Verilog and VHDL

Hardware description languages help us formalizing and representing a digital system at hand. Hence, it can be implemented on a target FPGA platform. Two popular HDLs are Verilog and VHDL. This chapter introduces basics of both HDLs. We will explore these HDLs in detail in representing digital systems in the following chapters. Although we provide Verilog and VHDL in one chapter, we strongly suggest the reader to master one HDL first, then learn the other. Throughout the book, we give precedence to Verilog since it resembles C programming language. Therefore, we start with Verilog fundamentals next. Then, we introduce testbench formation in Verilog. Afterward, we handle VHDL concepts in the same order. We also consider adding an IP block to a project.

5.1 Verilog Fundamentals

Verilog is the first HDL we will be using to describe a digital system. Therefore, we will introduce Verilog fundamentals with basic keywords in this section.

5.1.1 Module Representation

A digital system can be represented as a `module` with the following structure in Verilog:

```
module {module_name}(port_list);
// Port definitions
// Description of the digital system
    statement 1
    statement 2
    statement 3
    ...
endmodule
```

Let's analyze this structure in detail. First, the `module` should have a unique name which should not be the same as any of the predefined Verilog keywords. In the above description, we set the name as `module_name`. Second, the module should have input and output ports assigned to it. We represent these ports as `port_list` in the above description. The port list does not have a specific order. Therefore, input and output ports can be represented in any order within the list. For convenience, we suggest representing output ports first. At this stage, definition of the module is done. Next comes internal structure of the module. Here, we first define port elements within the module. Each element can be `input`, `output`, or `inout`. As the name implies, the

47

input keyword declares that the related port will get data from outside world. The output keyword declares that the related port will feed data to outside world. The inout keyword declares that the related port can be used for both input and output purposes. Then, we describe the digital system. This is indicated by statement 1, statement 2, and statement 3 above. It is important to remember that order of statements is not important in the description since they will be represented by hardware elements in the FPGA. Afterward, we close the module by keyword endmodule. Note that we can use the symbol // to add a comment to the Verilog description.

To understand the module definition, let's consider the first Verilog description in Listing 4.1. As a reminder, circuit diagram of this digital system has been given in Fig. 4.11. As can be seen in this figure, the digital system has two input ports in1 and in2. It also has two output ports out1 and out2. Now, let's focus on the first part of the description in Listing 4.1 given below.

```
module first_system(out1, out2, in1, in2);

// Port definitions
input in1, in2;
output out1, out2;

// Description of the digital system
    statement 1
    statement 2
    statement 3
    ...
endmodule
```

As can be seen here, the module name for this description is first_system. The port list is composed of out1, out2, in1, in2. Ports in1 and in2 are defined as input in the following line. Similarly, ports out1 and out2 are defined as output in the next line.

The following part in Listing 4.1 is the description of digital system. There are three different methods of modeling, such as structural, dataflow, and behavioral, in describing a digital system in Verilog. We will introduce each modeling method next.

5.1.1.1 Structural Modeling

The first method in describing a digital system is using structural modeling. In this method, each element to be used in the description statement should have been defined under Verilog as a structure. Since logic gates are extensively used in Verilog descriptions, they have been defined beforehand. Therefore, this description method is also called gate-level modeling.

Each gate is represented by the following structure in this method. First, gate type is defined by the corresponding Verilog keyword. Then, a name for the gate is assigned. Note that name assignment is not mandatory. Finally, output and input ports for the gate are defined within parenthesis. Therefore, the structural model of a logic gate will be as gate_keyword name (port_list). The port list should be such that output of the structure is defined first.

Let's describe the digital system in Listing 4.1 using structural modeling. The reader can also consult Fig. 4.11 for this purpose. As can be seen in this figure, four gates are used in this system as AND, OR, NOT, and XOR. Corresponding Verilog keywords for

---

Listing 5.1   Structural Model of the First System in Verilog

---

```
module first_system(out1,out2,in1,in2);

// Port definitions
input in1,in2;
output out1,out2;

// Description of the digital system
// Structural modeling

wire and_out,or_out;

and gate_and(and_out,in1,in2);
or  gate_or(or_out,in1,in2);
xor gate_xor(out1,and_out,or_out);
not gate_not(out2,in2);

endmodule
```

---

these are and, or, not, and xor, respectively. Let's give a name to each logic gate to be used in the description as gate_and, gate_or, gate_not, and gate_xor, respectively. Using these, we can construct the structural model. There is one issue to be solved in describing the digital system. Inputs of the XOR gate are output of the AND and OR gates. We should define variables using the Verilog keyword wire to make this connection. In fact, the user can remember this easily as if we are adding a wire between logic gates. Based on these, we can form the structural model of the digital system as in Listing 5.1. As can be seen in this description, the first system is defined using only pre-defined logic elements. To emphasize again, these elements can be defined in any order in Listing 5.1.

5.1.1.2   *Dataflow Modeling*

The second method in describing a digital system in Verilog is using dataflow modeling. In this method, the relation between input and output ports is formed as a function. Therefore, this description method is also called functional modeling.

The main keyword in dataflow modeling is assign. The syntax here is assign output = function of inputs. Output in this representation must always be a scalar or vector. Here, the function may be formed by logic gate representations. As in structural modeling, we will only consider logic gates AND, OR, NOT, and XOR here. Corresponding operators to be used in dataflow modeling are {& , | , ~ , ^} respectively.

In fact, the digital system in Listing 4.1 has been described by dataflow modeling such that we represented each logic gate input and output as a function. Then, we formed dataflow model of the digital system as in Listing 5.2. As in structural modeling, we used the wire keyword in this description to connect input and output of logic gates.

Dataflow modeling allows merging functions, which leads to a more compact representation. Let's reconsider the description in Listing 5.2. We provide the merged form of this description in Listing 5.3. As can be seen here, output out1 is defined in one merged line. Therefore, wire definitions are discarded from the description.

Listing 5.2 Dataflow Model of the First System in Verilog

```verilog
module first_system(out1,out2,in1,in2);

// Port definitions
input in1,in2;
output out1,out2;

// Description of the digital system
// Dataflow modeling

wire and_out,or_out;

assign and_out = in1 & in2;
assign or_out  = in1 | in2;
assign out1    = and_out ^ or_out;
assign out2    = ~ in2;

endmodule
```

Listing 5.3 Dataflow Model of the First System in Merged Form

---

```verilog
module first_system_merged(out1,out2,in1,in2);

// Port definitions
input in1,in2;
output out1,out2;

// Description of the digital system
// Dataflow modeling in merged form

assign out1 = (in1 & in2) ^ (in1 | in2);
assign out2 = ~ in2;

endmodule
```

5.1.1.3 Behavioral Modeling

The third method in describing a digital system in Verilog is using behavioral modeling. In this method, digital system at hand is represented by its behavior. In other words, Verilog keywords corresponding to conditional and recursive statements can be used within the model.

In behavioral modeling, statement (or statements) to be executed should be triggered by a signal (or signals) to operate. The keyword always is used to indicate this triggering operation. Once the signal changes its state, the statement is executed. If there is more than one statement to be executed, then they should be encapsulated by begin and end keywords. Hence, syntax for this representation becomes as follows:

```verilog
always @ (sensitivity_list)
    begin
    // behavioral description
        statement 1
```

```
        statement 2
        statement 3
          ...
     end
```

Here, `sensitivity_list` stands for triggering signal(s). The sensitivity list can be formed of signals separated by comma or combined by `or` keyword. If the behavioral description is to be executed for any input changes, then * sign can be used instead of the sensitivity list. Here, whenever one of the signals in the sensitivity list changes its state, the behavioral description is executed. Again, order of statements is not important in behavioral modeling.

One other important Verilog keyword for behavioral modeling is `initial`. Via this keyword, an initial block can be formed which is executed at time zero. Syntax of the initial block is as follows:

```
initial
    begin
        statements
    end
```

Let's describe the digital system in Listing 4.1 using behavioral modeling. Behavior of the system will change when the first or second input changes. Therefore, at the beginning of the always block, the sensitivity list will consist of inputs `in1` and `in2`. We can represent the relation between input and output of the system as in dataflow modeling. However, the `assign` keyword will not be used in behavioral modeling. Since there is more than one statement to be executed, they are encapsulated within `begin` and `end` keywords. As a result, behavioral model of the first system will be as in Listing 5.4.

Listing 5.4 Behavioral Model of the First System in Verilog

```
module first_system(out1,out2,in1,in2);

// Port definitions
input in1,in2;
output out1,out2;

// Description of the digital system
// Behavioral modeling

reg   out1,out2;

initial
begin
out1 = 0;
out2 = 0;
end

always @ (in1, in2)
begin
out1 = (in1 & in2) ^ (in1 | in2);
out2 = ~ in2;
end

endmodule
```

We should take a closer look at the description in Listing 5.4. The `always` keyword executes the beneath description block (encapsulated by `begin` and `end` keywords) whenever `in1` or `in2` changes. If there is no change in these variables, output will not be provided by the system. Therefore, we have to save previous output values. This can be done by the Verilog keyword `reg`. We used this keyword to keep the previous value of `out1` and `out2` in Listing 5.4. We also initialized these variables to logic level zero using the `initial` keyword.

There are two assignment types in behavioral modeling. These are called blocking and nonblocking. Statements having blocking assignment are executed one by one in sequential order. Therefore, as the name implies, each assignment blocks the execution of the next in hierarchy. Operator for the blocking assignment is =. Statements having nonblocking assignment are executed concurrently. Therefore, they don't block each other. Operator for the nonblocking assignment is <=.

Let's consider a simple example for blocking and nonblocking assignments. Assume that there is a Verilog module with output array y having six elements. Input of the module is represented by x. Within the always block, let's describe assignments as follows:

```
y[0] =x;
y[1] =y[0];
y[2] =y[1];

y[3] <=x;
y[4] <=y[3];
y[5] <=y[4];
```

Here, the first three assignments are of blocking type. Next three assignments are of nonblocking type. When input x becomes logic level one, blocking assignments result as y[0]=1, y[1]=1, and y[2]=1. In other words, input first affects output y[0]. Then, outputs affect each other in sequential order. On the other hand, nonblocking assignments will be as y[3]=1, y[4]=0, and y[5]=0. Hence, input only affects the first output y[3]. Remaining outputs do not change their initial value. This is because of the concurrent operation such that all output values are assigned at once. Hence, the new value of output y[3] could not affect remaining outputs.

We provide the complete Verilog description of the above example in Listing 5.5. Final simulation results for this description will be as in Fig. 5.1. Blocking and nonblocking assignment results are clearly seen in this figure.

It is strongly suggested in literature that blocking assignments should be used in combinational circuits. Nonblocking assignments should be used in sequential circuits. Hence, Verilog descriptions till Chap. 9 will only use blocking assignments in behavioral models. Starting from Chap. 9, nonblocking assignments will be used in behavioral models. There is also a good reference by Cummins [24] on the usage of blocking and nonblocking assignments in Verilog. We strongly suggest the reader to check this reference for in-depth understanding of this concept.

5.1.2 Timing and Delays in Modeling

Vivado allows adding simulation timings in Verilog descriptions. Moreover, if a blank Verilog file is to be opened, Vivado adds the first line automatically as `'timescale 1ns / 1ps`. These are the default timing values such that the first one

Listing 5.5 An Example on Blocking and Nonblocking Assignments

```
module blocking_nonblocking(y,x,clk);

input x,clk;
output reg [5:0] y;

initial y=6'b000000;

always @ (posedge clk)
begin
y[0] = x;
y[1] = y[0];
y[2] = y[1];

y[3] <= x;
y[4] <= y[3];
y[5] <= y[4];
end

endmodule
```

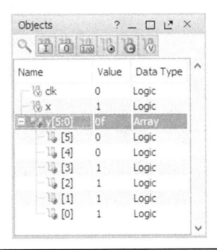

FIGURE 5.1 Simulation results for blocking and nonblocking assignments.

(1ns) indicates the reference time unit. Whenever a time value is added to the Verilog description, it will be in the order of one nanosecond. The second timing value (1ps) indicates the smallest precision that can be achieved. Hence, the default smallest precision in simulation is one picosecond. Again, these values will be of use during simulation. They will have no effect in the actual FPGA realization step.

Up to this point, we did not take physical characteristics of logic gates into account in simulation. In other words, we assumed all delay times to be zero within logic gates. If the user wants to obtain accurate results (especially in timing diagrams) of the implemented digital system, then delay values should be added to the Verilog description. These can be done in connection with the reference time unit.

There are three delay types that can be added to a digital device in Verilog. These are rise delay, fall delay, and turn-off delay. The rise delay indicates the transition time needed from any logic value to logic level one. The fall delay indicates the transition time needed from any logic value to logic level zero. The turn-off delay indicates the transition time needed from any logic value to high impedance. Next, we provide an example on the usage of these delay times in structural modeling.

```
and    #(5)       gate_and(and_out,in1,in1);
or     #(3, 4)    gate_or(or_out,in1,in2);
xor    #(3, 4, 5) gate_xor(out1,and_out,or_out);
```

In the first line, the delay value is specified as #(5). This indicates that all delay values are five time units. If the default reference time is used, this corresponds to 5 ns. In the second line, two delay values are specified as #(3, 4). Here, the rise delay is taken as three time units. The fall delay is taken as four time units. The turn-off delay is taken as the minimum of these two values. Hence, it becomes three time units. In terms of the default reference time, these values become 3 ns, 4 ns, and 3 ns, respectively. In the third line, three delay values are specified as #(3, 4, 5). Here, the rise delay is taken as three time units. The fall delay is taken as four time units. The turn-off delay is taken as five time units. Again, in terms of the reference time, these values will be as 3 ns, 4 ns, and 5 ns, respectively.

We can also apply delay values in dataflow modeling. Such an example is assign #10 and_out = in1 & in2. Here, #10 indicates that the assignment will be performed by a 10-time-unit delay. This will correspond to 10-ns delay with respect to the default reference time.

Let's apply delay to the dataflow model of the first system in Listing 5.3. Delay is applied such that out2 is calculated with a 20 time-unit lag. We provide the modified description in Listing 5.6.

We can simulate the Verilog description in Listing 5.6 using methods in Sec. 4.4. Obtained simulation result will be as in Fig. 5.2. As can be seen in this figure, the second output (out2) has a 20-ns delay.

Listing 5.6 Verilog Description of the First System After Adding a Delay

```
'timescale 1ns / 1ps

module first_system_delay(out1,out2,in1,in2);

// Port definitions
input in1,in2;
output out1,out2;

// Description of the digital system
// Dataflow modeling

assign out1 = in1 & in2 ^ in1 | in2;
assign #20 out2  = ~ in2;

endmodule
```

FIGURE 5.2   Simulation results after adding a delay of 20 ns to the second output.

5.1.3   Hierarchical Module Representation

Projects we have considered up to this point contain only one module. In larger projects, the number of modules may be more than one. In this section, we will show how a project with more than one module can be handled.

Let's reconsider dataflow model of the first system in Listing 5.2. We can represent the same description as a combination of three modules such that AND and OR gates are described in different modules. Let's call these as and_module and or_module, respectively. These should be formed as valid modules with their input/output ports and descriptions. We should instantiate the and_module and or_module in the top module first_system. This can be done as if structural modeling is used. In other words, the and_module should be represented within the first_system module as and_module instantiation_name (port_list).

There are two options in forming port list correspondence between module to be instantiated and the top module using it. The first one is using locations. Here, the port list order in the top module and instantiation should be the same. The second method in forming the port list correspondence is using the declaration .sub_module_name (top_module_name). Here, port in the module to be instantiated is declared as sub_module_name. The corresponding port in the top module is declared as (top_module_name). This operation should be done for all input/output ports. We will use both declarations throughout the book, although the second one should be picked whenever possible.

Based on the first port list declaration, hierarchical representation of the first system will be as in Listing 5.7. Here, instantiation name for the and_module and or_module is U1 and U2, respectively.

Schematic view of the modular design (under the RTL analysis option) in Listing 5.7 will be as in Fig. 5.3a. As can be seen in this figure, the and_module and or_module are represented as black boxes. As the "+" sign is pressed on these boxes, the RTL representation will be as in Fig. 5.3b. In this figure, black boxes are represented by the actual description of each module. Therefore, it becomes easy to analyze the overall description.

Vivado allows hierarchical module representation to be composed of more than one source file. Therefore, larger projects can be composed of smaller source files merged in Vivado. We can show how this method works as follows. Let's reconsider modular description of the first system in Listing 5.7. This file can be partitioned into two parts such that the first one holds the top module (first_system). The second one holds and_module and or_module. We can represent these two source files as in Listings 5.8 and 5.9. These two should be added to the project as source files. Then, Vivado merges them and forms the final description.

Listing 5.7 Verilog Description of the First System in Hierarchical Module Representation

```verilog
module and_module(and_out,in1,in2);

input in1,in2;
output and_out;

assign and_out = in1 & in2;
endmodule

module or_module(or_out,in1,in2);

input in1,in2;
output or_out;

assign or_out = in1 | in2;
endmodule


module first_system(out1,out2,in1,in2);

input in1,in2;
output out1,out2;

wire and_out,or_out;

and_module U1(and_out,in1,in2);
or_module  U2(or_out,in1,in2);

assign out1 = and_out ^ or_out;
assign out2 = ~ in2;

endmodule
```

5.2 Testbench Formation in Verilog

Characteristics of a digital system can be analyzed in Vivado by using a testbench. Here, we will explain the structure of a testbench file, taking the one in Listing 4.3 as an example. Note that we provide the testbench file for each Verilog description considered in this book on a companion website, www.mhprofessional.com/1259837904. Therefore, we strongly suggest that the reader visit this website. Finally, more information on Verilog testbench formation can be found in [25].

5.2.1 Structure of a Verilog Testbench File

A Verilog testbench file is composed of five parts as follows:

- Testbench module declaration
- Input/output port declaration
- Instantiation of the unit under test (UUT)
- Providing input to the UUT
- Displaying test results

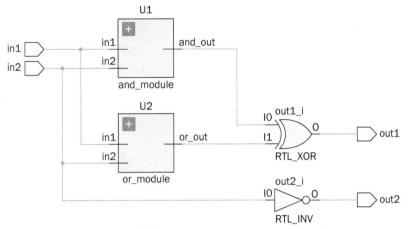

(A) Each module represented as a black box

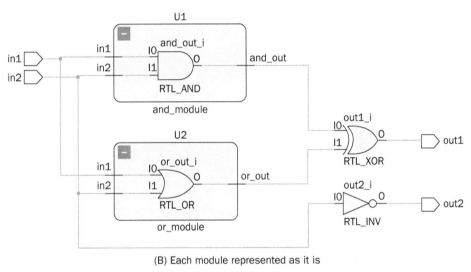

(B) Each module represented as it is

Figure 5.3   RTL schematic view of the first system in hierarchical module representation.

Let's explain these parts taking the testbench file in Listing 4.3 as an example.

The testbench is itself a Verilog module. Therefore, it needs valid module and input/ output port declarations. This is the first step in testbench formation. These declarations are done as follows in Listing 4.3.

```
'timescale 1ns / 1ps

module first_system_tb;

// Inputs
reg in1t, in2t;

// Outputs
wire out1t, out2t;
```

Listing 5.8 Verilog Description of the First System—the Top Module

```verilog
module first_system(out1,out2,in1,in2);

input in1,in2;
output out1,out2;

wire and_out,or_out;

and_module U1(and_out,in1,in2);
or_module U2(or_out,in1,in2);

assign out1 = and_out ^ or_out;
assign out2 = ~ in2;

endmodule
```

Listing 5.9 Verilog Description of the First System—the Supplement File

```verilog
module and_module(and_out,in1,in2);

input in1,in2;
output and_out;

assign and_out = in1 & in2;
endmodule

module or_module(or_out,in1,in2);

input in1,in2;
output or_out;

assign or_out = in1 | in2;
endmodule
```

Here, first the simulation timing value is declared by the `timescale` keyword. Then, the testbench module is declared as `module first_system_tb`. We specifically assigned such a name to the testbench module to associate it with the top module to be tested. The reader is free to choose any valid name here. Next, input and output ports of the testbench module are declared as `reg in1t`, `in2t` and `wire out1t`, `out2t`. Again, the reader can pick any valid name for each input or output port in the testbench module.

The second step in testbench formation is associating the module to be tested (unit under test) with the testbench module. This is done by instantiation. The related part in Listing 4.3 is as follows:

```verilog
// Instantiate the Unit Under Test (UUT)
first_system UUT (.out1(out1t),.out2(out2t),.in1(in1t),.in2(in2t));
```

Here, as in hierarchical module declaration, the module to be tested (for our case `first_system`) is instantiated in the testbench module with the name UUT. Then, each port in the testbench module and the module to be tested are associated (or connected) such as `.out1(out1t)`. Here, the port in the module to be tested is declared as `.out1`.

The corresponding port in the testbench module is declared as (out1t). This operation is done for all input/output ports.

The third step in testbench formation is providing input to the UUT. The related part in Listing 4.3 is as follows:

```
//Providing input to the UUT
initial begin
// Initialize Inputs
in1t = 0;
in2t = 0;

// Wait 100 ns for global reset to finish
#100;

// Add stimulus here
repeat (4)
#100 {in1t,in2t} = {in1t,in2t} + 1'b1;
end
```

Here, testbench input ports (in1t and in2t) are initialized first. Then, a delay of 100 ns is added by the command #100. This delay is added such that the module to be tested is reset properly. Otherwise, some undesired effects may occur during simulation. Next, input values are fed to the UUT. In Listing 4.3, this is done in two lines as follows. The first line contains the command repeat (4). This indicates that the following line will be repeated four times. The second line contains the command #100 {in1t,in2t} = {in1t,in2t} + 1'b1. This indicates that inputs will be incremented one by one sweeping the pattern 00, 01, 10, and 11. Transition between each input combination is done after a 100-ns delay. We will explain Verilog data formats in these lines in Chap. 6.

We can import input test signals from an existing text file. The testbench file in Listing 4.3 should be modified as in Listing 5.10 for this purpose. Here, a text file is opened by the attribute initial $readmemb. File entries are saved in ROM. Afterward, file entries are read and processed line by line from ROM. We will provide more information on this issue in Sec. 9.5.

5.2.2 Displaying Test Results

The testbench module is constructed following steps in previous section. The reader has two options to observe simulation results in Vivado. The first one is through input/output waveforms as explained in Sec. 4.4. This is a valid option and can be used in most tests.

The second option in observing output of the test is adding specific display commands such that output can be observed through Vivado's Tcl console. The related optional part in Listing 4.3 is as follows:

```
//Display the result on the Tcl console (Optional)
initial  begin
$display("   in1 in2 out1 out2");
$monitor("\t%b \t%b \t%b \t%b",in1t,in2t,out1t,out2t);
end
```

Here, the display function prints the string fed to it. The monitor function prints variables fed to it. The first part of this function handles formatting. Hence, \t%b stands for

---

Listing 5.10   The Testbench File Reading Input Signals from a Text File

---

```verilog
'timescale 1ns / 1ps

module first_system_file_read_tb;

// Inputs
reg in1t, in2t;

// Outputs
wire out1t, out2t;

// Instantiate the Unit Under Test (UUT)
first_system UUT (.out1(out1t),.out2(out2t),.in1(in1t),.in2(in2t));

reg [1:0] Testset [3:0];

integer count;

//load Testset content from file
initial $readmemb("H:/Xilinx_projects/first_project/Testset_entries_bin
    .txt", Testset);

//Providing input to the UUT
initial begin
#100;

count=0;
repeat (4)
begin
#50 {in1t,in2t}=Testset[count];
#50 count=count+1;
end

end
//Display the result on the Tcl console (Optional)
initial  begin
$display("   in1 in2 out1 out2");
$monitor("\t%b \t%b \t%b \t%b",in1t,in2t,out1t,out2t);
end

endmodule
```

"add tab and represent the value in binary form." The corresponding variable to be displayed is provided in the second part of the function as in1t. All input and output ports are tabulated this way. Therefore, whenever a change in input occurs, it is displayed on Vivado's Tcl console. The console output will be as in Fig. 5.4 for the testbench in Listing 4.3.

5.3   VHDL Fundamentals

VHDL is the second HDL we will be using to describe a digital system. Therefore, we will introduce VHDL fundamentals in this section. As in Verilog, we will introduce the

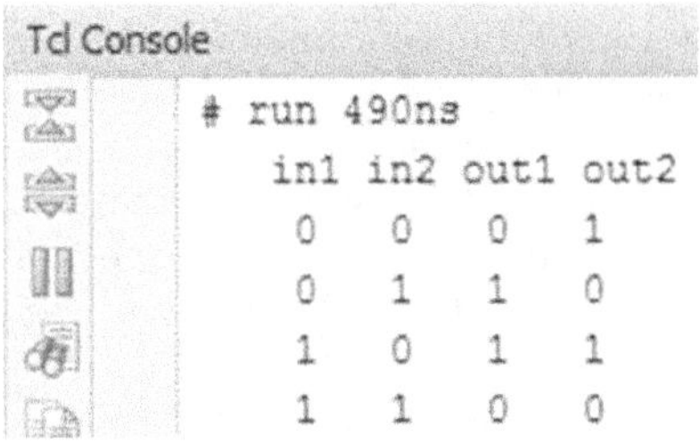

```
Tcl Console
    # run 490ns
        in1 in2 out1 out2
         0   0   0    1
         0   1   1    0
         1   0   1    1
         1   1   0    0
```

FIGURE 5.4 Simulation results observed in Vivado's Tcl console.

remaining VHDL keywords in connection with related digital design concepts in the following chapters.

5.3.1 Entity and Architecture Representations

A digital system should be declared in two parts in VHDL. The first part includes the entity declaration which defines input and output characteristics of the system to be implemented. The structure of the entity part will be as follows:

```
library library_name;
use library_elements;

entity system_name is
port (port_name : port_mode port_type;
      port_name : port_mode port_type;
      port_name : port_mode port_type)
end system_name;
```

Here, system_name is the name assigned to the system to be described. The keyword port defines actual ports of the device. Each port entry will have a unique name indicated by port_name. A port_mode can be in, out, or inout. As the name implies, the in keyword declares that the related port will get data from outside world. The out keyword declares that the related port will feed data to outside world. The inout keyword declares that the related port can be used for both input and output purposes. VHDL requires variable and port types to be used in entity declaration to be strongly defined. Therefore, port_type should be declared within library_elements included to the design by library and use keywords.

Second part of digital system declaration defines its architecture. This is done as follows:

```
architecture architecture_name of system_name is

--declarations part for
--variable, signal, constant, component

begin
-- Description of the digital system
    statement 1
    statement 2
    statement 3

    ...
end architecture_name;
```

Here, the user should give a specific name to architecture of the digital system as `architecture_name`. The `system_name` defined in the `entity` part should also be used in architecture definition. Then, `variable`, `signal`, `constant`, and `component` declarations should be made. The first three of these are related to data definitions and assignments within the design. These should have valid types defined in the included `library_elements` in the `entity` declaration. The `component` declaration allows hierarchical structural representation definition to be considered in detail in Sec. 5.3.5. Finally, system description is done within the architecture part. This is indicated by `statement 1`, `statement 2`, and `statement 3` above. It is important to remember that the order of statements is not important in the description since they will be represented by hardware elements in an FPGA. Note that we can use the symbol `--` to add comment to the VHDL description.

Next, we will consider the entity and architecture parts on an example. Therefore, let's revisit the VHDL description in Listing 4.2. The entity part of this declaration is as follows:

```vhdl
library ieee;
use ieee.std_logic_1164.all;

entity first_system is
port(in1 : in std_logic;
     in2 : in std_logic;
    out1 : out std_logic;
    out2 : out std_logic);
end first_system;
```

Here, the system has two input ports as `in1` and `in2`. It also has two outputs as `out1` and `out2`. We deliberately set the names of input and output ports as in the Verilog description in Listing 4.1. Hence, the reader can form a correspondence between them easily. In Listing 4.2, the library used in operation is picked as `ieee`. Within this library, all types defined under the `ieee.std_logic_1164` subset are imported. This allows using the `std_logic` type which can represent binary values such as logic level zero and one. We will evaluate this operation in detail in Chap. 6. As for now, please accept the provided representation as it is.

The only remaining part in the above VHDL description is representation of the digital system. In this book, we will only consider dataflow and behavioral models in VHDL. Note that some definitions in these models may overlap. We will introduce each modeling method next.

5.3.2 Dataflow Modeling

The first method to be considered in describing a digital system in VHDL is using dataflow modeling. In this method, the relation between input and output ports is formed by a function as in Verilog. Syntax in this function representation is `output <= function of inputs`.

The digital system described in Listing 4.2 has been formed in terms of dataflow modeling. There, we named the architecture as `dataflow_model`. The system name defined in the entity declaration has also been used in the architecture definition as `first_system`. We used logic gates AND, OR, NOT, and XOR within the architecture declaration. Corresponding VHDL keywords for these are `and`, `or`, `not`, `xor`, respectively. For completeness, let's provide the dataflow model of this system in Listing 5.11.

Listing 5.11 Dataflow Model of the First System in VHDL

```vhdl
library ieee;
use ieee.std_logic_1164.all;

entity first_system is
port(in1 : in std_logic;
     in2 : in std_logic;
     out1 : out std_logic;
     out2 : out std_logic);
end first_system;

architecture dataflow_model of first_system is
begin
out1 <= (in1 and in2) xor (in1 or in2);
out2 <= not in1;
end dataflow_model;
```

5.3.3 Behavioral Modeling

The second method in describing a digital system in VHDL is using behavioral modeling. As in Verilog, the digital system at hand is represented by its behavior in this method. In other words, VHDL keywords corresponding to conditional and recursive statements can be used within this model.

In behavioral modeling, statement(s) to be executed should be triggered by a signal (or signals) to operate. The keyword process is used to indicate this triggering operation. Once the signal changes its state, the statement(s) is executed. Syntax for this representation becomes as follows:

```vhdl
process (sensitivity_list)
-- process declarations
begin
-- system description
statement 1
statement 2
statement 3
    . . .
end process;
```

Here, the sensitivity_list stands for the triggering signal(s). In VHDL, the sensitivity list can be formed of signals separated by comma only. Whenever one of the signals in the sensitivity list changes its state, the behavioral description is executed. The process may have its own declarations which can be placed before the begin keyword. Then, the behavioral model is encapsulated by begin and end process keywords. To note again, the order of statements is not important in behavioral modeling.

Let's describe the digital system in Listing 4.2 using behavioral modeling. Behavior of the system will change when the first or second input changes its value. Therefore, the sensitivity list for the process will consist of inputs in1 and in2. We can represent the relation between inputs and outputs of the system similar to dataflow modeling. As a result, the behavioral model of the first system will be as in Listing 5.12. Here, architecture of the digital system is named as behavioral_model.

Dataflow and behavioral models share similar structures in VHDL. Their main difference is the process keyword used in dataflow modeling. Therefore, we will provide

---

Listing 5.12 Behavioral Model of the First System in VHDL

```vhdl
library ieee;
use ieee.std_logic_1164.all;

entity first_system is
port(in1 : in std_logic;
     in2 : in std_logic;
     out1 : out std_logic;
     out2 : out std_logic);
end first_system;

architecture behavioral_model of first_system is
begin
process (in1,in2)
begin
out1 <= (in1 and in2) xor (in1 or in2);
out2 <= not in1;
end process;
end behavioral_model;
```

---

either the dataflow or behavioral model from this point based on its appropriateness in describing the digital system at hand.

5.3.4 Timing and Delays in Modeling

As in Verilog, we can add delay times to descriptions in VHDL. This leads to precise simulation results especially in timing diagrams. Again, these values will be of use during simulation. They will have no effect in the actual FPGA realization step.

Different from Verilog, delay times can be added to a VHDL description using the keyword `after`. Let's assume that we want to add a 20-ns delay to the second output (out2) in Listing 4.2. The modified description line will be `out2 <= not in1 after 20 ns`. As this modification is done and simulation of the description is run, the same waveform in Fig. 5.2 should be observed.

5.3.5 Hierarchical Structural Representation

VHDL allows structural hierarchical representation for large projects. As in Verilog, the idea here is decomposing the project into subparts. Hence, it becomes manageable.

Let's reconsider dataflow model of the first system in Listing 5.11. As in Sec. 5.1.6, we can represent this description as a combination of three parts such that the AND and OR gates are represented separately. Let's call these as and_module and or_module, respectively. These should be formed with their valid entity and architecture descriptions. Then, we should instantiate the and_module and or_module in the top (main) module first_system. This can be done by using component declarations.

The component declaration should be made in architecture part of the top module with the following structure:

```vhdl
component component_name
 port (port_list);
end component;
```

This definition should be made at the beginning of the architecture part. Then, instantiation can be done by using the below structure:

```
instantiation_name: component_name port map(port_list correspondence);
```

There are two options in forming the port list correspondence. The first one is using locations. Here, the port list order in the main entity declaration and instantiation should be the same. Although this is a valid option, it may cause problems in implementation if the port order is not followed correctly. The second method in forming the port list correspondence is using the declaration `component_port_name => main_port_name`. Here, a correspondence is formed between each port in the component and main entity declarations. Based on these, hierarchical structural representation of the first system will be as in Listing 5.13. Here, the instantiation name for the `and_module` and `or_module` are `U1` and `U2`, respectively. Schematic view of this hierarchical representation (under the RTL analysis option) will be as in Fig. 5.3. Properties of this figure explained beforehand are also valid here.

Vivado allows hierarchical structural representation to be composed of more than one source file. Therefore, larger projects can be composed of smaller source files merged in Vivado. We can show how this method works as follows. Let's reconsider modular description of the first system in Listing 5.13. This file can be partitioned into two parts such that the first one holds the top module (`first_system`); the second one holds the `and_module` and `or_module`. We can represent these two files as in Listings 5.14 and 5.15. These two should be added to the project as source files. Then, Vivado merges them and forms the final description.

The supplement file in Listing 5.15 can be represented as a library in VHDL. This can be done by the keyword `package`. Afterward, the library can be called in the main file by the `library` and `use` keywords. For more detail on this issue, please see [26].

There are two more methods which can be used in hierarchical structural representation. These are `function` and `procedure` methods. For more information on these methods, please see [27].

5.4 Testbench Formation in VHDL

A VHDL description can be analyzed via its testbench in Vivado. Therefore, we will explore the structure of a testbench file, taking the one in Listing 4.4 as an example. We provide the testbench file for each VHDL description (as in Verilog) considered in this book on the companion website www.mhprofessional.com/1259837904. Therefore, we strongly suggest that the reader visit it. Finally, more information on VHDL testbench formation can be found in [25].

5.4.1 Structure of a VHDL Testbench File

A VHDL testbench file is composed of five parts as follows:

- Testbench entity and architecture declarations
- Input/output port declaration
- Instantiation of the unit under test (UUT)
- Providing input to the UUT
- Displaying test results

Listing 5.13 VHDL Description of the First System in Hierarchical Structural Representation

```vhdl
library ieee;
use ieee.std_logic_1164.all;

entity and_module is
port(ain1 : in std_logic;
     ain2 : in std_logic;
  and_out : out std_logic);
end and_module;

architecture dataflow_model of and_module is
begin
and_out <= (ain1 and ain2);
end dataflow_model;

library ieee;
use ieee.std_logic_1164.all;

entity or_module is
port(oin1 : in std_logic;
     oin2 : in std_logic;
   or_out : out std_logic);
end or_module;

architecture dataflow_model of or_module is
begin
or_out <= (oin1 or oin2);
end dataflow_model;

library ieee;
use ieee.std_logic_1164.all;

entity first_system is
port(in1 : in std_logic;
     in2 : in std_logic;
    out1 : out std_logic;
    out2 : out std_logic);
end first_system;

architecture dataflow_model of first_system is
signal and_out, or_out : std_logic;

component and_module
port(ain1 : in std_logic;
     ain2 : in std_logic;
  and_out : out std_logic);
end component;

component or_module
port(oin1 : in std_logic;
     oin2 : in std_logic;
   or_out : out std_logic);
end component;

begin

U1: and_module port map(ain1 =>in1,ain2 =>in2,and_out =>and_out);
U2:  or_module port map(oin1 =>in1,oin2 =>in2,or_out =>or_out);

out1 <= and_out xor or_out;
out2 <= not in1;
end dataflow_model;
```

Listing 5.14   VHDL Description of the First System—the Top Module

```vhdl
library ieee;
use ieee.std_logic_1164.all;

entity first_system is
port(in1 : in std_logic;
     in2 : in std_logic;
     out1 : out std_logic;
     out2 : out std_logic);
end first_system;

architecture dataflow_model of first_system is
signal and_out, or_out : std_logic;

component and_module
port(ain1 : in std_logic;
     ain2 : in std_logic;
   and_out : out std_logic);
end component;

component or_module
port(oin1 : in std_logic;
     oin2 : in std_logic;
   or_out : out std_logic);
end component;

begin

U1: and_module port map(ain1 =>in1,ain2 =>in2,and_out =>and_out);
U2:  or_module port map(oin1 =>in1,oin2 =>in2,or_out =>or_out);

out1 <= and_out xor or_out;
out2 <= not in1;
end dataflow_model;
```

These parts are almost the same as in Sec. 5.2. Let's explain them taking the testbench file in Listing 4.4 as an example.

The testbench is itself a VHDL description. Therefore, it needs valid entity and architecture declarations. This is the first step in testbench formation. These declarations are done as follows in Listing 4.4:

```vhdl
library ieee;
use ieee.std_logic_1164.all;

entity first_system_tb is
end first_system_tb;

architecture dataflow of first_system_tb is
signal  in1t : std_logic := '0';
signal  in2t : std_logic := '0';
signal out1t : std_logic := '0';
signal out2t : std_logic := '0';
```

Listing 5.15 VHDL Description of the First System—the Supplement File

```vhdl
library ieee;
use ieee.std_logic_1164.all;

entity and_module is
port(ain1 : in std_logic;
     ain2 : in std_logic;
  and_out : out std_logic);
end and_module;

architecture dataflow_model of and_module is
begin
and_out <= (ain1 and ain2);
end dataflow_model;

library ieee;
use ieee.std_logic_1164.all;

entity or_module is
port(oin1 : in std_logic;
     oin2 : in std_logic;
  or_out : out std_logic);
end or_module;

architecture dataflow_model of or_module is
begin
or_out <= (oin1 or oin2);
end dataflow_model;
```

Here, the testbench is declared as `first_system_tb`. We specifically assigned such a name to associate it with the architecture to be tested. The reader is free to choose any valid VHDL name here. Entity declaration of the testbench is empty since it will not get any input or feed output. Signals to be used within the testbench file are declared next. These are `in1t`, `in2t`, `out1t`, and `out2t`. Note that these signals are initialized while being declared. More information on this operation can be found in Chap. 6.

The second step in testbench formation is associating the description to be tested (unit under test) with the testbench module. This is done by instantiation. The related part in Listing 4.4 is as follows:

```vhdl
component first_system
port (in1 : in std_logic;
      in2 : in std_logic;
     out1 : out std_logic;
     out2 : out std_logic);
end component;

begin
UUT: first_system port map (in1 => in1t, in2 => in2t, out1 => out1t,
    out2 => out2t);
```

Here, as in hierarchical structural representation, unit to be tested (for our case `first_system`) is instantiated in testbench with the name UUT. Then, each port in the testbench and the unit to be tested are associated (or connected) such as `in1 => in1t`. Here, the port in unit to be tested is declared as `in1`. The corresponding port in the testbench is declared as `in1t`. This is done for all input/output ports.

The third step in testbench formation is providing input to the UUT. The related part in Listing 4.4 is as follows:

```
process
begin
wait for 100 ns;
in1t <= '0'; in2t <= '0';

wait for 100 ns;
in1t <= '0'; in2t <= '1';

wait for 100 ns;
in1t <= '1'; in2t <= '0';

wait for 100 ns;
in1t <= '1'; in2t <= '1';
wait;
end process;

end dataflow;
```

Here, testbench input ports (`in1t` and `in2t`) are set to zero first. Then, a delay of 100 ns is applied by the command line `wait for 100 ns`. This delay is added such that the description to be tested is reset properly. Otherwise, some undesired effects may occur during simulation. Afterward, different input combinations are fed to UUT. Transition between each input combination is done after a 100-ns delay. We will explain VHDL data formats in these lines in detail in Chap. 6.

VHDL allows receiving input signals from an existing text file. The testbench file in Listing 4.4 should be modified as in Listing 5.16 for this purpose. Here, a text file is opened by `file file_input: text open read_mode is`. Afterward, file entries are read line by line.

5.4.2   Displaying Test Results

The testbench in VHDL is constructed using steps in the previous section. The reader can observe simulation results through input/output waveforms as explained in Sec. 4.4. Waveforms for the testbench in Listing 4.4 will be as in Fig. 4.13.

Similar to Verilog, VHDL provides an explicit method to display results on Vivado's Tcl console. The related optional part in Listing 4.4 will be as follows:

```
//Display the result on the Tcl console (Optional)
process
begin
wait for 100 ns;
in1t <= '0'; in2t <= '0';
report "Outputs are "& std_logic'image(out1t) & std_logic'image(out2t);
wait for 100 ns;
in1t <= '0'; in2t <= '1';
report "Outputs are "& std_logic'image(out1t) & std_logic'image(out2t);
wait for 100 ns;
```

```
in1t <= '1'; in2t <= '0';
report "Outputs are "& std_logic'image(out1t) & std_logic'image(out2t);
wait for 100 ns;
in1t <= '1'; in2t <= '1';
report "Outputs are "& std_logic'image(out1t) & std_logic'image(out2t);
wait;
end process;
```

Here, the `report` attribute prints the string fed to it. The `std_logic'image` function prints the variable (in standard logic form) fed to it.

VHDL also allows writing simulation results to a text file. The testbench file in Listing 4.4 should be modified as in Listing 5.17 for this purpose. Operations here are similar to the ones in reading input data from a text file.

5.5   Adding an Existing IP to the Project

We can add an existing IP block to the project. The beauty of using IP blocks is that the HDL used for generating the IP is not important. In other words, we can use an IP generated by VHDL in a Verilog project or vice versa. Therefore, this option allows us merging Verilog and VHDL descriptions in the same project. Let's analyze how this can be done next.

5.5.1   Adding an Existing IP in Verilog

Let's start with the custom-generated IP block in Sec. 4.7. There, we have generated the IP block for the first system in Verilog. Now, let's add this IP to a new project. The first step here is adding the previously generated custom IP to IP catalog of the current project. To do so, we should first locate the custom IP files. Then, we should select the Interfaces tab in the IP Catalog. We should press the IP settings button (the last one) there. In the opened window, we should select the Repository_Manager in the IP tab. Here, we should add the IP repository by pressing the green + sign. Here, we should use location of the custom IP to be added. Then, the window should look like as in Fig. 5.5.

After adding the custom IP to the repository, it will be available in the IP catalog as in Fig. 4.23. To add it to the project, we should double click on it. A new window appears as in Fig. 5.6. Here, the first system is actually shown as a black box with input and output ports. As the OK button is pressed in Fig. 5.6, a new window appears summarizing which files will be generated. Here, we should select the "Synthesis Option" as "out of context per IP." As we press the Generate button in this window, the IP block will be added to the project.

We can observe the included files to the project from the sources → IP sources section as in Fig. 5.7. Here, there are two files of interest under the Instantiation Template section. These are `first_system_0.vho` and `first_system_0.veo`. These are instantiation blocks to be used in the top module. The first file is for use in a Verilog description. The second file is for use in a VHDL description.

The important step here is adding the IP to the top module of the project by instantiating it. Assume that we have generated a top module in Verilog and added it to the project. Then, we can add the instantiation template to the top module as in Listing 5.18. The RTL schematic of this description will be as in Fig. 5.8. As can be seen in this figure, the IP block is represented by a black box in the RTL schematic.

Listing 5.16 The Testbench File Reading Input Signals from a Text File

```vhdl
library ieee;
use ieee.std_logic_1164.all;
use ieee.std_logic_arith.all;
use ieee.std_logic_unsigned.all;
use ieee.std_logic_textio.all;
use std.textio.all;

entity first_system_file_read_tb is
end first_system_file_read_tb;

architecture dataflow of first_system_file_read_tb is

signal  in1t : std_logic := '0';
signal  in2t : std_logic := '0';
signal outlt : std_logic := '0';
signal out2t : std_logic := '0';


component first_system
port(in1 : in std_logic;
     in2 : in std_logic;
    out1 : out std_logic;
    out2 : out std_logic);
end component;

begin
UUT: first_system port map (in1 => in1t, in2 => in2t, out1 => out1t,
    out2 => out2t);

file_read : process

variable rdline : line;
variable r_data : std_logic_vector(0 to 1);
file file_input : text open read_mode is "H:\Xilinx_Projects\
    first_project\SimInputFile.txt";

begin

while not endfile(file_input) loop
    readline(file_input, rdline);
    read(rdline, r_data);
    in1t <= r_data(0);
    in2t <= r_data(1);
    wait for 5 ns;
end loop;
    wait;
end process;

end dataflow;
```

Listing 5.17 The Testbench File Reading Input Signals from a Text File and Writing Simulation Results to Another Text File

---

```vhdl
library ieee;
use ieee.std_logic_1164.all;
use ieee.std_logic_arith.all;
use ieee.std_logic_unsigned.all;
use ieee.std_logic_textio.all;
use std.textio.all;

entity first_system_file_read_write_tb is
end first_system_file_read_write_tb;

architecture dataflow of first_system_file_read_write_tb is

signal  in1t : std_logic := '0';
signal  in2t : std_logic := '0';
signal out1t : std_logic := '0';
signal out2t : std_logic := '0';
signal   clk : std_logic;

component first_system
port(in1 : in std_logic;
     in2 : in std_logic;
    out1 : out std_logic;
    out2 : out std_logic);
end component;

begin
UUT: first_system port map (in1 => in1t, in2 => in2t, out1 => out1t,
    out2 => out2t);

file_read : process

variable rdline : line;
variable r_data : std_logic_vector(0 to 1);
file file_input : text open read_mode is "H:\Xilinx_Projects\
    first_project\SimInputFile.txt";

begin

while not endfile(file_input) loop
    readline(file_input, rdline);
    read(rdline, r_data);
    in1t <= r_data(0);
    in2t <= r_data(1);
    wait for 5 ns;
end loop;
    wait;
end process;

file_write : process(clk)
variable cnt : integer := 0;
variable wrline : line;
```

```vhdl
variable w_data : std_logic_vector(0 to 1) := "00";
file file_output: text open write_mode is "H:\Xilinx_Projects\
    first_project\SimOutputFile.txt";

begin
    if (rising_edge(clk)) then
        w_data(0) := out1t;
        w_data(1) := out2t;
        write(wrline,w_data);
        writeline(file_output,wrline);
    end if;
end process;

clock: process
begin
    clk <= '0';
    wait for 1 ns;
    clk <= '1';
    wait for 1 ns;
end process;

end dataflow;
```

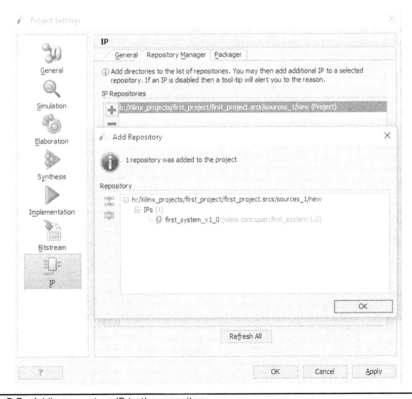

FIGURE 5.5 Adding a custom IP to the repository.

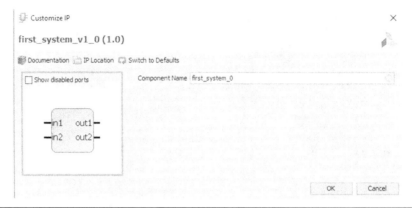

FIGURE 5.6 IP block representation of the first system.

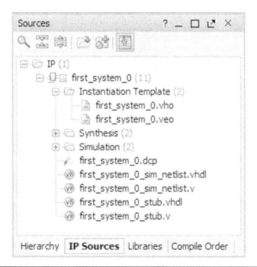

FIGURE 5.7 IP block representation in the IP sources section.

Listing 5.18 Adding the IP Block of the First System to the Top Module in Verilog

```
module top_module_IP(y1,y2,x1,x2);

// Port definitions
input x1,x2;
output y1,y2;

// Generated IP block

first_system_0 FS(.out1(y1),.out2(y2),.in1(x1),.in2(x2));

endmodule
```

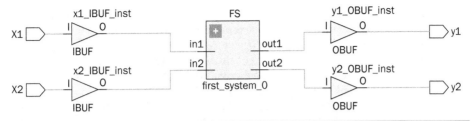

FIGURE 5.8   RTL schematic of the top module after adding the IP block.

Listing 5.19   Adding the IP Block of the First System to the Top Module in VHDL

```vhdl
library ieee;
use ieee.std_logic_1164.all;

entity top_module_IP is
port(x1 : in std_logic;
     x2 : in std_logic;
     y1 : out std_logic;
     y2 : out std_logic);

end top_module_IP;

architecture dataflow_model of top_module_IP is

component first_system_0
port(out1 : out std_logic;
     out2 : out std_logic;
      in1 : in std_logic;
      in2 : in std_logic);
end component;

begin

FS : first_system_0 port map (out1 => y1, out2 => y2, in1 => x1, in2 =>
     x2);

end dataflow_model;
```

5.5.2   Adding an Existing IP in VHDL

Next, we add the IP block of the first system to a VHDL description. We will follow the same steps as in the previous section. The new top module in VHDL will be as in Listing 5.19. As can be seen here, the IP block generated in Verilog can be directly used in the VHDL description.

5.6   Summary

Verilog and VHDL are the HDLs to be used throughout the book. We explored the fundamental properties of both HDLs through examples in this chapter. Basically, we explored the module representation in Verilog. Then, we introduced three modeling methods related to it. Afterward, we considered the effect of timing and delays in modeling. We also considered hierarchical module representation in Verilog. We finally analyzed how a testbench can be formed in Verilog. We followed the same methodology in exploring VHDL fundamentals next. We also considered adding an IP block to a Verilog and VHDL project. Here, we benefit from the generated IP block for the first system in Sec. 4.7. In all these sections, we benefit from examples introduced in Chap. 4. In the following chapters, we will expand our knowledge on Verilog and VHDL with digital system properties to be introduced. However, using fundamentals introduced in this chapter is a must to implement them. Therefore, topics in this chapter can be taken as basis for the following chapters.

5.7 Exercises

5.1 Check whether the structural, dataflow, and behavioral Verilog modeling of the first system evaluated in Sec. 5.1 require similar (or same) FPGA building blocks in Vivado.

5.2 Repeat Exercise 5.1 when VHDL is used in describing the first system evaluated in Sec. 5.3.

5.3 Does hierarchical module representations in Secs. 5.1 and 5.3 add any extra FPGA building blocks in implementation? Check in Vivado.

5.4 Does adding the first system as an IP block add any extra FPGA building blocks in implementation? Check in Vivado.

CHAPTER **6**

Data Types and Operators

This chapter is on basic data types and operators in digital systems. We will explore these concepts in two parts. In the first part of the chapter, we will handle data types and operators from a generic point of view without using any HDL description. Therefore, we will first consider binary, octal, and hexadecimal number representations. Then, we will explore methods to represent a negative number in a digital system. We will next introduce methods to represent a binary number with fractional parts. Here, we will use fixed- and floating-point representations. We will also consider the ASCII code to represent characters in a digital system. Then, we will evaluate arithmetic operations on binary numbers. In the second part of the chapter, we will explore data types and operators defined in Verilog and VHDL. Therefore, we will review most of the concepts introduced in the first part of the chapter using HDLs. Moreover, we will also refer to data types used in previous chapters. Finally, we will analyze how all these concepts can be realized in an FPGA.

6.1 Number Representations

We use the decimal number system in our daily life. This representation provides weights (powers of 10 here) of a digit with respect to its location. Here, the least significant integer digit gets weight 10^0, the next one gets 10^1, and so on. Using this form, we can represent an entity in a systematic way. Therefore, a decimal number 255 in fact means $2 \times 10^2 + 5 \times 10^1 + 5 \times 10^0$. A decimal number with fractional part can also be represented in a similar way. Now, weight of the digits in fractional part become 10^{-1}, 10^{-2}, and so on starting from the dot (separating integer and fractional parts) from left to right. As an example, the decimal number 1.25 corresponds to $1 \times 10^0 + 2 \times 10^{-1} + 5 \times 10^{-2}$.

6.1.1 Binary Numbers

A digit in binary number system (called bit) can take two values as 0 or 1. This perfectly matches with the digital system having two voltage levels as explained in Chap. 2. Therefore, binary numbers are used in digital systems instead of decimal representation.

Binary number representation has weights in powers of two as $2^0, 2^1, 2^2, \cdots, 2^N$. For the fractional part, weights become $2^{-1}, 2^{-2}, 2^{-3}$, and so on starting from the dot separating integer and fractional parts. In a binary number, bits with the highest and

77

lowest weight are specifically called the most significant bit (MSB) and least significant bit (LSB), respectively. Binary digits are grouped slightly different than decimal numbers. Eight bits correspond to one byte; 1024 bytes to one kilobyte (kB); 1024 kilobytes to one megabyte (MB); and 1024 megabytes to one gigabyte (GB).

6.1.1.1 Decimal to Binary Conversion

Integer part of a decimal number can be converted to binary form by iteratively dividing it by two. Iteration ends either when the dividend becomes less than two or number of iterations reach a predefined limit. Let's give a simple example on this operation. If we want to convert the decimal number 14 to binary, we divide it by two iteratively till we reach the remainder 0 or 1. This operation is tabulated in Table 6.1. As can be seen in this table, the division operation reaches remainder 1 after three iterations. We can construct the binary number by forming an array starting from this remainder and going backwards from the last division to the first in the list. Therefore, binary representation of the decimal number 14 will be 1110.

Fractional part of a decimal number can be converted to binary form by iteratively multiplying it by two. After each multiplication, integer part of the product is separated and multiplication continues from the fractional part. Iteration ends either when the fractional part becomes zero or number of iterations reach a predefined limit. Let's give a simple example on this operation. If we want to convert the decimal number 0.125 to binary, we multiply it iteratively till we reach the product 1.00. This operation is tabulated in Table 6.2. As can be seen in this table, the multiplication operation reaches the product 1.00 after three iterations. Since the fractional part becomes zero, iteration ends. We can construct the binary number by forming an array starting from the integer part of the first product to the last in the list. Therefore, binary representation of the decimal number 0.125 will be 0.001.

We can convert a decimal number with integer and fractional parts by applying the above procedures separately to the number. As an example, binary representation of the decimal number 14.125 will be 1110.001.

Iteration	Number	Division	Remainder
i	14	7	**0**
ii	7	3	**1**
iii	3	**1**	**1**

TABLE 6.1 Decimal to Binary Conversion Example, Integer Part

Iteration	Number	Product	Integer Part
i	0.125	0.25	**0**
ii	0.250	0.50	**0**
iii	0.500	1.00	**1**

TABLE 6.2 Decimal to Binary Conversion Example, Fractional Part

6.1.1.2 Binary to Decimal Conversion

We can convert a binary number to decimal by weighting each digit by its value and summing the result. Let's explain this operation on an example. To convert the binary number 1110.001 to decimal form, we apply the following operation: $1 \times 2^3 + 1 \times 2^2 + 1 \times 2^1 + 0 \times 2^0 + 0 \times 2^{-1} + 0 \times 2^{-2} + 1 \times 2^{-3}$. Summing these, we will obtain 14.125 in decimal form.

6.1.2 Octal Numbers

Although binary numbers are suitable for digital systems, their representation may not be compact. Octal numbers can be used instead to have a more compact representation. Here, there are eight digits as (0, 1, 2, 3, 4, 5, 6, 7). Next, we consider how conversions can be made between binary and octal numbers.

6.1.2.1 Binary to Octal Conversion

We can convert a binary number to octal by grouping bits in blocks of three. Then, each group can be represented by the corresponding octal digit. As a result, we will obtain the octal representation. If the number groups do not form blocks of three, then we append zeros to the integer part as a prefix and fractional part as a suffix.

Let's convert the binary number 1110.001 to octal. Since the total number of bits in the integer part of number is not a multiple of three, we should represent it by appending zeros as 001110.001. Then, we can group these digits as 001=1, 110=6, and 001=1. As a result, octal representation of the binary number 1110.001 will be 16.1. As can be seen in this example, the octal number is more compact compared to its binary counterpart.

6.1.2.2 Octal to Binary Conversion

We can convert an octal number to binary by applying the reverse operation. Hence, we represent each octal digit by three bits and form the final binary number. Let's take the octal number 16.1. We can represent each octal digit by three binary digits as 1=001, 6=110, and 1=001. As a result, binary representation of the octal number 16.1 will be 001110.001. Since the two leftmost zero bits do not change the value of number, it can be represented as 1110.001.

6.1.3 Hexadecimal Numbers

While representing binary numbers in compact form, hexadecimal numbers will be more useful compared to octal numbers. A hexadecimal number has 16 digits as (0, 1, 2, 3, 4, 5, 6, 7, 8, 9, A, B, C, D, E, F). Next, we consider how conversions can be made between binary and hexadecimal numbers.

6.1.3.1 Binary to Hexadecimal Conversion

We can convert a binary number to hexadecimal by grouping bits in blocks of four. Then, each group can be represented by the corresponding hexadecimal digit. If bit groups do not form blocks of four, then we append zeros to the integer part of the binary number as a prefix and fractional part as a suffix. As a result, we will obtain the hexadecimal representation.

Let's convert the binary number 1110.001 to hexadecimal form. Since the total number of bits in the fractional part of the number is not a multiple of four, we should represent it by appending zero as a suffix as 1110.0010. Then, we can group these digits as 1110=E and 0010=2. As a result, hexadecimal representation of the binary number

`1110.001` will be `E.2`. As can be seen in this example, the hexadecimal number is more compact compared to its binary (and octal) form.

6.1.3.2 *Hexadecimal to Binary Conversion*

We can convert a hexadecimal number to binary by applying the reverse operation. Hence, we represent each hexadecimal digit by four bits and form the final binary number. Let's take the hexadecimal number `E.2`. We can represent each hexadecimal digit by four binary digits as `E=1110` and `2=0010`. As a result, binary representation of the hexadecimal number `E.2` will be `1110.0010`. Since the rightmost zero bit does not affect the value of the number, it can also be represented as `1110.001`.

6.2   Negative Numbers

There may be negative binary numbers in operation. Although in daily life we put a negative sign in front of the number, this is not the case in a digital system. Instead, there are three methods to represent both positive and negative binary numbers. These are the signed bit, one's complement, and two's complement representation.

6.2.1   Signed Bit Representation

The first representation mimics the daily life practice (negative sign in front of number) by a sign bit in the MSB of number. In this representation, a positive number will have the sign bit as zero. A negative number will have the sign bit as one. Hence, the name signed bit representation. Although this method seems straightforward, it is not very effective since addition and subtraction may need extra operations as will be seen in Sec. 6.5.

Let's give two examples on signed bit representation. Assume that we have decimal number 14. We know that binary representation of this number is `1110`. As can be seen here, the MSB represents the number value. Therefore, it is not possible to assign it as the sign bit. To overcome this problem, let's append four more zeroes to the number. Then, it becomes `0000 1110`. In this representation, we can use the MSB as sign bit. Remaining bits will serve as value bits. Since the number 14 is positive, its sign bit representation will be `0000 1110`. Now, let's represent the decimal number −14 using signed bit. Corresponding binary number will become `1000 1110`. Therefore, only the MSB has changed to show that the number is negative.

6.2.2   One's Complement Representation

The second representation is based on the bit complement (NOT) operation. Here, the negative number is represented by the bit complement of the corresponding positive number. Therefore, this representation is called one's complement. In this representation, no extra bit is assigned to sign. However, arithmetic operations are not straightforward in this representation.

Let's give two examples on one's complement representation. As in the previous section, let's first take the decimal number 14. Based on the previous format, it will be represented as `0000 1110`. Now, let's represent the decimal number −14 in one's complement form. To do so, we take the complement of each bit and obtain `1111 0001`.

6.2.3   Two's Complement Representation

The third representation is based on two's complement. Here, the negative number is first represented in one's complement form. Then, the result is incremented by one. Two's complement has a major advantage compared to the previous representations. Subtracting two binary numbers can be formulated as adding the first number with two's complement of the second. The result also keeps the sign information. Therefore, need for an extra sign bit is eliminated. We will see this operation in Sec. 6.5.

Let's continue with the example given in one's complement form. There, the decimal number −14 was represented as 1111 0001 in one's complement form. To obtain the two's complement form of −14, we should add one to the LSB of one's complement representation. Hence, we obtain 1111 0010 as the two's complement representation of decimal number −14.

6.3 Fixed- and Floating-Point Representations

Binary number to be processed in a digital system may have a fractional part. We distinguished the integer and fractional parts of such numbers by a dot in the previous section. This is not possible in a digital system. Instead, there are two methods to represent a binary number with integer and fractional parts. These are fixed- and floating-point representations.

6.3.1 Fixed-Point Representation

The number of bits assigned to the integer and fractional parts is fixed in this representation. Hence the name fixed-point representation. This method is easy to implement since the number of bits assigned to the integer and fractional parts is fixed.

We can show an unsigned fixed-point number (without a sign bit) as UQp.q. Here, U indicates the unsigned bit notation; pq represents the number, p being the integer and q being the fractional part. We provide some fixed-point representation formats in Table 6.3. Note that we are not limited by these formats in an FPGA implementation since the user is free to assign any number of bits to the integer and fractional parts. We will see such examples in Secs. 6.7 and 6.9.

Let's reconsider the decimal number 14.125. We know that binary representation of this number is 1110.001. Assume that we would like to represent this number in UQ16. form. Therefore, there will be no fractional part. The number of bits to be assigned to the integer part will be 16. Hence, the resulting number in hexadecimal form will be 000E. Zeros appended to the left of the number will not affect its value. They will only satisfy the fixed-point representation format. If the UQ16.16 fixed-point representation is used for the same number, then the integer part of 14.125 will be the same in hexadecimal form as 000E. The fractional part will be in hexadecimal form as 0200. Here, zeros are appended to the right of the number. Therefore, the value of the fractional part will not be affected. As a result, fixed-point representation of the number will be 000E0200. As can be seen here, there is no separator between the integer and fractional parts of the number. Knowing that the number is in UQ16.16 form, we can easily extract the integer and fractional parts (since we know the number of bits assigned to each).

Format	Minimum	Maximum	Resolution	# bits for p	# bits for q	# total bits
UQ16.	0	$2^{16}-1$	1	16	0	16
UQ.16	0	$1-2^{-16}$	2^{-16}	0	16	16
UQ16.16	0	$2^{16}-1$	2^{-16}	16	16	32

TABLE 6.3 Fixed-Point Unsigned Number Representation Formats

Format	Minimum	Maximum	Resolution	# bits for p	# bits for q	# total bits
Q15.	-2^{15}	$2^{15} - 1$	1	15	0	16
Q.15	-1	$1 - 2^{-15}$	2^{-15}	0	15	16
Q15.16	-2^{15}	$2^{15} - 2^{-16}$	2^{-16}	15	16	32

TABLE 6.4   Fixed-Point Signed Number Representation Formats

In a similar way, we can represent signed numbers. In this form, the MSB is reserved for the sign bit. Therefore, we use the sign bit representation here. We provide three signed bit formats for the fixed-point representation in Table 6.4. Similar to the unsigned bit representation, fixed-point number will be in the form Qp.q.

Let's consider the decimal number -14.125. Assume that we would like to represent this number in Q15. form. The resulting number in hexadecimal form will be 800E. Here, the MSB is set to 1 as the sign bit to represent that the number is negative. If the Q15.16 fixed-point representation is used for the same number, then hexadecimal form of the number will be 800E0200. Again, the MSB is kept as the sign bit in this representation.

6.3.2   Floating-Point Representation

Fixed-point representation is easy to implement and process. However, it has a major drawback. The number of bits assigned to integer and fractional parts is always fixed in this representation. This causes limitations both in the range of numbers to be represented and their resolution. Floating-point representation can be used to overcome these problems. As the name implies, the number of bits assigned to integer and fractional parts is not fixed in this representation. Instead, the assigned number of bits differ for each number depending on its significant digits. Therefore, a much wider range of values can be represented in this form.

In floating-point representation, a binary number with fractional part will be shown as $N = (-1)^S \times 2^E \times F$. Here, S stands for the sign bit, E represents the exponent value, and F stands for the fractional part. Then, floating-point number N is kept in memory as $X = SEF$.

To represent a floating-point number as $N = (-1)^S \times 2^E \times F$, the number should be normalized such that the integer part will have one digit. For ease of binary representation, the exponent will be biased by $2^{(e-1)} - 1$, where e is the number of bits to be used for E in the given format. Finally, certain number of bits will be assigned to S, E, and F depending on the standard format used for representation. The IEEE 754 standard is used by most digital systems in floating-point representation. This standard is summarized in Table 6.5.

Let's take the decimal number 14.125 and represent it in floating-point form. We will follow the below itemized procedure for this purpose:

- Decide on the format: Let's pick the "half" format for this example.
- Represent the integer and fractional parts of the decimal number in binary form: The number becomes 1110.001.
- Decide on the sign bit S: Since the number is positive, $(-1)^0 = 1$, S=0.
- Normalize the number such that the integer part will have one digit: The number becomes 1.110001×2^3.

Format	Exponent bias	# bits for S	# bits for E	# bits for F	# total bits
Half	15	1	5	10	16
Single	127	1	8	23	32
Double	1023	1	11	52	64
Quad	16383	1	15	112	128

TABLE 6.5 The IEEE 754 Standard for Floating-Point Representation

- Find the exponent value: For the half format, the exponent bias is 15. Therefore, the exponent will become $E = 15 + 3 = 18$ with bias. Or, in binary form E=10010.
- Find the fractional part: The fractional part (after normalization) was 110001. Since 10 bits should be used to represent the fractional part of the number in half format, F=1100010000. Remember, since this is the fractional part, we append extra zeros to its right so that the value of the number is not affected.
- Construct $X = SEF$: Finally, X = 0 10010 1100010000. Or in hexadecimal form, X=4B10.

Next, let's represent the decimal number -14.125 in floating-point form. As in the previous example, let's use the half format. Then, the only change will be in the sign bit. As a result, the number will become X = 1 10010 1100010000. Or in hexadecimal form, X=CB10.

6.4   ASCII Code

We do not only process numbers in digital systems. For some applications, we may need to handle characters and symbols as well. We know that everything in a digital system is represented in binary form. Therefore, characters and symbols should also be represented as such. One way of representing characters and symbols in binary form is using the ASCII code. ASCII stands for the American Standard Code for Information Interchange. The ASCII code for characters and symbols are given in Table 6.6. In this table, LSB stands for least significant byte and MSB stands for most significant byte. To represent a specific character (or symbol), its corresponding code should be given. Let's

		\multicolumn{16}{c}{LSB}
		0	**1**	**2**	**3**	**4**	**5**	**6**	**7**	**8**	**9**	**A**	**B**	**C**	**D**	**E**	**F**
	0	NUL	SOH	STX	ETX	EOT	ENQ	ACK	BEL	BS	HT	LF	VT	FF	CR	SO	SI
	1	DLE	DC1	DC2	DC3	DC4	NAK	SYN	ETB	CAN	EM	SUB	ESC	FS	GS	RS	US
	2		!	"	#	$	%	&	'	(	)	*	+	`	-	.	/
M	3	0	1	2	3	4	5	6	7	8	9	:	;	<	=	>	?
S	4	@	A	B	C	D	E	F	G	H	I	J	K	L	M	N	O
B	5	P	Q	R	S	T	U	V	W	X	Y	Z	[	\	]	^	_
	6	`	a	b	c	d	e	f	g	h	i	j	k	l	m	n	o
	7	p	q	r	s	t	u	v	w	x	y	z	{			}	~	DEL

TABLE 6.6 ASCII Code Table

assume that we would like to represent the @ symbol. Using Table 6.6, the corresponding ASCII code in hexadecimal form will be 40.

6.5 Arithmetic Operations on Binary Numbers

We will consider arithmetic operations on binary numbers from a generic point of view in this section. Therefore, we will first analyze each arithmetic operation based on binary numbers having only integer part. Then, we will consider arithmetic operations on numbers with fractional part (represented by fixed- and floating-point forms).

6.5.1 Addition

Adding two binary numbers is not different than adding two decimal numbers. The only condition the reader should remember is that a binary number can take only two values as zero or one. Therefore, adding two binary digits will produce a carry bit whenever two digits with value one are added.

Let's give an example on adding two binary numbers represented by eight bits as 0000 1110 and 0010 0111. We can also call these numbers as fixed-point with format UQ8.0. The sum will be 0011 0101.

There may be cases where adding two N bit numbers result in a $N+1$ bit number. For such cases, the MSB ($N+1$th bit) is called overflow. This bit should be handled separately if the number of bits assigned to the sum is N.

Adding two binary numbers with fractional part is also the same as in its decimal counterpart. Here, the important point is that the two numbers should be represented in the same format. If this is not the case, the first step is making formats the same. Next, let's consider the binary addition operation on fixed- and floating-point numbers.

6.5.1.1 Fixed-Point Addition

Let's start with adding two binary numbers represented by the same unsigned fixed-point format. Since both numbers will have the same number of integer and fractional bits, addition will be straightforward for this case. As an example, let's take two binary numbers represented in UQ8.4 format as 0000 1110 0010 and 0010 0111 0110. The sum will be 0011 0101 1000. The first and second numbers are 14.125 and 39.375 in decimal form with the sum 53.5. The sum obtained in UQ8.4 format is also the same as this number is in binary form.

Adding two fixed-point signed numbers with common format is the same as adding two numbers with unsigned fixed-point format. The only difference is that the sign bit in each number should not be taken into account in the addition operation. At this step, we assume that the two fixed-point signed numbers have the same sign. We will see adding two numbers with different sign bits in the next section under subtraction.

6.5.1.2 Floating-Point Addition

Adding two binary numbers represented by floating-point format is more complicated. As a reminder, a binary number is represented as $N = (-1)^S \times 2^E \times F$ in floating-point form. Moreover, the number is saved as $X = SEF$. To make the addition operation, the first constraint is that the two numbers should have the same floating-point format such as half, single, double, or quad. Moreover, the exponent value (E) should be the same for both numbers. If they are not the same, then fractional parts should be adjusted accordingly. Then, addition can be done. After addition, the fractional part and exponent should be adjusted such that a valid floating-point representation is obtained. Here, we assume that the sign bit of two numbers to be added are the same. We will handle adding two numbers with different sign bits in the next section under subtraction.

Let's give an example on adding two decimal numbers 14.125 and 39.375. Assume that the half floating-point form is selected such that first and second numbers are represented as X_1 = 0100 1011 0001 0000 and X_2 = 0101 0000 1110 1100. These numbers can be represented as 1.110001 $\times 2^3$ and 1.00111011 $\times 2^5$ after discarding the exponent bias. We can equate the exponential value for these such that the second number becomes 100.111011 $\times 2^3$. Then, we can add these two numbers as (1.110001 + 100.111011) $\times 2^3$. Here, the addition operation on two numbers can be done as if they are in a fixed-point form. The result becomes 110.101100 $\times 2^3$. This number can be represented as 1.10101100 $\times 2^5$. Hence, half floating-point representation of the result becomes X_3=0101 0010 1011 0000. As can be seen in this example, adding two floating-point numbers require format changes and condition checks. Therefore, it is not straightforward to add two numbers represented in floating-point form.

6.5.2   Subtraction

Two binary numbers can be subtracted in two different ways. The first method is plain subtraction as in decimal numbers. There is nothing specific about this operation. The second method is using two's complement representation. Here, the negative number is represented in two's complement form. This provides a clear advantage such that subtraction is performed by addition. Hence, no second circuitry is needed for the subtraction operation. Moreover, if the result of subtraction is negative it is automatically represented in two's complement form as well. Therefore, this method is used in most digital systems.

Let's give two examples on subtracting two binary numbers using two's complement representation. In the first example, let's subtract 0000 1110 from 0010 0111. First, we obtain the two's complement of 0000 1110 as 1111 0010. Adding 1111 0010 to 0010 0111 gives 1 0001 1001. As can be seen, the result is represented by nine bits. In other words, an overflow occurred. If overflow occurs, we should discard it and the result is final. In other words, subtraction results in 0001 1001. In the second example, let's subtract 0010 0111 from 0000 1110. Here, we obtain the two's complement of 0010 0111 and add it to 0000 1110. The result becomes 1110 0111. In this operation, no overflow occurs. This indicates that the result is negative and represented in two's complement form. We can check it by obtaining the two's complement of the first subtraction result which gives 1110 0111. As can be seen here, two's complement representation handles sign of the result after operation.

Subtraction operation can also be applied to two binary numbers with fractional part. As in addition, the important point here is that the two numbers should be represented in the same format. If this is not the case, the first step is making formats the same. Next, let's consider binary subtraction operation on fixed- and floating-point numbers.

6.5.2.1   *Fixed-Point Subtraction*

Let's start with subtracting two binary numbers represented by the same unsigned fixed-point format. To explain subtraction, let's take two binary numbers represented in UQ8.4 format as 0000 1110 0010 and 0010 0111 0110. In the first example, let's subtract 0000 1110 0010 from 0010 0111 0110. We can apply two's complement method as in the previous example. Therefore, we first obtain the two's complement of 0000 1110 0010 as 1111 0001 1110. Adding 1111 0001 1110 to 0010 0111 0110 gives 1 0001 1001 0100. As can be seen here, the result should be represented by

13 bits but the original format had 12 bits. Since overflow occurred, we discard the MSB and obtain the final result as `0001 1001 0100`. Here, the first and second numbers were 14.125 and 39.375, respectively in decimal form. Subtracting 14.125 from 39.375 results in 25.25. Binary subtraction result obtained in UQ8.4 form is also the same as this number. In the second example, let's subtract `0010 0111 0110` from `0000 1110 0010`. Applying the same steps as in the previous example, we will obtain the subtraction result as `1110 0110 1100`. In this operation, no overflow occurs. This indicates that the result is negative and represented in two's complement form.

Subtracting two fixed-point signed numbers with the same format is the same as in subtracting two numbers with unsigned fixed-point format. The only difference is that the sign bit should be taken into account such that if the number is negative, it should be represented as such in the subtraction operation.

6.5.2.2   *Floating-Point Subtraction*

As in addition, subtracting two binary numbers represented by floating-point format is more complicated. To subtract numbers, the first constraint is their having the same floating-point format as half, single, double, or quad. Moreover, exponent (E) should be the same for both numbers. If they are not the same, then fractional parts should be adjusted accordingly. Then, subtraction can be done. Afterward, the fractional part, exponent, and sign bit should be adjusted such that a valid floating-point representation is obtained. While subtracting numbers, the sign bit should be taken into account such that if the number is negative, it should be represented as such in operations.

Let's take two examples on subtracting two floating-point numbers. For these let's pick two decimal numbers as 14.125 and 39.375 (which we have been using up to now). Assume that the half floating-point form is selected. Hence, the first and second numbers are represented as $X_1 = $ `0100 1011 0001 0000` and $X_2 = $ `0101 0000 1110 1100`. These numbers can be represented as `1.110001` $\times 2^3$ and `1.00111011` $\times 2^5$ after discarding the exponent bias. We can equate the exponential value for these such that the second number becomes `100.111011` $\times 2^3$. As first example, let's subtract `1.110001` $\times 2^3$ from `100.111011` $\times 2^3$. We can represent the subtraction as (`100.111011` $-$ `1.110001`) $\times 2^3$. Here, subtraction can be done as if they are in fixed-point form. The result becomes `11.001010` $\times 2^3$. This number can be represented as `1.1001010` $\times 2^4$. Hence, the half floating-point form of the result becomes $X_3 = $`0100 1110 0101 0000`. As second example, let's subtract `100.111011` $\times 2^3$ from `1.110001` $\times 2^3$ which can be shown as (`1.110001` $-$ `100.111011`) $\times 2^3$. The result of this operation becomes $-$`1.1001010` $\times 2^4$. Hence, the half floating-point form of the result becomes $X_3 = $`1100 1110 0101 0000`.

6.5.3   Multiplication

Multiplying two binary numbers is also the same as multiplying two decimal numbers. The reader should be aware that the product term requires more bits for representation compared to multiplied numbers. Let's give an example on multiplying two binary numbers represented by eight bits as `0000 1110` and `0010 0111`. Their product will be `10 0010 0010`. As can be seen here, the product term requires 10 bits for representation. The multiplication operation can also be applied on two binary numbers with fractional part. Next, let's consider the binary multiplication operation on fixed- and floating-point numbers.

6.5.3.1 Fixed-Point Multiplication

Let's start with multiplying two binary numbers represented by the same unsigned fixed-point format. To explain the multiplication operation, let's take two binary numbers represented in UQ8.4 format as `0000 1110 0010` and `0010 0111 0110`. Here, the reader can represent these two numbers as 11100010×2^{-4} and 1001110110×2^{-4}. Product of these two numbers will be $11100010 \times 1001110110 \times 2^{-8}$. Hence, the result becomes $100010110000101100 \times 2^{-8}$. We can represent this number in UQ8.4 format as `0010 1100 0010`. As can be seen in this example, an overflow with two and four bits occurred in integer and fractional parts, respectively. Therefore, a larger format should be used in representing the result. Multiplying two fixed-point signed numbers is the same as in unsigned numbers. However, the sign bit should be taken into account in deciding the sign of the product.

6.5.3.2 Floating-Point Multiplication

Multiplying two binary numbers represented by floating-point format is more complicated as in addition and subtraction. Let's give an example on multiplying two decimal numbers 14.125 and 39.375. Assume that these numbers are represented by the half floating-point form. From previous examples we know that these numbers can be represented as 1.110001×2^3 and 1.00111011×2^5 or in simplified form as 1110001×2^{-3} and 100111011×2^{-3}, respectively. Hence, their product will be $1110001 \times 100111011 \times 2^{-6}$. The result becomes $1000101100001011 \times 2^{-6}$. This number can be represented as $1.000101100001011 \times 2^9$. Hence, half floating-point form of the result will be $X_3 = 0110\ 0000\ 0101\ 1000$. In this representation, least significant five bits are discarded due to the half floating-point format. However, the effect of these bits are minor compared to the overflow in fixed-point representation. In this example, the two floating-point numbers had the same sign bit as positive. For floating-point numbers having negative sign bit, this should be taken into account in operations.

6.5.4 Division

Dividing two binary numbers is also the same as dividing two decimal numbers. The reader should be aware that the division of two integer numbers may result in a number with extra fractional part. Let's give an example on dividing two binary numbers represented by eight bits as `0010 0111` and `0000 1110`. Let's divide the first number by the second. Integer part of the division will be `10`. Besides, there is also a fractional part of the division. For ease of demonstration, we can represent this fractional part by four bits as `1100`. The division operation can also be applied on two binary numbers with fractional part. Next, let's consider the binary division operation on fixed- and floating-point numbers.

6.5.4.1 Fixed-Point Division

To explain the division operation, let's take two binary numbers in UQ8.4 format as `0000 1110 0010` and `0010 0111 0110`. Here, the reader can represent these two numbers as 11100010×2^{-4} and 1001110110×2^{-4}. Let's divide the second number by the first which can be represented as $1001110110 \div 11100010 \times 2^0$. The division results in a fractional number with overflow. Therefore, it should be truncated. Then, the result becomes 101100×2^{-4}. We can represent this number in UQ8.4 format as `0000 0010 1100`.

6.5.4.2   *Floating-Point Division*

Let's finally give an example on dividing the decimal number 39.375 by 14.125 represented by half floating-point form. We know that these numbers can be represented as 1.00111011×2^5 and 1.110001×2^3, respectively, from previous sections. We can represent these numbers as 100111011×2^{-3} and 1110001×2^{-3}. Therefore, their division can be represented as $100111011 \div 1110001 \times 2^0$. The result of division will be 1.0110010011×2^1. Hence, half floating-point form of the result becomes X_3=0100 0001 1001 0011. In this representation, least significant bits lower than digit 10 are discarded due to half floating-point format. However, effect of these bits are minor compared to the overflow in fixed-point representation. In this example, two floating-point numbers had the same sign bit as positive. For floating-point numbers having negative sign bit, this should be taken into account in operations.

6.6   Data Types in Verilog

We introduced number representations and related concepts from a generic point of view in previous sections. Starting from this section, we will handle these concepts using HDLs. Therefore, we will start exploring data types in Verilog in this section. Then, we will consider constants and parameters. Afterward, we will introduce vectors. We will analyze the FPGA implementation details of these in Sec. 6.11.

6.6.1   Net and Variable Data Types

A value in a digital system can basically be represented either as net or variable in Verilog. The net data type is specific for connecting two elements. For us, the most important net data type is wire. As the name implies, this data type acts simply as a wire connecting two elements. The variable data type can be used to represent a generated data till it changes. Useful variable data types are reg and integer in Verilog. A reg variable can be used to represent one-bit data. An integer variable typically represents 32-bit long data. We can define a net or variable data type in Verilog by the structure data_type data_name. For example, we can define wire in1 to indicate a variable in1 of type wire.

6.6.2   Data Values

A net or variable data type can get one of four predefined values. These are as follows:

0 corresponds to logic level zero.

1 corresponds to logic level one.

x represents the undefined logic level.

z represents high impedance.

We are familiar with logic level zero and one from previous chapters. The undefined logic level x is used in logical operations when the corresponding value is unknown or it does not affect the operation. For the second case, x is most of the times called "don't care" condition. The high impedance value z indicates that connection at that point is disabled. In other words, it indicates an open circuit at the given location.

6.6.3   Naming a Net or Variable

While describing a digital system in Verilog, one may want to name a net or variable. Here, the reader is free to choose among many options. The only constraint here is that the name should not begin with a digit and it should not be any of Verilog keywords.

Besides, Verilog is case sensitive. Hence, an uppercase and lowercase character is not the same. This should be taken into account while assigning a name. More importantly, meaningful and representative names should be picked for assignment to increase the readability of Verilog description.

6.6.4   Defining Constants and Parameters

We can represent binary, octal, hexadecimal, and decimal constant values (besides others) in Verilog. General structure of representing a constant for these types is `bit_width 'radix constant_value`. Here, `bit_width` indicates the number of bits to represent the constant value. If this is not set, the default value is 16 bits. The `radix` can be binary (b), octal (o), hexadecimal (h), or decimal (d). The `constant_value` is the actual constant to be represented.

Let's give some examples on constants. `1'b0` indicates the binary number `0`. `2'b10` indicates the binary number `10`. `4'b10` indicates the binary number `0010`. `6'o75` indicates the octal number `75`. `8'hCA` indicates the hexadecimal number `CA`. Finally, `8'd251` indicates the decimal number `251` which can be represented by eight bits.

6.6.5   Defining Vectors

A net or variable need not be composed of one bit in Verilog. Instead, it can be represented as a vector. This allows us to represent data in compact form. The vector format for representation will be the same as a net or variable definition with an extra `[N-1:0]` prefix which indicates that there will be N net variable entries packed as a vector. Here, MSB and LSB are located at the N-1th and zeroth entries, respectively.

As an example, we can define `wire[7:0] in1` to indicate a variable `in1` of type `wire` with eight entries. Here, `in1` represents all eight-bit values at once. `in1[7]` represents the most significant entry. `in1[0]` represents the least significant entry. We can select a subpart of the vector as `in1[5:3]` such that the fourth, fifth, and sixth entries are selected.

We can also change the order of bits in representing a vector. Continuing from the above representation, we can redefine `wire[0:7] in1` to indicate a variable `in1` of type `wire` with eight entries. Now, the most significant bit will be represented by `in1[0]`. The least significant entry will be represented by `in1[7]`.

We next provide Verilog description as an example of vector operations in Listing 6.1. Here, first a specific vector entry is selected. Then, subpart of the vector is selected. Finally, the vector bit order is reversed. We provide the RTL schematic of these vector operations in Fig. 6.1. As can be seen in this figure, vector operations are performed by wiring input and output ports only.

To explain working principles of vector operations, we provide a testbench file in Listing 6.2. Here, the input vector to be processed is taken as FA. We provide the results obtained from the testbench file in Fig. 6.2. These results indicate that vector entries can be processed as desired in Verilog.

6.7   Operators in Verilog

There are basically six operator groups in Verilog. These are logical, arithmetic, shift, concatenate, replicate, and condition. We will introduce arithmetic, concatenation, and replication operators in this chapter. The rest will be introduced in the following chapters. We will analyze the FPGA implementation details of operations considered here in Sec. 6.11.

Listing 6.1 Basic Vector Operations in Verilog

```verilog
module vector_defn(num1,res1,res2,res3);

input [7:0] num1;
output res1;
output [3:0] res2;
output [0:7] res3;

//selecting a specific vector entry
assign res1=num1[2];

//selecting specific vector entries
assign res2=num1[7:4];

//changing the order of bits
assign res3=num1;

endmodule
```

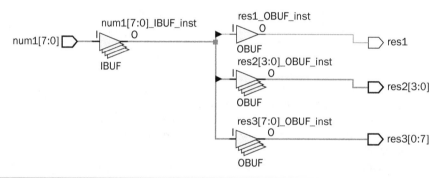

FIGURE 6.1 RTL schematic of basic vector operations.

Listing 6.2 Testbench File for Basic Vector Operations in Verilog

---

```verilog
'timescale 1ns / 1ps

module vector_defn_tb;

reg [7:0] in1;
wire out1;
wire [3:0] out2;
wire [0:7] out3;

vector_defn UUT(.num1(in1),.res1(out1),.res2(out2),.res3(out3));

initial begin
in1 = 8'hFA;
#100;
end

endmodule
```

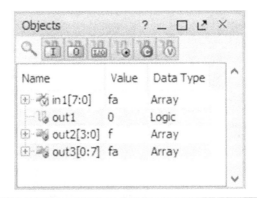

FIGURE 6.2   Basic Vector Operation Results in Verilog.

6.7.1   Arithmetic Operators

Verilog has five arithmetic operations as addition (+), subtraction (-), multiplication (*), division (/), and modulus (%). All these operations can be performed on vectors with user-defined size. Hence, these can be called fixed-point operations with user-defined format. When result of an operation becomes negative, it is represented in two's complement form.

Let's give basic examples on the usage of arithmetic operations. The first example is on arithmetic operations by using a vector input and constant defined as a parameter. We provide the corresponding Verilog description in Listing 6.3. Here, five arithmetic operations (addition, subtraction, multiplication, division, and modulus) are applied on the constant coef=8'h02 and input vector num. Dataflow modeling is used in describing these operations. We provide the RTL schematic of the description in Fig. 6.3. As can be seen in this figure, constant values are taken as fixed voltage levels in the schematic.

To explain working principles of arithmetic operations including a constant, we provide the testbench file in Listing 6.4. Here, input vector to be processed is taken as 8'h07. Arithmetic operation results are provided (in hexadecimal form) in Fig. 6.4. As can be seen in this figure, only the integer part of the division operation is kept. Besides, obtained results are as expected.

The second example on arithmetic operations is based on examples (on two eight-bit numbers) in Sec. 6.5. Here, again five arithmetic operations are applied on two eight-bit input vectors num1 and num2. We provide the corresponding Verilog description in Listing 6.5. Dataflow modeling is used in describing these operations. Note that the multiplication result is represented by a 14-bit vector in the description. The reader can also use a 16-bit vector as well. We provide the RTL schematic of the description in Fig. 6.5. As can be seen in this figure, all arithmetic operations are represented as basic blocks.

To be consistent with the examples in Sec. 6.5, we construct the testbench file in Listing 6.6. Here, the two vectors are taken as 8'b00001110 and 8'b00100111. Arithmetic operation results are provided in Fig. 6.6. As can be seen in this figure, the negative result is represented in two's complement form. Also, only integer part of the division operation is given. Besides, the reader can observe that results obtained here are the same as in Sec. 6.5.

The third example on arithmetic operations is based on examples (on fixed-point numbers with UQ8.4 format) in Sec. 6.5. Here, fixed-point numbers are represented by two 12-bit input vectors num1 and num2. We provide the corresponding Verilog

---

Listing 6.3 Arithmetic Operations on a Constant and Vector in Verilog

---

```verilog
module arithmetic_constant(num,res1,res2,res3,res4,res5);

input [7:0] num;
output [7:0] res1;
output [7:0] res2;
output [7:0] res3;
output [7:0] res4;
output [7:0] res5;

parameter coef=8'h02;

//addition
assign res1=coef+num;

//subtraction
assign res2=num-coef;

//multiplication
assign res3=coef*num;

//division
assign res4=num/coef;

//modulus
assign res5=num%coef;

endmodule
```

---

description in Listing 6.7. Note that the multiplication result is represented by a 24-bit vector in the description. The RTL schematic of this description is the same as in Fig. 6.5. Only the number of wires used in operations differ.

To be consistent with the fixed-point arithmetic examples in Sec. 6.5, we construct the testbench file in Listing 6.8. Here, the two vectors are taken as 12'b000011100010 and 12'b001001110110. Arithmetic operation results are provided in Fig. 6.7. As in the previous example, the reader can observe that results obtained here are the same as in Sec. 6.5.

As can be seen in all these examples, arithmetic operations can be performed without any difficulty in Verilog. Therefore, we will not explore dedicated arithmetic operation circuits in the following chapters. We should warn the reader about multiplication and division operations at this point. Although these operations can be performed, they heavily dissipate the FPGA resources. We will see this resource dissipation by actual examples in Sec. 6.11. Therefore, multiplication and division operations should be avoided whenever possible.

6.7.2 Concatenation and Replication Operators

The concatenation operator in Verilog allows merging two or more vectors. This is done by the curly bracket. Let's give an example. Assume that we want to merge two vectors num1 and num2. We can do this by {num1, num2}. The replication operation can be used to copy a vector multiple times to generate a new vector. This can be done by n{num1} where n is the duplication number.

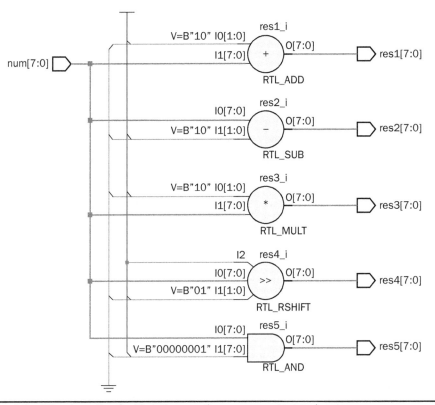

FIGURE 6.3   RTL schematic of arithmetic operations on a constant and vector.

Listing 6.4   Testbench File for Arithmetic Operations on a Constant and Vector in Verilog

```
'timescale 1ns / 1ps

module arithmetic_constant_tb;

reg  [7:0] in1;
wire [7:0] out1;
wire [7:0] out2;
wire [7:0] out3;
wire [7:0] out4;
wire [7:0] out5;

arithmetic_constant UUT (.num(in1),.res1(out1),.res2(out2),.res3
    (out3),.res4(out4),.res5(out5));

initial begin
in1 = 8'h07;

#100;
end

endmodule
```

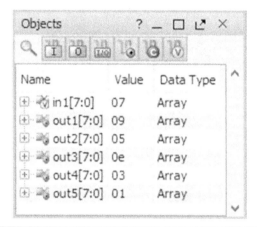

FIGURE 6.4   Result of arithmetic operations on a constant and vector in Verilog.

Listing 6.5   Arithmetic Operations on Two Eight-bit Vectors in Verilog

```
module arithmetic_operations(num1,num2,res1,res2,res3,res4,res5,res6);

input  [7:0]  num1;
input  [7:0]  num2;
output [7:0]  res1;
output [7:0]  res2;
output [7:0]  res3;
output [13:0] res4;
output [7:0]  res5;
output [7:0]  res6;

//addition
assign res1=num1+num2;

//subtraction
assign res2=num2-num1;

assign res3=num1-num2;

//multiplication
assign res4=num1*num2;

//division
assign res5=num2/num1;

//modulus
assign res6=num2%num1;

endmodule
```

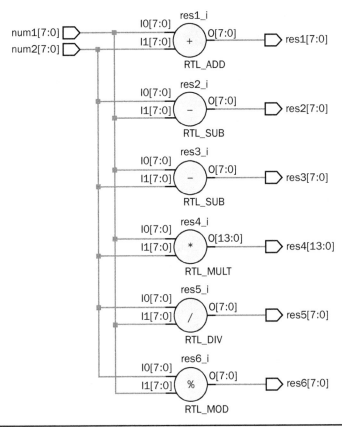

FIGURE 6.5 RTL schematic of arithmetic operations on two vectors in Verilog.

Listing 6.6 Testbench File for Arithmetic Operations on Two Eight-bit Vectors in Verilog

```
module arithmetic_operations_tb;

reg   [7:0] in1;
reg   [7:0] in2;
wire  [7:0] out1;
wire  [7:0] out2;
wire  [7:0] out3;
wire  [13:0] out4;
wire  [7:0] out5;
wire  [7:0] out6;

arithmetic_operations UUT (.num1(in1),.num2(in2),.res1(out1),.res2
        (out2),.res3(out3),.res4(out4),.res5(out5),.res6(out6));

initial begin
in1 = 8'b00001110;
in2 = 8'b00100111;

#100;
end

endmodule
```

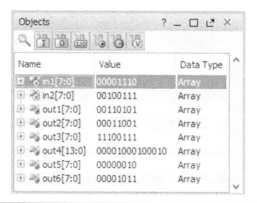

FIGURE 6.6 Result of arithmetic operations on two eight-bit vectors in Verilog.

Listing 6.7 Arithmetic Operations on Two 12-bit Vectors in Verilog

```verilog
module arithmetic_operations_UQ8_4(num1,num2,res1,res2,res3,res4,res5,
    res6);

input   [11:0] num1;
input   [11:0] num2;
output  [11:0] res1;
output  [11:0] res2;
output  [11:0] res3;
output  [23:0] res4;
output  [11:0] res5;
output  [11:0] res6;

//addition
assign res1=num1+num2;

//subtraction
assign res2=num2-num1;

assign res3=num1-num2;

//multiplication
assign res4=num1*num2;

//division
assign res5=num2/num1;

//modulus
assign res6=num2%num1;

endmodule
```

We provide dataflow model of concatenation and replication operations on vectors in Listing 6.9. Here, first two vectors num1 and num2 are concatenated. Then, the replicate of the vector num1 is generated twice. The RTL schematic of these operations are as in Fig. 6.8. As can be seen in this figure, concatenation and replication operations are implemented by using wiring between input and output ports.

Listing 6.8 Testbench File for Arithmetic Operations on Two 12-bit Vectors in Verilog

```verilog
module arithmetic_operations_UQ8_4_tb;

reg   [11:0] in1;
reg   [11:0] in2;
wire  [11:0] out1;
wire  [11:0] out2;
wire  [11:0] out3;
wire  [23:0] out4;
wire  [11:0] out5;
wire  [11:0] out6;

arithmetic_operations_UQ8_4 UUT (.num1(in1),.num2(in2),.res1(out1),.
    res2(out2),.res3(out3),.res4(out4),.res5(out5),.res6(out6));

initial begin
in1 = 12'b000011100010;
in2 = 12'b001001110110;

#100;
end

endmodule
```

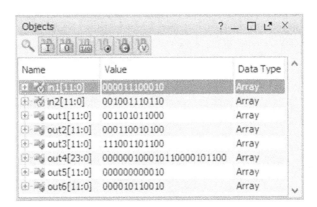

Figure 6.7 Result of arithmetic operations on two 12-bit vectors in Verilog.

We provide the testbench file in Listing 6.10 to explain concatenation and replication operations on an example. Here, the two vectors are taken as 8'hFA and 8'h0F. Concatenation and replication operation results are provided in Fig. 6.9. The reader can see how both operations resulted there.

6.8 Data Types in VHDL

As in Verilog, we should know data types in VHDL. Hence, they can be used in processing data in digital systems. In this section, we will introduce data types and their usage for this purpose.

Listing 6.9   Concatenation and Replication Operations in Verilog

```verilog
module concatenate_replicate(num1,num2,res1,res2);

input   [7:0] num1;
input   [7:0] num2;
output [15:0] res1;
output [15:0] res2;

//concatenate
assign res1={num1,num2};

//replicate
assign res2={2{num1}};

endmodule
```

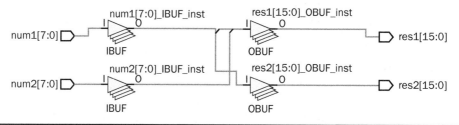

Figure 6.8   RTL schematic of concatenation and replication operations.

Listing 6.10   Testbench File for Concatenation and Replication Operations in Verilog

```verilog
'timescale 1ns / 1ps

module concatenate_replicate_tb;

reg  [7:0] in1;
reg  [7:0] in2;
wire [15:0] out1;
wire [15:0] out2;

concatenate_replicate UUT(.num1(in1),.num2(in2),.res1(out1),.res2(out2)
    );

initial begin
in1 = 8'hFA;
in2 = 8'h0F;

#100;
end

endmodule
```

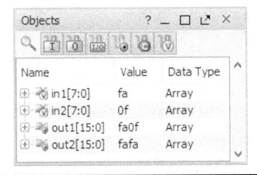

Figure 6.9 Result of concatenation and replication operations in Verilog.

6.8.1 Signal and Variable Data Types

A value in a digital system can be basically represented either as a `signal` or `variable` in VHDL. The signal data type is similar to the `wire` in Verilog. Hence, it can be used to connect two elements. The signal (with its assigned type) can be defined as `signal signal_name : signal_type`. The `variable` data type in VHDL is similar to the one in Verilog. However, it is generally used in storing intermediate values and loop counters. Therefore, we will provide its usage in the following chapters when needed.

The signal should have an associated type which defines values that can be taken by it. Although there are several signal types in VHDL, we will use four of them at this level as `std_logic`, `std_logic_vector`, `signed`, and `unsigned`. We may introduce new types in the following chapters if needed.

The `std_logic` type is for bitwise representations. Related to it, the `std_logic_vector` type is for an array of bits to be explored in detail in Sec. 6.8.5. To use `std_logic` and `std_logic_vector` types, we should include the `ieee` library in the description. We should also add the `use ieee.std_logic_1164.all` line to the description.

`Signed` and `unsigned` types have the same properties as `std_logic_vector`. However, they are specifically used in arithmetic operations to be introduced in Sec. 6.9. To use signed and unsigned types, we should include the `ieee` library in the description. We should also add the `use ieee.numeric_std.all` line to the description.

6.8.2 Data Values

`Std_logic` data type has nine different values. We will use the following four values throughout the book:

0 corresponds to logic level zero.

1 corresponds to logic level one.

- represents the undefined logic level.

z represents high impedance.

We are familiar with logic level zero and one from previous chapters. The undefined logic level, -, is used in logical operations when the corresponding value is unknown or it does not affect the operation. For the second case, - is most of the times called "don't care" condition. The high impedance value z indicates that connection at that point is disabled. In other words, an open circuit is present at that location. `Std_logic_vector`, `signed`, and `unsigned` types also use the mentioned data values.

6.8.3 Naming a Signal or Variable

As in Verilog, the user can select a wide range of names for a signal or variable in VHDL. However, a VHDL keyword cannot be used as a name. Besides, the name should begin with a letter. It cannot end with an underscore or it cannot have two successive underscores. Unlike Verilog, VHDL is not case sensitive. Therefore, the reader should take this into account while defining a name. Meaningful and representative names should be picked for assignment to increase the readability of a VHDL description.

6.8.4 Defining Constants

A constant can be defined to represent a value in VHDL. This is done to improve the readability of description. Structure of a constant declaration is `constant constant_name : type_name := value`. Here, if the `value` is one bit, then it should be represented between apostrophes as `` `0' `` or `` `1' ``. If the `value` has more than one bit, then it should be represented between double quotes as `"0101"`. Moreover, we can use the format `x"value"` or `o"value"` to represent the hexadecimal and octal values, respectively. For example, the binary value `"0101"` can also be represented as `x"5"` as hexadecimal.

6.8.5 Defining Arrays

In VHDL, we can use `std_logic_vector`, `signed`, and `unsigned` types to represent bit arrays. The signal array (with its assigned type) can be defined as `signal array_name : array_type (low to high)` or `signal array_name : array_type (high downto low)`. Here, `low` and `high` values indicate the array's first and last index values.

Each array entry can be reached in VHDL. Let's give an example for this operation. Assume we define an array `in1` as `signal in1 : std_logic_vector (7 downto 0)`. Here, `in1` represents all eight bits at once. `in1(7)` represents the MSB. `in1(0)` represents the LSB. We can also change the order of bits in representing an array. To do so, we should redefine the array `in1` as `signal in1 : std_logic_vector (0 to 7)`. Now, the MSB will be represented by `in1(0)`. The LSB will be represented by `in1(7)`.

We next provide the VHDL description as an example of array operations in Listing 6.11. Here, first a specific array entry is selected. Then, subpart of the array is selected. Finally, the array bit order is reversed. Dataflow modeling is used in describing these operations. The RTL schematic of this description is the same as in Fig. 6.1. As can be seen in this figure, array operations are performed by wiring input and output ports only.

To explain working principles of array operations, we provide the testbench file in Listing 6.12. Here, input array to be processed is taken as `"11111010"`. Array operation

---

Listing 6.11 Basic Array Operations in VHDL

```
library ieee;
use ieee.std_logic_1164.all;

entity vector_defn is
port(num1 : in std_logic_vector (7 downto 0);
     res1 : out std_logic;
     res2 : out std_logic_vector (3 downto 0);
```

```vhdl
      res3 : out std_logic_vector (0 to 7));
end vector_defn;

architecture dataflow_model of vector_defn is
begin

--selecting a specific vector entry
res1 <= num1(2);

--selecting specific vector entries
res2<= num1(7 downto 4);

--changing the order of bits
res3 <= num1;

end dataflow_model;
```

Listing 6.12 Testbench File for Basic Array Operations in VHDL

```vhdl
library ieee;
use ieee.std_logic_1164.all;

entity vector_defn_tb is
end vector_defn_tb;

architecture dataflow of vector_defn_tb is

component vector_defn
port(num1 : in std_logic_vector (7 downto 0);
     res1 : out std_logic;
     res2 : out std_logic_vector (3 downto 0);
     res3 : out std_logic_vector (0 to 7));
end component;

signal  in1 : std_logic_vector (7 downto 0);
signal out1 : std_logic;
signal out2 : std_logic_vector (3 downto 0);
signal out3 : std_logic_vector (0 to 7);

begin
UUT: vector_defn port map (num1 => in1,res1 => out1,res2 => out2,
          res3 => out3);

process
begin
wait for 5 ns;

in1 <= "11111010";

--wait;
end process;

end dataflow;
```

results will be as in Fig. 6.2. These results indicate that array entries can be processed as desired in VHDL.

6.9 Operators in VHDL

There are basically five operator groups in VHDL. These are arithmetic, relational, shift and rotate, concatenation, and logical operators. We will introduce arithmetic and concatenation operators in this chapter. The rest will be introduced in the following chapters.

6.9.1 Arithmetic Operators

We will use seven arithmetic operators in VHDL throughout the book. These are absolute value (abs), multiplication (*), division (/), modulus (mod), remainder (rem), addition (+), and subtraction (-). Except abs, all arithmetic operations are performed on signed or unsigned numbers. Obtained result from these operations will also be either a signed or unsigned number. The abs needs a signed number to operate. As in Verilog, when the result of an operation is negative, it is represented in two's complement form in VHDL. Note that addition and subtraction operations can also be applied to signals defined by std_logic_vector.

Let's give three examples on the usage of arithmetic operations. The first example is on arithmetic operations using an array input and constant. We provide the VHDL description in Listing 6.13. Here, three arithmetic operations (addition, subtraction, and

Listing 6.13 Arithmetic Operations on a Constant and Array in VHDL

```vhdl
library ieee;
use ieee.numeric_std.all;

entity arithmetic_constant is
port(num : in signed (7 downto 0);
    res1 : out signed (7 downto 0);
    res2 : out signed (7 downto 0);
    res3 : out signed (15 downto 0));

constant coef : signed (7 downto 0) :="00000010";
end arithmetic_constant;

architecture dataflow_model of arithmetic_constant is
begin

--addition
res1 <= num + coef;

--subtraction
res2 <= num - coef;

--multiplication
res3 <= coef * num;

end dataflow_model;
```

multiplication) are applied on a constant `coef` and input array num. Here, the constant is defined as `"00000010"`. Dataflow modeling is used in describing these operations. This description is the VDHL version of the one given in Listing 6.3.

To explain working principles of arithmetic operations including a constant, we provide the testbench file in Listing 6.14. Here, the input array to be processed is taken as `"0000111"`. Arithmetic operation results obtained will be the same as in Fig. 6.4.

The second example is arithmetic operations based on examples (on two eight-bit numbers) in Sec. 6.5. Here, six arithmetic operations (addition, subtraction, multiplication, division, modulus, and remainder) are applied on two eight-bit input arrays num1 and num2. We provide the corresponding VHDL description in Listing 6.15. Dataflow modeling is used in describing these operations. This description is the VDHL version of the one given in Listing 6.5. We provide the RTL schematic of the VHDL description in Fig. 6.10. As can be seen in this figure, all arithmetic operations are represented as basic blocks.

Listing 6.14 Testbench File for Arithmetic Operations on a Constant and Array in VHDL

```vhdl
library ieee;
use ieee.numeric_std.all;

entity arithmetic_constant_tb is
end arithmetic_constant_tb;

architecture dataflow of arithmetic_constant_tb is

component arithmetic_constant
port(num : in signed (7 downto 0);
     res1 : out signed(7 downto 0);
     res2 : out signed (7 downto 0);
     res3 : out signed (15 downto 0));
end component;

signal  in1 : signed (7 downto 0);
signal out1 : signed (7 downto 0);
signal out2 : signed (7 downto 0);
signal out3 : signed (15 downto 0);

begin
UUT:  arithmetic_constant port map (num => in1,res1 => out1,res2 =>
      out2,res3 => out3);

process
begin
wait for 5 ns;

in1 <= "00000111";

--wait;
end process;

end dataflow;
```

Listing 6.15 Arithmetic Operations on Two Eight-bit Arrays in VHDL

```vhdl
library ieee;
use ieee.numeric_std.all;

entity arithmetic_operations is
port(num1 : in unsigned (7 downto 0);
     num2 : in unsigned (7 downto 0);
     res1 : out unsigned (7 downto 0);
     res2 : out unsigned (7 downto 0);
     res3 : out unsigned (7 downto 0);
     res4 : out unsigned (15 downto 0);
     res5 : out unsigned (7 downto 0);
     res6 : out unsigned (7 downto 0);
     res7 : out unsigned (7 downto 0));

end arithmetic_operations;

architecture dataflow_model of arithmetic_operations is
begin

--addition
res1 <= num1 + num2;

--subtraction
res2 <= num2 - num1;

res3 <= num1 - num2;

--multiplication
res4 <= num1 * num2;

--division
res5 <= num2 / num1;

--modulus
res6 <= num2 mod num1;

--remainder
res7 <= num2 rem num1;

end dataflow_model;
```

To be consistent with examples in Sec. 6.5, we construct the testbench file in Listing 6.16. Here, the two eight-bit arrays are taken as "00001110" and "00100111". Arithmetic operation results are provided in Fig. 6.11. These are the same as in Fig. 6.6. Besides, the reader can observe that results obtained here are the same as in Sec. 6.5.

The third example is arithmetic operations based on examples (on fixed-point numbers with UQ8.4 format) in Sec. 6.5. Here, fixed-point numbers are represented by two 12-bit input arrays num1 and num2. We provide the corresponding VHDL description in Listing 6.17. This description is VDHL version of the one in Listing 6.7. The RTL schematic of this description is the same as in Fig. 6.10. Only number of wires used in operations differ.

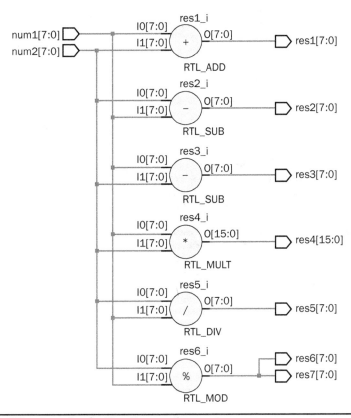

FIGURE 6.10 RTL schematic of arithmetic operations on two arrays.

To be consistent with fixed-point arithmetic operation examples in Sec. 6.5, we construct the testbench file in Listing 6.18. Here, two arrays are taken as "000011100010" and "001001110110". Arithmetic operation results are provided in Fig. 6.12. As in the previous example, the reader can observe that results obtained here are the same as in Sec. 6.5.

Similar to Verilog, all arithmetic operations can be performed without any difficulty in VHDL as can be seen in Listing 6.15. Therefore, we will not explore dedicated arithmetic operation circuits in the following chapters. We should warn the reader about multiplication and division operations at this point. Although these operations can be performed, they heavily dissipate the FPGA resources. Therefore, multiplication and division operations should be avoided whenever possible.

6.9.2 Concatenation Operator

The concatenation operator in VHDL allows merging two or more arrays. This is done by the & operator. Let's give an example. Assume that we want to merge two arrays num1 and num2. We can do this by num1&num2.

We provide dataflow model of concatenation operation on arrays in Listing 6.19. Here, two arrays num1 and num2 are concatenated. The RTL schematic of this description is a part of Fig. 6.8.

Listing 6.16 Testbench File for Arithmetic Operations on Two Eight-bit Arrays in VHDL

```vhdl
library ieee;
use ieee.numeric_std.all;

entity arithmetic_operations_tb is
end arithmetic_operations_tb;

architecture dataflow of arithmetic_operations_tb is

component arithmetic_operations
port(num1 : in unsigned (7 downto 0);
     num2 : in unsigned (7 downto 0);
     res1 : out unsigned (7 downto 0);
     res2 : out unsigned (7 downto 0);
     res3 : out unsigned (7 downto 0);
     res4 : out unsigned (15 downto 0);
     res5 : out unsigned (7 downto 0);
     res6 : out unsigned (7 downto 0);
     res7 : out unsigned (7 downto 0));
end component;

signal  in1 : unsigned (7 downto 0);
signal  in2 : unsigned (7 downto 0);
signal out1 : unsigned (7 downto 0);
signal out2 : unsigned (7 downto 0);
signal out3 : unsigned (7 downto 0);
signal out4 : unsigned (15 downto 0);
signal out5 : unsigned (7 downto 0);
signal out6 : unsigned (7 downto 0);
signal out7 : unsigned (7 downto 0);

begin
UUT:  arithmetic_operations port map (num1 => in1, num2 => in2, res1 =>
      out1, res2 => out2,     res3 => out3, res4 => out4,     res5 =>
      out5, res6 => out6,     res7 => out7);

process
begin
wait for 5 ns;

in1 <= "00001110";
in2 <= "00100111";

--wait;
end process;

end dataflow;
```

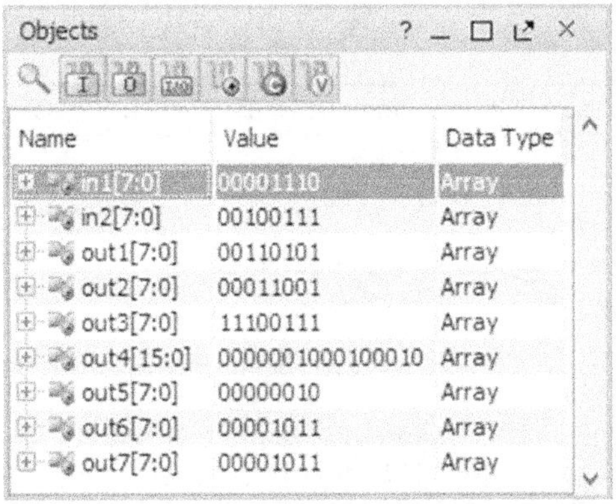

FIGURE 6.11 Result of arithmetic operations on two eight-bit arrays in VHDL.

Listing 6.17 Arithmetic Operations on Two 12-bit Arrays in VHDL

```vhdl
library ieee;
use ieee.numeric_std.all;

entity arithmetic_operations_UQ8_4 is
port(num1 : in unsigned (11 downto 0);
     num2 : in unsigned (11 downto 0);
     res1 : out unsigned (11 downto 0);
     res2 : out unsigned (11 downto 0);
     res3 : out unsigned (11 downto 0);
     res4 : out unsigned (23 downto 0);
     res5 : out unsigned (11 downto 0);
     res6 : out unsigned (11 downto 0);
     res7 : out unsigned (11 downto 0));
end arithmetic_operations_UQ8_4;

architecture dataflow_model of arithmetic_operations_UQ8_4 is
begin

--addition
res1 <= num1 + num2;

--subtraction
res2 <= num2 - num1;

res3 <= num1 - num2;

--multiplication
res4 <= num1 * num2;

--division
res5 <= num2 / num1;

--modulus
res6 <= num2 mod num1;

--remainder
res7 <= num2 rem num1;

end dataflow_model;
```

Listing 6.18   Testbench File for Arithmetic Operations on Two 12-bit Arrays in VHDL

```vhdl
library ieee;
use ieee.numeric_std.all;

entity arithmetic_operations_UQ8_4_tb is
end arithmetic_operations_UQ8_4_tb;

architecture dataflow of arithmetic_operations_UQ8_4_tb is

component arithmetic_operations_UQ8_4
port(num1 : in unsigned (11 downto 0);
     num2 : in unsigned (11 downto 0);
     res1 : out unsigned (11 downto 0);
     res2 : out unsigned (11 downto 0);
     res3 : out unsigned (11 downto 0);
     res4 : out unsigned (23 downto 0);
     res5 : out unsigned (11 downto 0);
     res6 : out unsigned (11 downto 0);
     res7 : out unsigned (11 downto 0));
end component;

signal  in1 : unsigned (11 downto 0);
signal  in2 : unsigned (11 downto 0);
signal out1 : unsigned (11 downto 0);
signal out2 : unsigned (11 downto 0);
signal out3 : unsigned (11 downto 0);
signal out4 : unsigned (23 downto 0);
signal out5 : unsigned (11 downto 0);
signal out6 : unsigned (11 downto 0);
signal out7 : unsigned (11 downto 0);

begin
UUT: arithmetic_operations_UQ8_4 port map (num1 => in1, num2 => in2,
        res1 => out1, res2 => out2,     res3 => out3, res4 => out4,
         res5 => out5, res6 => out6,     res7 => out7);

process
begin
wait for 5 ns;

in1 <= "000011100010";
in2 <= "001001110110";

--wait;
end process;

end dataflow;
```

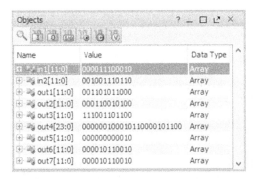

FIGURE **6.12**   Result of arithmetic operations on two 12-bit arrays in VHDL.

Listing **6.19**   Concatenation Operation in VHDL

```
library ieee;
use ieee.std_logic_1164.all;

entity concatenate is
port(num1 : in std_logic_vector (7 downto 0);
     num2 : in std_logic_vector (7 downto 0);
     res1 : out std_logic_vector (15 downto 0));
end concatenate;

architecture dataflow_model of concatenate is
begin
res1 <= num1 & num2;
end dataflow_model;
```

We provide the testbench file in Listing 6.20 to explain the concatenation operation on an example. Here, two arrays are taken as "11111010" and "00001111". The obtained result will be the same as in Fig. 6.9.

6.10   Application on Data Types and Operators

In this section, we will construct a primitive calculator to add, subtract, multiply, and divide two four-bit numbers on the Basys3 board. Input bits and the operation type is represented by switches on the board. Output bit values are represented by LEDs on the board. The reader can consult Sec. 4.8 related to this setup. In Listing 6.21, we provide Verilog description of the calculator.

6.11   FPGA Building Blocks Used in Data Types and Operators

We introduced several operators to process data in this chapter. The aim here is trying to show the reader how these are implemented in an FPGA. Therefore, he or she can grasp the fundamental idea in using this device. Note that the FPGA implementations provided in this section are not unique. They are the ones provided by Vivado. In other words, we are bound by Vivado's optimization tools in generating these implementations.

Listing 6.20 Testbench File for Concatenation Operation in VHDL

```vhdl
library ieee;
use ieee.std_logic_1164.all;

entity concatenate_tb is
end concatenate_tb;

architecture dataflow of concatenate_tb is

component concatenate
port(num1 : in std_logic_vector (7 downto 0);
     num2 : in std_logic_vector (7 downto 0);
     res1 : out std_logic_vector (15 downto 0));
end component;

signal  in1 : std_logic_vector (7 downto 0);
signal  in2 : std_logic_vector (7 downto 0);
signal  out1 : std_logic_vector (15 downto 0);

begin
UUT: concatenate port map (num1 => in1, num2 => in2, res1 => out1);

process
begin
wait for 5 ns;
in1 <= "11111010";
in2 <= "00001111";

end process;

end dataflow;
```

In this section, we picked Verilog descriptions used in the chapter. The reader may also test VHDL descriptions. However, we do not expect them to be totally different than the ones given here.

6.11.1 Implementation Details of Vector Operations

We first focus on vector operations in Listing 6.1. To show implementation details on this description, let's set the input vector length to four as input [3:0] num1. With this new form, a specific vector entry is selected (assign res1=num1[2]), subpart of a vector is selected (assign res2=num1[3:2]), and vector bit order is reversed. After synthesizing the modified description in Vivado, its schematic will be as in Fig. 6.13.

As can be seen in Fig. 6.13, 11 input/output ports are used in the implementation. Besides, each input or output port has an associated buffer with it. Moreover, only wiring is done between input and output ports. Therefore, this implementation only uses input/ output blocks and interconnect resources from the FPGA building blocks introduced in Sec. 2.2.

Listing 6.21 Calculator Implemented on the Basys3 Board in Verilog

```verilog
module calculator(led,sw);

//sw[7:4] and sw[3:0] represent numbers,
//sw[9:8] represents the arithmetic operation (+, - , *, /)
input [9:0] sw;
output [7:0] led;

wire [7:0] addition;
wire [7:0] subtraction;
wire [7:0] multiplication;
wire [7:0] division;

assign addition = sw[7:4] + sw[3:0];
assign subtraction = sw[7:4] - sw[3:0];
assign multiplication = sw[7:4] * sw[3:0];
assign division = sw[7:4] / sw[3:0];

assign led = ({8{~sw[8]}} & {8{~sw[9]}} & addition) +
({8{sw[8]}} & {8{~sw[9]}} & subtraction) +
({8{~sw[8]}} & {8{sw[9]}} & multiplication) +
({8{sw[8]}} & {8{sw[9]}} & division);

endmodule
```

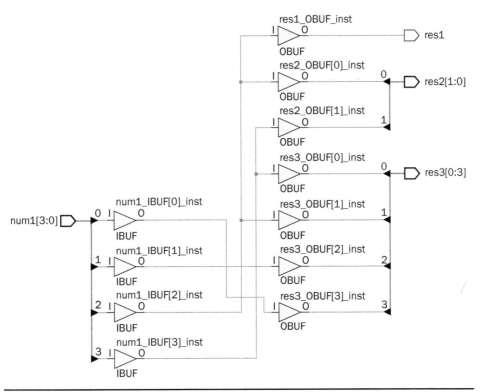

FIGURE 6.13 FPGA implementation of vector operations.

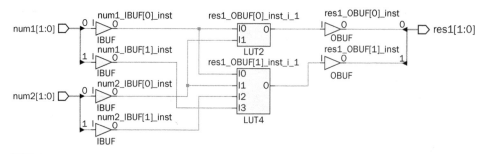

FIGURE 6.14 FPGA implementation of addition operation.

Listing 6.22 Enforcing Vivado to Use DSP Block in Arithmetic Operations in Verilog

```verilog
(* use_dsp48="yes" *)

module addition_operation_dsp48(num1,num2,res1);

input [7:0] num1;
input [7:0] num2;
output [7:0] res1;

assign res1=num1+num2;

endmodule
```

Listing 6.23 Enforcing Vivado to Use DSP Block in Arithmetic Operations in VHDL

```vhdl
library ieee;
use ieee.numeric_std.all;

entity addition_operation_dsp48 is
port(num1 : in unsigned (7 downto 0);
     num2 : in unsigned (7 downto 0);
     res1 : out unsigned (7 downto 0));

attribute use_dsp48 : string;
attribute use_dsp48 of addition_operation_dsp48 : entity is "yes";
end addition_operation_dsp48;

architecture dataflow_model of addition_operation_dsp48 is
begin

res1 <= num1 + num2;

end dataflow_model;
```

Implementation schematic should emphasize that no variable or memory element is used in the design as in a programming language. Only wires and ports are used. This is also the case for concatenation and replication operations in Listing 6.9.

6.11.2 Implementation Details of Arithmetic Operations

Implementing arithmetic operations in the FPGA is an important topic by itself. Therefore, let's closely analyze implementation details of the description in Listing 6.5. To understand how Vivado implements arithmetic operations, let's first focus on the addition operation. As in the previous section, let's apply addition on two two-bit vectors. Schematic of the description after synthesis will be as in Fig. 6.14. As can be seen in this figure, the addition operation is implemented by two LUTs in the FPGA.

The architecture in Fig. 6.14 is kept when subtraction, multiplication, and division operations are implemented. However, the reader should remember that these operations are done on two vectors each having two bits. If the vector length is increased, resource usage difference between arithmetic operations become more apparent. For example, when eight-bit addition, subtraction, multiplication, and division operations in Listing 6.5 are implemented separately, addition and subtraction operations will need eight LUTs. However, multiplication and division operations will need 67 and 69 LUTs, respectively. Moreover, if the bit length is increased to 12 as in Listing 6.7, then addition and subtraction operations will need 12 LUTs. The multiplication operation will need one DSP block. The division operation will need 155 LUTs. Hence, the multiplication and division operation implementations need extensive number of LUTs or DSP blocks. Note that LUT and DSP usage numbers are obtained using tools in Sec. 4.3.

We can enforce Vivado to synthesize arithmetic operations using DSP blocks. The way to do this in Verilog is adding attribute `(* use_dsp48="yes" *)` in front of the module to be handled this way. We provide such an example in Listing 6.22. Here, the addition operation is implemented using the DSP block. In VHDL, the same operation can be done by adding an attribute in the port definition part. We provide such an addition example in Listing 6.23. More information on this topic can be found in [28].

We can summarize basic findings in this section as follows. Arithmetic operations are implemented either using CLBs or DSP blocks in the FPGA. Besides, interconnect resources and input/output blocks are needed during implementation. There is one important issue. Size of data to be processed directly affects the resource usage. Related to this, multiplication and division operations may require heavy resource usage when data size increases.

6.12 Summary

We introduced key data type and operator concepts in this chapter. While doing this, we first explored number representations, negative numbers, and fixed- and floating-point numbers from a generic point of view. Then, we explored binary arithmetic operations. We next explored all of these concepts using HDLs. We postponed floating-point operation implementation in HDL descriptions till Chap. 13 since it requires advanced tools. We also analyzed HDL descriptions introduced in this chapter from an FPGA implementation perspective. The idea here was to give an insight how these descriptions are implemented in the FPGA. We will also apply the same methodology in the following chapters.

6.13 Exercises

6.1 Find the fixed-point representation of number 315.2342 in formats
  a. UQ16.
  b. UQ.16
  c. UQ16.16

6.2 Find the fixed-point representation of numbers −315.2342 and 315.2342 in formats
  a. Q15.
  b. Q.15
  c. Q15.16.

6.3 You have four numbers as 13.25, 15.50, 17.50, and 19.25. Find the hexadecimal representation of these numbers in fixed-point UQ16.16 format.

6.4 Find the floating-point representation of numbers −315.2342 and 315.2342 in formats
  a. half
  b. single
  c. double

6.5 We will only have an approximation in representing the number 8751.135 in half floating-point form. What is the difference between the actual number and this approximation?

6.6 Find the floating-point representation of the number 8751.135 in single form. Will there be an approximation here?

6.7 Find the floating-point representation of π in half form.

6.8 The ASCII codes given in Table 6.6 are called regular. What happens if we want to represent regional characters like ü, ı, and ç?

6.9 Two 16-bit numbers are taken as FFFF and 0005 in hexadecimal form. Write a Verilog or VHDL description and its testbench to implement and simulate below operations.
  a. FFFF+0005
  b. FFFF-0005
  c. 0005-FFFF

6.10 We know that only lowercase characters enter a system. Write a Verilog or VHDL module to convert each entry to uppercase form. Simulate the result by forming a testbench file.

6.11 Vivado offers an IP block called Adder/Subtracter in its IP Catalog. Use it to implement addition and subtraction operations in previous exercises in Verilog or VHDL.

6.12 What will be the value of y2, y1, y0 when the below Verilog description is simulated? The input is set as x=8h'4F for simulation.

```
module question12(y2, y1, y0, x);

output reg [7:0] y2, y1, y0;
input [7:0] x;

initial
begin
y0=8'h00;
y1=8'h00;
y2=8'h00;
end

always @ (x)
begin
y0 <= x/2;
y1 <= y0 + x;
y2 <= y1 + x*2;
end

endmodule
```

6.13 Form a Verilog description in behavioral modeling to calculate cube of a given number. Only one multiplication operation can be used at once.
   a. use nonblocking assignments
   b. use blocking assignments

6.14 **(Joystick application.)** A two-axis joystick provides analog voltage values corresponding to its horizontal (x-axis) and vertical (y-axis) position when an analog interfacing is done. These analog voltage values can be converted to digital form by an analog-to-digital converter (ADC) module. Assume that the analog interfacing is done and the ADC module is set to work. Hence, you get two vectors as xp and yp each with 12 bits each. We will take the most significant eight bits for xp

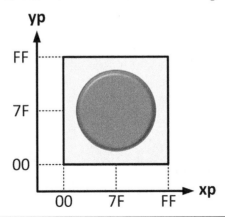

FIGURE 6.15 Sample readings from the joystick.

and yp. Hence, sample hexadecimal values of these vectors with respect to joystick position are as in Fig. 6.15.

We will use LEDs and switches on the Basys3 board for our operation. Therefore, LEDs and switches 15 to 8 are assigned to the vertical position (yp array) reading. LEDs and switches 7 to 0 are assigned to the horizontal position (xp array) reading.

a. Form a Verilog or VHDL description to display the values of joystick axes directly via designated LEDs.

b. Let's design a simple game using our setup. The first user forms a 16-bit pattern with setting each switch as on or off. The second user (without seeing this pattern) tries to match this pattern by moving the joystick in x- and y-axes. When the second user matches the pattern with the joystick position, all LEDs will turn off. Form a Verilog or VHDL description to realize this game.

CHAPTER 7

Combinational Circuits

A digital system can be implemented in two forms. In the first one, output depends on current input only. This form can be realized by combinational circuits, which is the main topic of this chapter. In the second form, output depends on past input or output values besides the current input. This form can be realized by sequential circuits, which will be introduced in Chap. 10.

A combinational circuit is composed of logic gates to perform a specific task. To understand the working principles of a combinational circuit, we will start with basic definitions. Then, we will review logic gates from a combinational circuit perspective. Afterward, we will introduce tools to analyze combinational circuits. Related to this, we will explore how a combinational circuit can be implemented in an field-programmable gate array (FPGA). Then, we will evaluate combinational circuit design steps. We will also provide sample designs so that the reader can grasp the idea in designing such a circuit. We will finally summarize how FPGA building blocks are used in combinational circuit implementation.

7.1 Basic Definitions

Before going further, we should make basic definitions which will be used throughout the book. Let's start with defining binary variable.

7.1.1 Binary Variable

While analyzing or designing a combinational circuit, logic level at certain location may be needed. To represent this value in generic form, we will assign a binary variable at that location. This variable can only take either logic level zero or one by its definition. Since we will be extensively using these logic levels for binary variables, we will call them as 0 and 1 from this point on.

The customary way to represent a binary variable is using characters. Throughout the book, we will adopt the same methodology by using characters such as x, y, and z for this purpose. Therefore, we can represent value of a binary variable as $x = 1$ or $x = 0$.

7.1.2 Logic Function

A logic function by its definition is formed of logic gates operating on binary variables. To be more specific, inputs of a logic function are defined as binary variables. Then, logic gates operating on these produce output, again as a binary variable. This will allow us to represent a combinational circuit in formal way.

Inputs	Outputs
Binary variables	**Binary variables**
Input	Corresponding
combinations	outputs

TABLE 7.1 Generic Truth Table

We will represent a logic function by capital letter throughout the book. One such example is $z = F(x, y)$. Here, the logic function F is defined on two binary variables x and y. Output of the function is another binary variable z. Depending on the definition of the logic function F, z will be represented in terms of binary variables x and y.

7.1.3 Truth Table

One way of describing input/output characteristics of a logic function is by forming its truth table which will tabulate all input combinations on its left-hand side. For each input combination, corresponding output will be provided on the right-hand side of the table. Hence, a generic truth table will be as in Table 7.1.

Let's assume that the logic function (or corresponding combinational circuit) has N input variables. Since each binary variable can take two values, total number of input combinations will be 2^N. The truth table should tabulate all these combinations. Output of the logic function for each input combination is either 0 or 1. Therefore, the truth table describes combinational circuit characteristics precisely. In other words, we know how the combinational circuit behaves for any given input. Hence, the truth table will be our main tool in analyzing and designing combinational circuits.

7.2 Logic Gates

We have introduced logic gates as digital electronic devices in Sec. 2.1. Here, we review them by focusing on their combinational characteristics. Moreover, we provide hardware description language (HDL) description of all logic gates considered here.

7.2.1 The NOT Gate

NOT is the first logic gate to be considered. It is actually an inverter with single input and output. Let's assume that input to the NOT gate is represented by binary variable x; and let output of the gate be binary variable y. Then, the NOT gate can be represented by the logic function $y = \bar{x}$. Truth table of the NOT gate based on this logic function will be as in Table 7.2. Symbol of the NOT gate for this logic function is as in Fig. 7.1.

Input	Output
x	$\mathbf{z = \bar{x}}$
0	1
1	0

TABLE 7.2 Truth Table of the NOT Gate

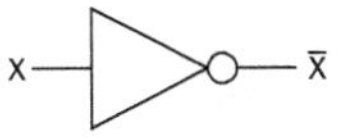

FIGURE 7.1 Symbol of the NOT gate.

7.2.1.1 The NOT Gate in Verilog

The NOT gate has a specific keyword not for structural modeling in Verilog. For dataflow and behavioral modeling, operator for the NOT gate is " $\sim$ ". Using these, we can describe the logic function $y = \bar{x}$ in Verilog as follows:

```
// Structural modeling
not not_gate(y,x);

// Dataflow modeling
assign y = ~x;

// Behavioral modeling
y = ~x;
```

Here, y and x correspond to output and input of the NOT gate, respectively. We named the NOT gate as not_gate in structural modeling.

7.2.1.2 The NOT Gate in VHDL

The VHDL keyword for the NOT gate is not. Using it, we can describe the logic function $y = \bar{x}$ as follows:

```
library ieee;
use ieee.std_logic_1164.all;

entity not_gate is
port (x : in std_logic;
          y : out std_logic);
end not_gate;

architecture dataflow_model of not_gate is
begin
y<= not x;
end dataflow_model;
```

7.2.2 The OR Gate

OR is the second logic gate to be considered. It may have two or more inputs. However, the gate has one output. The working principles of the OR gate are as follows. Whenever any of its inputs has value 1, output will be 1. Output will be 0 if and only if all inputs have value 0. To represent the input/output characteristics of the OR gate, let's assume it has two inputs as binary variables x and y; and let output of the gate be binary variable z. The operator to represent the OR gate is " $+$ ". Based on these, the two-input OR gate can be represented by the logic function $z = x + y$. The truth table of the OR gate based on this logic function will be as in Table 7.3. The symbol of the OR gate for this logic function is as in Fig. 7.2.

Inputs		Output
x	**y**	**z = x + y**
0	0	0
0	1	1
1	0	1
1	1	1

TABLE 7.3 Truth Table of the OR Gate

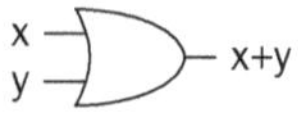

FIGURE 7.2 Symbol of the OR gate.

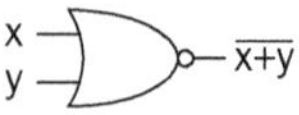

FIGURE 7.3 Symbol of the NOR gate.

A NOT gate can be connected to output of the OR gate. This combination forms the NOR (NOT-OR) gate. As in the OR gate, let's assume inputs of this gate be represented by binary variables x and y; and let output of the gate be binary variable z. Then, the two-input NOR gate can be represented by the logic function $z = \overline{x + y}$. The truth table of the NOR gate based on this logic function will be as in Table 7.3. Only output values will be inverted. The symbol of the NOR gate for this logic function is as in Fig. 7.3.

7.2.2.1 The OR Gate in Verilog

The OR gate has a specific keyword `or` for structural modeling in Verilog. For dataflow and behavioral modeling, the operator of the OR gate is " | ". Using these, we can describe the logic function $z = x + y$ in Verilog as follows:

```
// Structural modeling
or or_gate(z, x, y);

// Dataflow modeling
assign z = x | y;

// Behavioral modeling
z = x | y;
```

Here, z corresponds to output of the OR gate. x and y correspond to inputs of the gate. Note that we can increase the number of inputs as we like. In structural modeling, we named the OR gate as `or_gate`.

7.2.2.2 *The OR Gate in VHDL*

The VHDL keyword for OR gate is `or`. Using it, we can describe the logic function $z = x + y$ as follows:

```
library ieee;
use ieee.std_logic_1164.all;

entity or_gate is
port (x : in std_logic;
         y : in std_logic;
         z : out std_logic);
end or_gate;

architecture dataflow_model of or_gate is
begin
z <= x or y;
end dataflow_model;
```

7.2.3   The AND Gate

AND is the third logic gate to be considered. As in the OR gate, it may have two or more inputs. However, it has one output. The working principles of the AND gate are as follows. Whenever all of its inputs have value 1, output will be 1. Output will be 0 if any of the inputs has value 0. To represent input/output characteristics of the AND gate, let's assume two inputs as binary variables x and y; and let output of the gate be binary variable z. Operator to represent the AND gate is " $\cdot$ ". Based on these, the two-input AND gate can be represented by the logic function $z = x \cdot y$. The truth table of the AND gate based on this logic function will be as in Table 7.4. The symbol of the AND gate for this logic function is as in Fig. 7.4.

A NOT gate can be connected to output of the AND gate. This combination forms the NAND gate. As in the AND gate, let's assume input to this gate be binary variables x and y; and output of the gate be binary variable z. Then, the two-input NAND gate can be represented by the logic function $z = \overline{x \cdot y}$. The truth table of the NAND gate based on this logic function will be as in Table 7.4. Only output values will be inverted. The symbol of the NAND gate for this logic function is as in Fig. 7.5.

Inputs		Output
x	**y**	**z = x · y**
0	0	0
0	1	0
1	0	0
1	1	1

TABLE 7.4   Truth Table of the AND Gate

FIGURE 7.4   Symbol of the AND gate.

FIGURE 7.5   Symbol of the NAND gate.

7.2.3.1   The AND Gate in Verilog

The AND gate has a specific keyword and for structural modeling in Verilog. For dataflow and behavioral modeling, the operator for the AND gate is " & ". Using these, we can describe the logic function $z = x \cdot y$ in Verilog as follows:

```
// Structural modeling
and and_gate(z, x, y);

// Dataflow modeling
assign z = x & y;

// Behavioral modeling
z = x & y;
```

Here, z corresponds to output of the AND gate. x and y correspond to inputs of the gate. Note that we can increase the number of inputs as we like. In structural modeling, we named the AND gate as and_gate.

7.2.3.2   The AND Gate in VHDL

The VHDL keyword for the AND gate is and. Using it, we can describe the logic function $z = x \cdot y$ as follows:

```
library ieee;
use ieee.std_logic_1164.all;

entity and_gate is
port (x : in std_logic;
          y : in std_logic;
          z : out std_logic);
end and_gate;

architecture dataflow_model of and_gate is
begin
z <= x and y;
end dataflow_model;
```

7.2.4   The XOR Gate

The fourth and final logic gate to be considered is XOR (Exclusive-OR). This gate can be constructed by using AND, OR, and NOT gates. Therefore, it may or may not be taken as a fundamental logic gate. However, XOR is used in combinational circuit representation. Therefore, we explore it in this section. The working principles of the XOR gate are as follows. When two inputs of the gate have the same logic level (either 0 or 1), its output will be 0. Whenever the two inputs of the gate have different logic levels, its output will be 1. To represent input/output characteristics of the XOR gate based on this definition, let's assume two inputs as binary variables x and y. Let output of the gate be binary variable z. Then, two-input XOR gate can be represented by the logic function $z = (x \cdot \overline{y}) + (\overline{x} \cdot y)$. This logic function can be simplified by using the " $\oplus$ " operator to represent

Inputs		Output
x	**y**	**z = x ⊕ y**
0	0	0
0	1	1
1	0	1
1	1	0

TABLE 7.5 Truth Table of the XOR Gate

FIGURE 7.6 Symbol of the XOR gate.

the XOR gate as $z = x \oplus y$. The truth table of the XOR gate based on this logic function will be as in Table 7.5. The symbol of the XOR gate for this logic function is as in Fig. 7.6.

7.2.4.1 The XOR Gate in Verilog

The XOR gate has a specific keyword xor for structural modeling in Verilog. For dataflow and behavioral modeling, the operator for the XOR gate is "^". Using these, we can describe the logic function $z = x \oplus y$ in Verilog as follows:

```verilog
// Structural modeling
xor xor_gate(z, x, y);

// Dataflow modeling
assign z = x ^ y;

// Behavioral modeling
z = x ^ y;
```

Here, z corresponds to output of the XOR gate. x and y correspond to inputs of the gate. In structural modeling, we named the XOR gate as xor_gate.

7.2.4.2 The XOR Gate in VHDL

The VHDL keyword for the XOR gate is xor. Using it, we can describe the logic function $z = x \oplus y$ as follows:

```vhdl
library ieee;
use ieee.std_logic_1164.all;

entity xor_gate is
port (x : in std_logic;
         y : in std_logic;
         z : out std_logic);
end xor_gate;

architecture dataflow_model of xor_gate is
begin
z <= x xor y;
end dataflow_model;
```

7.3 Combinational Circuit Analysis

Logic gates introduced in the previous section can be used to construct combinational circuits. To understand the working principles of a combinational circuit, we should analyze it. Therefore, we should first form a logic function between its inputs and output(s). If needed, we can also form the truth table of combinational circuit based on this representation. The final step in analysis is representing the combinational circuit by less (or simpler) elements, which is called gate-level minimization.

7.3.1 Logic Function Formation between Input and Output

The first step in analyzing a combinational circuit is forming the logic function between its inputs and output(s). Here, we assume that the corresponding circuit diagram is at hand. Then, we should "read" this diagram. Let's give a simple example. Assume that a combinational circuit has been designed beforehand by discrete logic gates as in Fig. 7.7. We would like to form the corresponding logic function.

In Fig. 7.7, we specifically labeled output of each logic gate by a binary variable. Based on these, we can represent input/output characteristics of the combinational circuit. To do so, we first obtain output of each logic gate separately as follows:

$$z1 = x \cdot y$$
$$z2 = x + y$$
$$z3 = \overline{x \cdot y}$$
$$z4 = \overline{y}$$
$$z5 = z1 \oplus z2$$
$$z = \overline{z5 + z3 + z4}$$

These lead to input/output characteristics of the combinational circuit as follows:

$$z = \overline{((x \cdot y) \oplus (x + y)) + \overline{x \cdot y} + \overline{y}}$$

This logic function can be implemented by an HDL in an FPGA. However, some simplifications can be done on it before its implementation. Next, we will consider how this can be done.

7.3.2 Boolean Algebra

We can benefit from Boolean algebra for gate-level minimization. Boolean algebra is the framework to represent and analyze logic functions formed by binary variables and

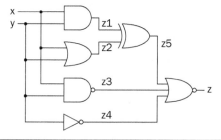

FIGURE 7.7   Circuit diagram of a combinational circuit.

logic gates. Boolean algebra can be explained in a rigorous way using mathematical definitions. However, we will take a simpler approach in this book. The idea is to cover basic definitions of Boolean algebra necessary for our purposes.

First, we will review basic identities by Boolean algebra. Let's assume two binary variables x and y. We can define identities on AND and OR gates as in Table 7.6. Although these identities can be justified by using a truth table, the reader can consult the mentioned reference for more rigorous proof [29].

We can describe Boolean algebra identities in Verilog as in Listing 7.1. Here, the output of identities are represented by two arrays y_or[3:0] and y_and[3:0]. Corresponding VHDL description will be as in Listing 7.2. Synthesis result of the Verilog description is as in Fig. 7.8. As can be seen in this figure, Vivado's optimization tool actually applied Boolean identities such that outputs are simplified accordingly. Note that ground and supply voltage levels are represented by special signs in this figure.

Next, we will review basic Boolean algebra properties on AND, OR, and NOT gates (or operations corresponding to them). These are involution, commutative, associative, distributive, and absorption properties and DeMorgan's theorem as summarized in Table 7.7. Involution property tells us that applying NOT on a binary variable twice gives its original value. Commutative property tells us that the order of variables in logic gates

The OR gate	The AND gate
$x + 0 = x$	$x \cdot 0 = 0$
$x + 1 = 1$	$x \cdot 1 = x$
$x + x = x$	$x \cdot x = x$
$x + \overline{x} = 1$	$x \cdot \overline{x} = 0$

TABLE 7.6 Boolean Algebra Identities

Listing 7.1 Boolean Identity Operations in Verilog

```verilog
module Boolean_identity(y_or,y_and,x);

parameter one=1'b1;
parameter zero=1'b0;

input x;
output [3:0] y_or;
output [3:0] y_and;

assign y_or[0] = x | zero;
assign y_or[1] = x | one;
assign y_or[2] = x | x;
assign y_or[3] = x | ~x;

assign y_and[0] = x & zero;
assign y_and[1] = x & one;
assign y_and[2] = x & x;
assign y_and[3] = x & ~x;

endmodule
```

Listing 7.2 Boolean Identity Operations in VHDL

```vhdl
library ieee;
use ieee.std_logic_1164.all;

entity Boolean_identity is
port(x : in std_logic;
  y_or : out std_logic_vector (3 downto 0);
  y_and : out std_logic_vector (3 downto 0));
end Boolean_identity;

architecture dataflow_model of Boolean_identity is

constant one  : std_logic :='1';
constant zero : std_logic :='0';

begin
y_or(0) <= x or zero;
y_or(1) <= x or one;
y_or(2) <= x or x;
y_or(3) <= x or not x;

y_and(0) <= x and zero;
y_and(1) <= x and one;
y_and(2) <= x and x;
y_and(3) <= x and not x;

end dataflow_model;
```

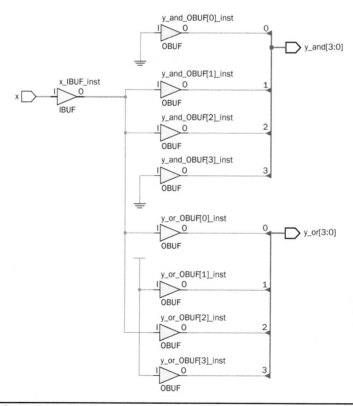

FIGURE 7.8 Synthesis result of Boolean identities.

Property		
Involution	$\bar{\bar{x}} = x$	
Commutative	$x + y = y + x$	$x \cdot y = y \cdot x$
Associative	$x + (y + z) = (x + y) + z$	$x \cdot (y \cdot z) = (x \cdot y) \cdot z$
Distributive	$x \cdot (y + z) = (x \cdot y) + (x \cdot z)$	$x + (y \cdot z) = (x + y) \cdot (x + z)$
Absorption	$x + (x \cdot y) = x$	$x \cdot (x + y) = x$
DeMorgan's theorem	$\overline{x + y} = \bar{x} \cdot \bar{y}$	$\overline{x \cdot y} = \bar{x} + \bar{y}$

TABLE 7.7   Boolean Algebra Properties on AND, OR, and NOT Operations

Inputs		Output
x	**y**	**z**
0	0	0
0	1	0
1	0	0
1	1	1

TABLE 7.8   Truth Table of the Combinational Circuit in Fig. 7.7

is not important. Associative property tells us that if more than one operation is done, then the order is not important. Distributive property tells us that AND and OR operations are distributive on each other. As the name implies, absorption property discards unnecessary variables. The reader can remember DeMorgan's theorem as follows. If the NOT operation is applied on an AND or OR operation, inputs will be inverted. Moreover, the operation will be changed from AND to OR or vice versa. Again, rigorous proof of these properties can be found in [29].

7.3.3   Gate-Level Minimization

Gate-level minimization aims to simplify input/output characteristics of a combinational circuit. The idea here is obtaining the same truth table with less number of logic gates. This operation can be done using Boolean algebra identities and properties introduced in the previous section. However, this requires expertise. There are also very effective methods for gate-level minimization. In this book, we will depend on Vivado's optimization tool for gate-level minimization since it can handle most cases very effectively. This does not mean that the reader should not know basics of gate-level minimization.

Let's see how gate-level minimization can be done on two examples. The first combinational circuit to be minimized is the one in Fig. 7.7. As can be seen there, the circuit is composed of six logic gates. The truth table of this combinational circuit is as in Table 7.8. This truth table corresponds to the AND gate. Hence, the combinational circuit can be represented by the logic function $z = x \cdot y$. Therefore, one logic gate is sufficient to implement it instead of using six gates.

As second example, let's take the combinational circuit with the logic function $z = \bar{x} \cdot y + x \cdot y$. Boolean algebra identity and properties given in Tables 7.6 and 7.7 can be used to simplify this logic function such that the end result will be $z = y$. In other words, input x does not have any effect on the output of the combinational circuit. We provide Verilog and VHDL descriptions of this combinational circuit in Listings 7.3 and 7.4.

Let's take the Verilog description in Listing 7.3. We provide initial form of the combinational circuit in Vivado (the RTL design) in Fig. 7.9a. As can be seen in this figure, the combinational circuit is constructed exactly as represented by the Verilog description. We also provide the synthesization result in Fig. 7.9b. As can be seen in this figure,

Listing 7.3 Verilog Description of the Combinational Circuit to be Minimized

```
module minimization_example(z,x,y);

input x,y;
output z;

assign z=(~x&y)|(x&y);

endmodule
```

Listing 7.4 VHDL Description of the Combinational Circuit to be Minimized

```
library ieee;
use ieee.std_logic_1164.all;

entity minimization_example is
port(x : in std_logic;
     y : in std_logic;
     z : out std_logic);
end minimization_example;

architecture dataflow_model of minimization_example is
begin
z <= (not x and y) or (x and y);
end dataflow_model;
```

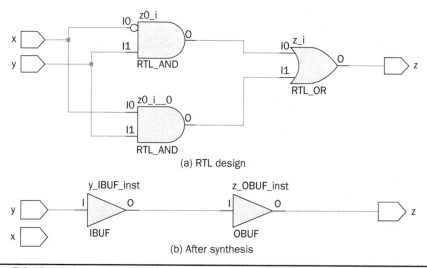

(a) RTL design

(b) After synthesis

Figure 7.9 Gate-level minimization example in Vivado.

input x is not connected to any logic block. Therefore, Vivado's optimization tool worked as expected.

7.4 Combinational Circuit Implementation

We can implement a combinational circuit using several methods as explained in Sec. 2.3. Since the main focus of this book is on the FPGA-based implementation, we will use the look-up table (LUT) representation for combinational circuits here. To do so, we will start with the truth table–based implementation next. Then, we will consider implementing combinational circuits with different number of inputs.

7.4.1 Truth Table-Based Implementation

A combinational circuit can be implemented when its truth table is available. The idea here is focusing on input combinations producing output 0 or 1 separately. Each input combination can be represented by a standard logic function. This leads to the overall logic function of the combinational circuit.

To explain the truth table–based implementation methodology, let's first focus on input combinations producing output 1. Assume that the truth table of a two-input combinational circuit is as in Table 7.9. As can be seen in this table, the output z will be 1 when $x = 0$ and $y = 1$ or $x = 1$ and $y = 1$. This helps us forming the logic function for the combinational circuit as follows. First, z should be 1 when $x = 0$ and $y = 1$. We can satisfy this constraint by the logic function $z = \bar{x} \cdot y$. Second, z should be 1 when $x = 1$ and $y = 1$. Using the same reasoning, we can form the logic function $z = x \cdot y$. Now, z will be 1 when either the first or second constraint is satisfied. Therefore, we can form the final logic function as $z = \bar{x} \cdot y + x \cdot y$.

The logic function $z = \bar{x} \cdot y + x \cdot y$ can be described by only mentioning which input combinations produce output 1. This representation is called sum of products (SOP). As the name implies, each constraint is represented by an AND gate. The final logic function is formed by applying OR gate to all constraints. Hence, the name sum of products. For our example, the SOP form will be as $z = \sum(1,3)$. Here, the sum sign represents the SOP form. The numbers within the parentheses stand for which input combinations produce the output 1.

The truth table–based implementation can also be done by focusing on input combinations producing the output 0 as the second case. Here, we can modify the truth table by taking the inverse of the output. Then, it becomes as in Table 7.10. We can form

Inputs		Output
x	y	z
0	0	0
0	1	1
1	0	0
1	1	1

TABLE 7.9 Truth Table of the Example Two-Input Combinational Circuit

Inputs		Output
x	y	$\bar{z}$
0	0	1
0	1	0
1	0	1
1	1	0

TABLE 7.10 Modified Truth Table of the Example Two-Input Combinational Circuit

Input	Output
x	y
0	F(0)
1	F(1)

TABLE 7.11 Generic Truth Table of a One-Input Combinational Circuit

the logic function for $\bar{z}$ using the SOP representation as $\bar{z} = \sum(0, 2)$. Or, as a logic function it becomes $\bar{z} = \bar{x} \cdot \bar{y} + x \cdot \bar{y}$. Using Boolean algebra properties in Table 7.7, we can obtain $z = \overline{\bar{x} \cdot \bar{y} + x \cdot \bar{y}}$. This logic function can be represented in the simplified form as $z = (x + y) \cdot (\bar{x} + y)$. This representation is called product of sums (POS). Different from SOP, here each constraint is represented by an OR gate. The final logic function is formed by applying AND gate to all constraints. Hence, the name product of sums. The above example can be represented in POS form as $z = \prod(0, 2)$. Here, the product sign represents the POS form. Numbers within the parentheses stand for which input combinations produce the output 0.

The reader is free to choose the SOP or POS form in implementation. However, it is advisable to choose the one which requires the less number of logic operations (gates) in implementation. Next, we will focus on the multiplexer-based implementation methodology for SOP and POS forms.

7.4.2 Implementing One-Input Combinational Circuits

Combinational circuits are implemented by LUTs in an FPGA. As explained in Sec. 2.2.3, a generic LUT is composed of a multiplexer and memory elements in its basic form. Therefore, we will explore how a logic function (corresponding to a combinational circuit) can be implemented by memory elements and multiplexers in this and the following sections.

The first group of combinational circuits to be explored has one input. We can represent a generic logic function for such a combinational circuit as $y = F(x)$. Here, x and y are the input and output variables, respectively. To implement this logic function by a multiplexer, we should first form its truth table. The generic truth table will be as in Table 7.11.

Based on the truth table in Table 7.11, we can construct an implementation using a two-to-one multiplexer and memory elements as in Fig. 7.10. Here, the select pin of

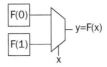

FIGURE 7.10   Generic implementation of a one-input combinational circuit.

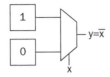

FIGURE 7.11   NOT gate implemented by a two-to-one multiplexer.

Inputs		Output
x	y	z
0	0	F(0,0)
0	1	F(0,1)
1	0	F(1,0)
1	1	F(1,1)

TABLE 7.12   Generic Truth Table of a
Two-Input Combinational Circuit

the multiplexer is set as the input variable x. The multiplexer input pins are connected
to memory elements which are set according to output values of the logic function to be
implemented as indicated in its truth table. The multiplexer output corresponds to the
output of the logic function y.

We can take the NOT gate as an example of one-input combinational circuit. Based
on its truth table in Table 7.2, implementation of this gate will be as in Fig. 7.11. As can
be seen in this figure, characteristics of the setup can be changed just by changing input
values of the multiplexer (set as memory elements).

7.4.3   Implementing Two-Input Combinational Circuits

The second group of combinational circuits to be explored has two inputs. We can form
a generic logic function to represent such a combinational circuit as $z = F(x, y)$. Here,
x and y are input variables and z is the output of the logic function. The truth table of
this function will be as in Table 7.12.

We can implement the logic function $z = F(x, y)$ in two different ways. First, the
truth table in Table 7.12 leads to the structure in Fig. 7.12 as in the previous section.
Here, a four-to-one multiplexer and memory elements are used. The select pins of the
multiplexer are set as input variables x and y. The multiplexer input pins are connected
to memory elements which are set according to output values of the logic function to be
implemented as indicated in its truth table. The multiplexer output corresponds to the
output of the logic function z.

Let's consider the two-input OR, AND, and XOR gates as examples. These can be
implemented using the structure in Fig. 7.12 by their truth table as in Fig. 7.13.

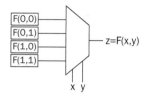

FIGURE 7.12 Generic implementation of a two-input combinational circuit.

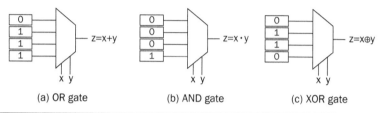

(a) OR gate   (b) AND gate   (c) XOR gate

FIGURE 7.13 OR, AND, and XOR gates implemented by a four-to-one multiplexer.

Inputs		Output	
x	y	z	
0	0	F(0,0)	*first*
0	1	F(0,1)	*part*
1	0	F(1,0)	*second*
1	1	F(1,1)	*part*

TABLE 7.13 Generic Truth Table of a Two-Input Combinational Circuit in Decomposed Form

The second implementation method for the logic function $z = F(x,y)$ is by using three separate two-to-one multiplexers. To explain this structure, let's closely look at Table 7.12. As can be seen in this table, the variable x will have the value 0 for the first two input combinations. It will have the value 1 for the last two input combinations. This allows us to decompose the truth table into two parts as in Table 7.13.

Let's consider a hierarchical implementation strategy based on Table 7.13. To do so, we should initially handle the first and second parts. Since the binary variable x is fixed for each subpart, we will consider only the binary variable y. Therefore, the first and second parts can be implemented by two two-to-one multiplexers. Input values of the first multiplexer will be $F(0,0)$ and $F(0,1)$. Input values of the second multiplexer will be $F(1,0)$ and $F(1,1)$. The select pin of these multiplexers will be set as the binary variable y. The output of these multiplexers is fed to another two-to-one multiplexer as input. The select pin of this multiplexer will be connected to the binary variable x. The output of this multiplexer corresponds to the output of the logic function $F(x,y)$. Therefore, this multiplexer will decide which part in Table 7.13 will be connected to the output. Generic structure of this hierarchical implementation will be as in Fig. 7.14. Logic gates in Fig. 7.13 can also be implemented this way.

As can be seen in Fig. 7.14, the hierarchical implementation is more complex compared to the one in Fig. 7.12. However, it has one main advantage. This structure allows

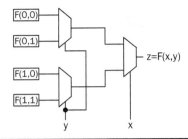

Figure 7.14   Generic implementation of a two-input combinational circuit using hierarchical structure.

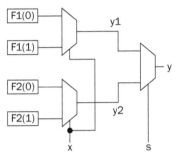

Figure 7.15   Generic implementation of two one-input combinational circuits using hierarchical structure.

Inputs			Output
x	**y**	**z**	**w**
0	0	0	F(0,0,0)
0	0	1	F(0,0,1)
0	1	0	F(0,1,0)
0	1	1	F(0,1,1)
1	0	0	F(1,0,0)
1	0	1	F(1,0,1)
1	1	0	F(1,1,0)
1	1	1	F(1,1,1)

Table 7.14   Generic Truth Table of a Three-Input Combinational Circuit

implementing two different one-input combinational circuits with the same input. Let's assume that we have two such logic functions as $y1 = F1(x)$ and $y2 = F2(x)$. We can implement these using the hierarchical structure as in Fig. 7.15. Here, the binary variable s decides on which logic function is active.

7.4.4   Implementing Three-Input Combinational Circuits

The third and final group of combinational circuits to be explored has three inputs. We can form a generic logic function to represent such a combinational circuit as $w = F(x, y, z)$. Here, $x, y,$ and z are the input variables and w is the output of the logic function. The truth table of this function will be as in Table 7.14.

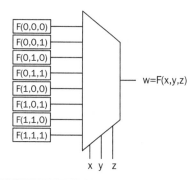

FIGURE 7.16   Generic implementation of a three-input combinational circuit.

Inputs			Output	
x	**y**	**z**	**w**	
0	0	0	F(0,0,0)	
0	0	1	F(0,0,1)	*first*
0	1	0	F(0,1,0)	*part*
0	1	1	F(0,1,1)	
1	0	0	F(1,0,0)	
1	0	1	F(1,0,1)	*second*
1	1	0	F(1,1,0)	*part*
1	1	1	F(1,1,1)	

TABLE 7.15   Generic Truth Table of a Three-Input Combinational Circuit Decomposed into Two Parts

We can implement the logic function $w = F(x, y, z)$ in three different ways. The first implementation method is based on a single eight-to-one multiplexer as in Fig. 7.16. This is the straightforward method as introduced in the previous sections.

The second and third implementation methods for the logic function $w = F(x, y, z)$ are based on the hierarchical structure introduced in the previous section. Let's start with the second implementation method by decomposing the truth table of the logic function $w = F(x, y, z)$ into two parts as in Table 7.15. We can implement the first and second parts separately using four-to-one multiplexers. The final form of this implementation will be as in Fig. 7.17. Similar to the previous section, this structure can also be used to implement two different two-input combinational circuits as $z1 = F1(x, y)$ and $z2 = F2(x, y)$. We can implement these using the hierarchical structure as in Fig. 7.18. Here, the binary variable s decides on which logic function is active.

The third implementation method for the logic function $w = F(x, y, z)$ is based on the hierarchical structure using two-to-one multiplexers. To do so, we should decompose the truth table of the logic function $w = F(x, y, z)$ into four parts as in Table 7.16. This leads to the implementation as in Fig. 7.19. This structure can also be used to implement three one-input combinational circuits.

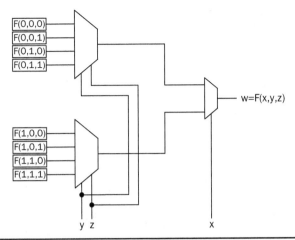

FIGURE 7.17 Generic implementation of a three-input combinational circuit using four-to-one multiplexers.

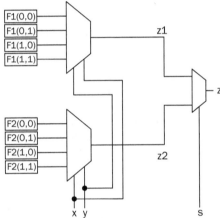

FIGURE 7.18 Generic implementation of two two-input combinational circuits using hierarchical structure.

Inputs			Output	
x	**y**	**z**	**w**	
0	0	0	F(0,0,0)	*first*
0	0	1	F(0,0,1)	*part*
0	1	0	F(0,1,0)	*second*
0	1	1	F(0,1,1)	*part*
1	0	0	F(1,0,0)	*third*
1	0	1	F(1,0,1)	*part*
1	1	0	F(1,1,0)	*fourth*
1	1	1	F(1,1,1)	*part*

TABLE 7.16 Generic Truth Table of a Three-Input Combinational Circuit Decomposed into Four Parts

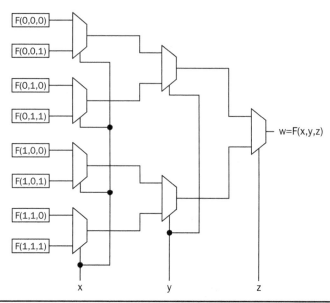

FIGURE 7.19 Generic implementation of a three-input combinational circuit using two-to-one multiplexers.

The hierarchical implementation strategy can be generalized to combinational circuits with more than three inputs. In fact, a similar idea has been applied to the LUT formation in FPGAs as mentioned in Sec. 2.2.3. There, it is mentioned that in the Artix-7 FPGA, each CLB slice has four six-input LUTs. This allows two seven-input LUT or one eight-input LUT formation.

7.5 Combinational Circuit Design

Designing a combinational circuit requires expertise. Moreover, this is a topic of its own. On the other hand, there are some standard steps to be followed for any design. In this section, we will introduce these steps such that they can be applied throughout the book.

7.5.1 Analyzing the Problem to Be Solved

The first and most important step in designing a combinational circuit is understanding the problem to be solved. In other words, the problem to be solved should be well-defined. This leads to forming the solution. At this step, design constraints should also be taken into account. Related to this, the input data to be processed and the output to be fed by the circuit should be set. This will allow defining input and output binary variables to be processed. Here, the reader should remember that a combinational circuit gets input as logic levels 0 or 1 (or voltage values corresponding to these). Therefore, if an input is to be received from a sensor, it should be adjusted accordingly. The output of the combinational circuit will also be in the form of logic levels 0 or 1. Therefore, if an actuator is to be driven by output of the combinational circuit, a suitable interface should be established between the combinational circuit and actuator.

7.5.2   Selecting a Solution Method

After analyzing the problem, the next step is forming a method or algorithm to solve it. Since we are dealing with combinational circuits, the solution will be in terms of a logic function between the inputs and output of the circuit. The formed logic function should satisfy all design constraints specified in the previous step.

7.5.3   Implementing the Solution

The final step in the design process is the implementation. Since the main focus of this book is on the FPGA, we will implement the design on it. Therefore, the corresponding HDL for the designed combinational circuit should be formed first. Afterward, we can benefit from the Vivado's optimization tool for gate-level minimization. It is also advisable to simulate the designed system before implementation. If it satisfies all design constraints, then the corresponding bitstream can be generated and embedded on the FPGA chip. Hence, the design is concluded.

7.6   Sample Designs

We can apply the previous design steps on designing combinational circuits to solve real-life problems. Here, we pick three such cases as home alarm, digital safe, and car park occupied slot counting system. We will discuss each design next.

7.6.1   Home Alarm System

We can design a basic home alarm system using tools introduced in this and previous chapters. To do so, let's first define the problem. Assume that the alarm system to be designed is to be applied on three windows and a door. Each window and the door has a sensor such that when it is opened, it will give logic level 1. There should be an on/off switch for the alarm. If we want to activate the alarm, the switch will give logic level 1. Otherwise, it will give logic level 0. At this point, the problem is defined and design constraints are set.

To implement the combinational circuit for the design, let's assign binary variables $s0, s1, s2,$ and $s3$ to each sensor output. Let the on/off switch be represented by the binary variable m. Let's define the binary variable a as an output. This variable will have logic value 1 when an intruder triggers the alarm. Otherwise, the output of the system will be logic level 0. Based on all these constraints, the logic function between the input and output will be $a = (s0 + s1 + s2 + s3) \cdot m$. The corresponding circuit diagram will be as in Fig. 7.20.

We can form Verilog description of the circuit in Fig. 7.20 as in Listing 7.5. The VHDL description of the same circuit will be as in Listing 7.6. Vivado synthesizes the Verilog or VHDL description as in Fig. 7.21. As can be seen in this figure, one five-input LUT is sufficient for implementation.

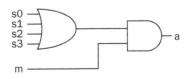

FIGURE 7.20   Circuit diagram of the home alarm system.

Listing 7.5 Verilog Description of the Home Alarm System

```verilog
module home_alarm(a,s,m);

input [3:0] s;
input m;
output a;

assign a=(s[0]|s[1]|s[2]|s[3])&m;

endmodule
```

Listing 7.6 VHDL Description of the Home Alarm System

```vhdl
library ieee;
use ieee.std_logic_1164.all;

entity home_alarm is
port(s : in std_logic_vector (3 downto 0);
     m : in std_logic;
     a : out std_logic);
end home_alarm;

architecture dataflow_model of home_alarm is
begin
a <= (s(0) or s(1) or s(2) or s(3)) and m;
end dataflow_model;
```

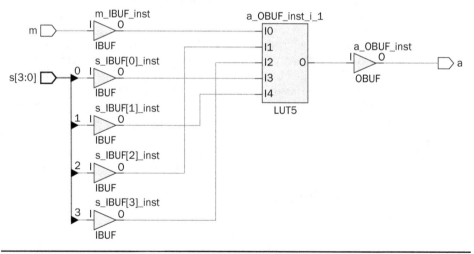

FIGURE 7.21 Synthesization result of the home alarm system.

7.6.2 Digital Safe System

We can design a simple digital safe using combinational circuits. Assume that the system has a four-bit predefined password. We will use four switches as the input to the system. If the input matches the predefined password, then the first output will have logic level 1. Otherwise, the second output will have logic level 1.

We can implement the corresponding combinational circuit using an XOR gate followed by a NOT gate for each bit to be tested. Therefore, if input bit matches the corresponding password bit, then the XOR gate followed by NOT will give logic level 1. If all input bits match corresponding predefined password bits this way, the first output will have logic level 1 and the second output will have logic level 0. The second output will simply be inverse of the first output.

To implement the combinational circuit for the design, let's assign binary variables $s0$, $s1$, $s2$, and $s3$ as input. Predefined password can be represented as $p[0] \cdots p[3]$. Let's define the first and second outputs as binary variables $l1$ and $l2$, respectively. The logic function between the inputs and first output variable will be $l0 = \overline{s0 \oplus p[0]} \cdot \overline{s1 \oplus p[1]} \cdot \overline{s2 \oplus p[2]} \cdot \overline{s3 \oplus p[3]}$. The second output will be $l1 = \overline{l0}$. The corresponding circuit diagram will be as in Fig. 7.22.

We can form Verilog description of the circuit in Fig. 7.22 as in Listing 7.7. The VHDL description of the same circuit will be as in Listing 7.8. Vivado synthesizes the Verilog description as in Fig. 7.23. As can be seen in this figure, two four-input LUTs are sufficient for implementation.

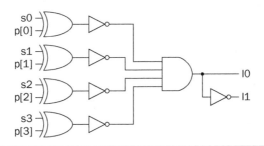

FIGURE 7.22 Circuit diagram of the digital safe system.

Listing 7.7 Verilog Description of the Digital Safe System

```verilog
module digital_safe(l,s);

input [3:0] s;
output [1:0] l;

parameter p=4'b0101;

assign l[0] = ~(s[0]^p[0]) & ~(s[1]^p[1]) & ~(s[2]^p[2]) & ~(s[3]^p[3])
    ;
assign l[1] = ~l[0];

endmodule
```

Listing 7.8 VHDL Description of the Digital Safe System

```
library ieee;
use ieee.std_logic_1164.all;

entity digital_safe is
port(s : in std_logic_vector (3 downto 0);
     lo : inout std_logic;
     l1 : out std_logic);

constant p : std_logic_vector (0 to 3) :="0101";

end digital_safe;

architecture dataflow_model of digital_safe is
begin

lo <= not (s(0) xor p(0)) and not (s(1) xor p(1)) and not (s(2) xor p
    (2)) and not (s(3) xor p(3));
l1 <= not lo;

end dataflow_model;
```

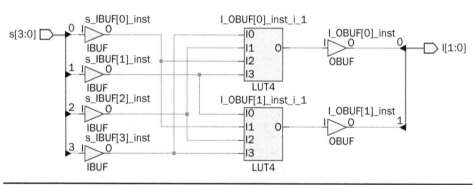

FIGURE 7.23 Synthesization result of the digital safe system.

7.6.3 Car Park Occupied Slot Counting System

Our last real-life problem is as follows. There is a car park with three slots and we would like to know how many of its slots are occupied at a given time. Within the design, occupied slot locations are not important. We can design a combinational circuit for this purpose. Assume that we placed a sensor over each slot which gives output logic level 1 when the slot is occupied. If the slot is empty, sensor gives output logic level 0. Let's label output of sensors as binary variables $s0$, $s1$, and $s2$. The designed combinational circuit will provide the output as a two-bit binary number $c1$ (MSB) and $c0$ (LSB). Therefore, we should cover all input combinations in terms of a truth table as in Table 7.17.

Using Table 7.17, we can form logic functions for $c0$ and $c1$ in the SOP form as follows:

$$c0 = \overline{s0} \cdot \overline{s1} \cdot s2 + \overline{s0} \cdot s1 \cdot \overline{s2} + s0 \cdot \overline{s1} \cdot \overline{s2} + s0 \cdot s1 \cdot s2$$
$$c1 = \overline{s0} \cdot s1 \cdot s2 + s0 \cdot \overline{s1} \cdot s2 + s0 \cdot s1 \cdot \overline{s2} + s0 \cdot s1 \cdot s2$$

Inputs			Outputs	
s0	**s1**	**s2**	**c1**	**c0**
0	0	0	0	0
0	0	1	0	1
0	1	0	0	1
0	1	1	1	0
1	0	0	0	1
1	0	1	1	0
1	1	0	1	0
1	1	1	1	1

TABLE 7.17 Truth Table of the Car Park Occupied Slot Counting System

---

Listing 7.9 Verilog Description of the Car Park Occupied Slot Counting System

```
module car_park(c,s);

input [0:2] s;
output [1:0] c;

assign c[0]=(~s[0]&~s[1]&s[2])+(~s[0]&s[1]&~s[2])+(s[0]&~s[1]&~s[2])+
    (s[0]&s[1]&s[2]);
assign c[1]=(~s[0]&s[1]&s[2])+(s[0]&~s[1]&s[2])+(s[0]&s[1]&~s[2])+
    (s[0]&s[1]&s[2]);

endmodule
```

Listing 7.10 VHDL Description of the Car Park Occupied Slot Counting System

```
library ieee;
use ieee.std_logic_1164.all;

entity car_park is
port(s : in std_logic_vector (0 to 2);
    c : out std_logic_vector (1 downto 0));
end car_park;

architecture dataflow_model of car_park is
begin

c(0)<=(not s(0) and not s(1) and s(2)) or (not s(0) and s(1) and not
    s(2)) or (s(0) and not s(1) and not s(2)) or (s(0) and s(1)
    and s(2));
c(1)<=(not s(0) and s(1) and s(2)) or (s(0) and not s(1) and s(2)) or
    (s(0) and s(1) and not s(2)) or (s(0) and s(1) and s(2));

end dataflow_model;
```

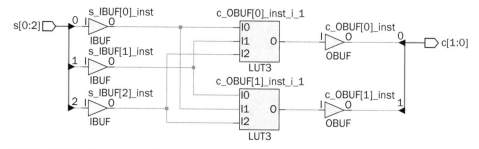

Figure 7.24 Synthesization result of the car park occupied slot counting system.

Listing 7.11 Home Alarm System Implemented on the Basys3 Board in Verilog

```verilog
module topmodule(sw,led);

input  [4:0] sw;
output [0:0] led;

// Generated IP block
home_alarm_0 HA(.a(led),.s(sw[3:0]),.m(sw[4]));

endmodule
```

We can implement these logic functions in Verilog and VHDL as in Listings 7.9 and 7.10. Vivado synthesizes the Verilog description as in Fig. 7.24. As can be seen in this figure, two LUTs each with three inputs are sufficient for implementation.

7.7 Applications on Combinational Circuits

In this section, we will implement sample designs in Sec. 7.6 on the Basys3 board. Therefore, we will cover home alarm, digital safe, and car park occupied slot counting systems. For all three applications, we will get input bit values from switches on the board. Output bit values are represented by LEDs on the board. The reader can consult Sec. 4.8 related to this setup.

7.7.1 Implementing the Home Alarm System

We can implement the home alarm system on the Basys3 board. Therefore, we provide the Verilog description in which LEDs and switches on the board are connected as the input and output in Listing 7.11. Here, we use the Verilog description of the system in Listing 7.5 as an IP block. Therefore, we expect the reader has generated the corresponding IP block.

7.7.2 Implementing the Digital Safe System

As in previous application, we can implement the digital safe system on the Basys3 board. In Listing 7.12, we provide the Verilog description in which LEDs and switches on the board are connected as the input and output. Here, we use the Verilog description of the system in Listing 7.7 as an IP block. Therefore, we expect the reader has generated the corresponding IP block.

Listing 7.12 Digital Safe System Implemented on the Basys3 Board in Verilog

```verilog
module topmodule(sw,led);

input   [3:0] sw;
output  [1:0] led;

// Generated IP block
digital_safe_0 DS(.l(led),.s(sw));

endmodule
```

Listing 7.13 Car Park Occupied Slot Counting System Implemented on the Basys3 Board in Verilog

```verilog
module topmodule(sw,led);

input   [2:0] sw;
output  [1:0] led;

// Generated IP block
car_park_0 CP(.c(led),.s(sw));

endmodule
```

7.7.3 Implementing the Car Park Occupied Slot Counting System

We can also implement the car park occupied slot counting system on Basys3 board. As in previous applications, we provide Verilog description in which LEDs and switches on the board are connected in Listing 7.13. Here, we use Verilog description of the system in Listing 7.9 as an IP block. Therefore, we expect the reader has generated the corresponding IP block.

7.8 FPGA Building Blocks Used in Combinational Circuits

LUTs are extensively used in the combinational circuit implementation as explained in detail in Sec. 7.4. Hence, CLBs will be the main blocks to be used in this chapter. Besides, interconnect resources and input/output blocks are needed while implementing a combinational circuit.

7.9 Summary

Combinational circuits and their properties were the main focus of this chapter. Therefore, we started with analyzing basic logic gates NOT, OR, AND, and XOR. Then, we introduced tools to analyze combinational circuits formed by these basic logic gates. At this step, we benefited from Vivado extensively. Hence, we did not cover mathematical derivations and methods. Instead, we directed the reader to related references. We then explored how combinational circuits can be designed. Related to this, we provided sample designs to show how real-life problems can be solved using combinational circuits. We also provided sample designs on real-life problems in exercises. We believe that solving these will let the reader grasp digital design principles at least from the combinational circuit perspective.

7.10   Exercises

7.1 Form the truth table of a three-input
  a. AND gate.
  b. OR gate.

7.2 Construct three- and four-input AND gates using two-input AND gates.

7.3 Construct three- and four-input OR gates using two-input OR gates.

7.4 A combinational circuit is represented by logic function $F(x, y, z) = x \cdot y + y \cdot z + z \cdot x$. Implement this circuit using
  a. an eight-to-one multiplexer.
  b. four-to-one and two-to-one multiplexers.
  c. two-to-one multiplexers.

7.5 Describe the combinational circuit in Exercise 7.4 in Verilog or VHDL.

7.6 A combinational circuit is represented by logic function $F(x, y, z) = \bar{x} \cdot \bar{z} + x \cdot y$. Implement this circuit using
  a. an eight-to-one multiplexer.
  b. four-to-one and two-to-one multiplexers.
  c. two-to-one multiplexers.

7.7 Describe the combinational circuit in Exercise 7.6 in Verilog or VHDL.

7.8 A combinational circuit is represented in the SOP form $F(x, y, z) = \sum(0, 2, 4, 6)$.
  a. Describe this circuit in Verilog or VHDL using dataflow modeling.
  b. Obtain the simplest form of this circuit.

7.9 Construct the truth table of a three-input XOR gate. Describe the POS form of this gate in Verilog or VHDL using the dataflow modeling.

7.10 **(Two's complement calculator.)** Design a combinational circuit with the following specifications. Input to the circuit is a three-bit unsigned number. Output of the circuit is two's complement of input. Implement the designed combinational circuit either in Verilog or VHDL.

7.11 **(Two's complement calculator with sign bit.)** Design a combinational circuit with the following specifications. Input to the circuit is a four-bit signed number. Output of the circuit is the three value bits. If the number is negative, then it is represented in two's complement form at output. Implement the designed combinational circuit either in Verilog or VHDL.

7.12 **(Arithmetic operations.)** Arithmetic operations introduced in Chap. 6 can be implemented by combinational circuits. Let's take two two-bit numbers $x[1]x[0]$ and $y[1]y[0]$.
  a. Design combinational circuits for arithmetic operations on these numbers as addition, subtraction, multiplication, and division.
  b. Implement the designed combinational circuits either in Verilog or VHDL.
  c. Compare the implemented design with the ones provided in Chap. 6 in terms of the FPGA resource usage.

7.13 **(Fire alarm system.)** Design a fire alarm system with the following specifications. The system has an on/off switch. The system works only if the switch is on. There is a smoke detector giving the output in three bits. When the smoke density is maximum, the output of the sensor is seven in the binary form. When there is no smoke

FIGURE 7.25 Seven-segment display.

Displayed	Segment						
number	A	B	C	D	E	F	G
0	0	0	0	0	0	0	1
1	1	0	0	1	1	1	1
2	0	0	1	0	0	1	0
3	0	0	0	0	1	1	0
4	1	0	0	1	1	0	0
5	0	1	0	0	1	0	0
6	0	1	0	0	0	0	0
7	0	0	0	1	1	1	1
8	0	0	0	0	0	0	0
9	0	0	0	0	1	0	0

TABLE 7.18 Seven-Segment Display Patterns

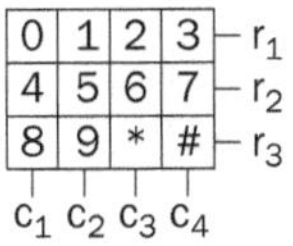

```
 0  1  2  3 ─ r₁
 4  5  6  7 ─ r₂
 8  9  *  # ─ r₃
 │  │  │  │
 c₁ c₂ c₃ c₄
```

FIGURE 7.26 Simple keypad.

detected, the output of the sensor is zero in the binary form. The alarm will be active if the output of the smoke detector exceeds four in the binary form. Implement the designed combinational circuit either in Verilog or VHDL.

7.14 (Seven-segment display decoder.) In digital systems, seven-segment displays are used extensively. The display has seven independent segments (A, B, C, D, E, F, G) as in Fig. 7.25.

Design a decoder circuitry with a four-bit input representing a decimal number. The decoder converts this number to corresponding seven-segment pin pattern as in Table 7.18. Implement the designed combinational circuit either in Verilog or VHDL.

7.15 (Keypad decoder.) A simple keypad can be represented as in Fig. 7.26. As can be seen in this figure, the keypad has seven output lines, three for row and four for column locations, respectively. When a key is pressed, corresponding row and column lines will produce logic level 1. Design a combinational circuit working as a keypad decoder. The input of the circuit will be the output lines of the keypad. The output of the circuit will be the corresponding binary number in three bits.

If * or # key is pressed, the output of the circuit will be zero. Implement the designed combinational circuit either in Verilog or VHDL.

7.16 Merge the designs in Exercises 7.14 and 7.15 such that when a number is pressed on the keypad, it is shown in the seven-segment display. Implement the designed combinational circuit either in Verilog or VHDL.

7.17 (Remote controller—key pattern generator.) Design a simple remote controller key pattern generator system with the following specifications. Only the key pattern part is handled in the design. There are three buttons on the controller. When the first one is pressed, the combinational circuit should produce pattern 001. For the second and third buttons this pattern will become 010 and 100, respectively. When more than one button is pressed, the output of the combinational circuit will be the pattern 000. This pattern will also be used when no button is pressed. Implement the designed combinational circuit either in Verilog or VHDL.

7.18 (Even/odd number detector.) Design an even/odd number detector with the following specifications. Input to the system is a four-bit number. If the number is even, the first output will be logic level 1. Otherwise, the second output will be logic level 1. Implement the designed combinational circuit either in Verilog or VHDL.

7.19 (Simple safety belt alarm system for cars.) Design a simple safety belt alarm system for cars. Only the front seat safety belts are of focus. The alarm system works as follows. If the car engine has started, the passenger has seated, and the passenger has not plugged in the belt, then alarm signal starts till the belt has been plugged in. The engine status (started or not) is provided by a digital signal. If the engine has started and operating, logic level 1 is fed. Otherwise, logic level 0 is fed. Pressure sensor attached to the driver and passenger seats provide a digital signal with logic level 1 when a mass produces pressure. Otherwise, the sensor provides logic level 0. The safety belt plug-in apparatus has a digital sensor such that when the belt is plugged in, it produces logic level 1. Otherwise, it produces logic level 0. Although an audio alarm signal is desirable, in this question we will use two LEDs to indicate the alarm. If the driver has seated, started the engine, and not plugged the belt, the alarm will turn on till the belt is plugged in. The same settings in the driver seat apply to the passenger seat. Please note that the two seat alarms operate independently. Implement the designed combinational circuit either in Verilog or VHDL.

7.20 (Joystick application.) Use the joystick setup in Exercise 6.14 to form a new Verilog or VHDL description. Here, when the joystick goes to its four limits (two for x-axis and two for y-axis) four separate LEDs on the Basys3 board (led[0], led[3], led[6], and led[9]) will turn on separately. Otherwise, all LEDs will turn off.

CHAPTER 8

Combinational Circuit Blocks

We have introduced combinational circuits in the previous chapter. There, the focus was on general characteristics of these circuits. There are also well-known combinational circuit blocks used in digital systems. These can be counted as adders, comparators, decoders, encoders, multiplexers, parity generators, and checkers. This chapter discusses these combinational circuit blocks.

8.1 Adders

Although addition is performed using a different method in an FPGA, the basic adder circuit is still worth analyzing. Therefore, we will consider it in this section. There are two basic adder types: half and full.

8.1.1 Half Adder

The half adder (for one-bit addition) has two inputs and two outputs. It adds input bits and gives sum and carry-out bits as the output. The truth table of the one-bit half adder is presented in Table 8.1. In this table, binary variables x and y stand for input bits to be added. Binary variables s and co represent sum and carry-out values, respectively.

As can be seen in Table 8.1, the carry-out bit has logic level 1 when both input bits are at logic level 1. This corresponds to the AND operation. The sum bit (s) has logic level 1 when input bits have different logic levels. This corresponds to the XOR operation. Based on these observations, the half adder can be constructed as in Fig. 8.1.

Inputs		Outputs	
x	y	s	co
0	0	0	0
0	1	1	0
1	0	1	0
1	1	0	1

TABLE 8.1 Truth Table of the Half Adder (for One-Bit Addition)

Figure 8.1 Circuit diagram of half adder.

Inputs			Outputs	
x	y	ci	s	co
0	0	0	0	0
0	1	0	1	0
1	0	0	1	0
1	1	0	0	1
0	0	1	1	0
0	1	1	0	1
1	0	1	0	1
1	1	1	1	1

Table 8.2 Truth Table of Full Adder (for One-Bit Addition)

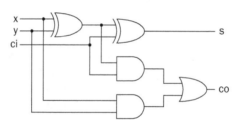

Figure 8.2 Circuit diagram of full adder.

8.1.2 Full Adder

The half adder does not take the input carry-in bit into account in operation. This causes problems when adding binary numbers with more than one digit. The full adder is introduced to overcome this problem. Besides having two input pins, the full adder also has a carry-in pin. The truth table of this device is presented in Table 8.2. In this table, binary variables x and y stand for input bits to be added. Binary variable ci stands for the carry-in bit. Binary variables s and co represent sum and carry-out bits, respectively. As in half adder, circuit diagram of full adder can be constructed by analyzing Table 8.2. The final constructed circuit diagram of the full adder is shown in Fig. 8.2.

8.1.3 Adders in Verilog

As mentioned in Chap. 6, addition is handled differently in an FPGA: either the DSP block is used for this operation or a LUT structure is formed. We have introduced the addition operation in Verilog in Sec. 6.7. Here, we will only provide half and full adders in the gate level. Let's start with the one-bit half adder. We provide the Verilog description for this circuit in Listing 8.1. Here, x and y represent input bits to be added. s and co stand for sum and carry-out bits, respectively.

We next provide the Verilog description of the one-bit full adder in Listing 8.2. The only difference here is that the device has an extra carry-in bit represented as ci.

8.1.4 Adders in VHDL

We have introduced the addition operation in VHDL in Sec. 6.9. Here, we will only provide half and full adders in the gate level. Let's start with the one-bit half adder.

Listing 8.1 Verilog Description of One-Bit Half Adder

```verilog
module one_bit_half_adder(s,co,x,y);

// Port definitions
input  x,y;
output s,co;

// Structural modeling
and g1(co,x,y);
xor g2(s,x,y);

// Dataflow modeling
assign co = x & y;
assign s = x ^ y;

endmodule
```

Listing 8.2 Verilog Description of One-Bit Full Adder

```verilog
module one_bit_full_adder(s,co,x,y,ci);

// Port definitions
input  x,y,ci;
// for structural and functional modeling
output s,co;

// for structural modeling
wire o1,o2,o3;

// Structural modeling
and g1(o1,x,y);
xor g2(o2,x,y);
xor g3(s,o1,ci);
and g4(o3,o1,ci);
or  g5(co,o2,o3);

// Dataflow modeling
assign co = (x & y) | (ci & (x^y));
assign s = x^y^ci;

endmodule
```

Listing 8.3 VHDL Description of One-Bit Half Adder

```
library ieee;
use ieee.std_logic_1164.all;

entity one_bit_half_adder is
port(x : in std_logic;
     y : in std_logic;
     s : out std_logic;
    co : out std_logic);
end one_bit_half_adder;

architecture dataflow of one_bit_half_adder is
begin
 s <= x xor y;
co <= x and y;
end dataflow;
```

Listing 8.4 VHDL Description of One-Bit Full Adder

```
library ieee;
use ieee.std_logic_1164.all;

entity one_bit_full_adder is
port(x : in std_logic;
     y : in std_logic;
    ci : out std_logic);
     s : out std_logic;
    co : out std_logic);
end one_bit_full_adder;

architecture dataflow of one_bit_full_adder is
begin
 s <= x xor y xor ci;
co <= (x and y) or (x and ci) or (y or ci);
end dataflow;
```

We provide the VHDL description for this circuit in Listing 8.3. Binary variables used in this description are the same as in the previous section.

We next provide the VHDL description of the one-bit full adder in Listing 8.4. As in the Verilog description, the only difference here is that the circuit has an extra carry-in bit represented as ci.

8.2 Comparators

We may need to compare the magnitude of two binary numbers to obtain their status. Here, the first number may be greater than the second. The two numbers may be equal. Or, the first number may be less than the second. To achieve this goal, we will need a comparator. We can explain the comparison operation on two binary variables x and y

Inputs		Outputs		
x	**y**	**g**	**e**	**l**
0	0	0	1	0
0	1	0	0	1
1	0	1	0	0
1	1	0	1	0

TABLE 8.3   Truth Table of the One-Bit Comparator

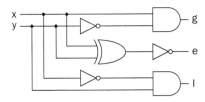

FIGURE 8.3   Circuit diagram of one-bit comparator.

(each being one bit) using the truth table presented in Table 8.3. Here, g, e, and l stand for greater, equal, and less, respectively.

Based on Table 8.3, we can obtain logic functions between inputs and outputs of the one-bit comparator as follows:

$$g = x \cdot \overline{y}$$
$$e = \overline{x \oplus y}$$
$$l = \overline{x} \cdot y$$

Obtained logic functions lead to the circuit diagram of the one-bit comparator as in Fig. 8.3.

8.2.1   Comparators in Verilog

We provide the Verilog description of the one-bit comparator in terms of structural and dataflow modeling forms in Listing 8.5. Here, we implemented the circuit in Fig. 8.3. Therefore, input bits to be compared are represented by binary variables x and y. Output values are represented by binary variables g, e, and l.

The Verilog description of an N-bit comparator to compare two N-bit numbers using dataflow and structural modeling will be complex. Therefore, behavioral modeling will be more appropriate for this case. To do so, we need to introduce relational operators and conditional statements in Verilog. Let's start with relational operators.

8.2.1.1   Relational Operators in Verilog

While constructing a Verilog description, we may need to compare two variables or vectors. Verilog has specific operators for this purpose. We provide operators to be used in this book and their explanation in Table 8.4 using two binary variables x and y. As these operations are executed, their result will be either logic level 0 or 1 based on whether the given condition is satisfied or not.

Listing 8.5 Verilog Description of One-Bit Comparator

```verilog
module one_bit_comparator(g,e,l,x,y);

// Port definitions
input   x,y;
// for structural and functional modeling
output  g,e,l;

// for structural modeling
wire o1,o2,o3;

// Structural modeling
not g1(o1,y);
and g2(g,o1,x);
xor g3(o2,x,y);
not g4(e,o2);
not g5(o3,x);
and g6(l,o3,y);

// Dataflow modeling
assign g = x & ~y;
assign e = ~(x ^ y);
assign l = ~x & y;

endmodule
```

Operation	Explanation
x==y	x is equal to y
x!=y	x is not equal to y
x>y	x is greater than y
x<y	x is less than y
x>=y	x is greater than or equal to y
x<=y	x is less than or equal to y

TABLE 8.4 Relational Operators in Verilog

8.2.1.2 Conditional Statements in Verilog

Verilog allows forming conditional statements using if keyword under behavioral modeling. Via this keyword, given statements can be executed if the condition is satisfied. The condition can be formed by a single variable, two or more variables combined with logical operators, or relational operators. The syntax of a conditional statement using the if keyword is as follows:

```verilog
if (condition)
statements;
```

The if keyword also allows using else if and else keywords. The syntax for their usage is as follows. The else if keyword allows adding a new condition (in a sequential manner). The `else` keyword is executed if none of the above conditions are satisfied.

```
if (condition)
statements;

else if (condition)
 statements;

else
statements;
```

An *N*-bit comparator can be constructed by the `if` keyword. We provide such a Verilog description only for behavioral modeling in Listing 8.6. Here, two vectors each with four-bits (x and y) are compared and the result is written to another vector `comp`. If the first vector is greater than the second one, then `comp[2]=1`. If the second vector is greater than the first one, then `comp[0]=1`. Finally, if the two vectors are equal, then `comp[1]=1`. We provide the RTL schematic of the four-bit comparator in Fig. 8.4. As can be seen in this figure, equality operators and multiplexers are used in synthesizing the Verilog description. We will analyze how the comparator is implemented in an FPGA in Sec. 8.8.

Listing 8.6 Verilog Description of Four-Bit Comparator Using `if` Keyword

```
module N_bit_comparator(comp,x,y);

parameter N = 4;

input [N-1:0] x, y;
output reg [2:0] comp;

initial
comp = 3'b0;

always @ (x or y)
if (x > y) comp = 3'b100;
else if (x == y) comp = 3'b010;
else if (x < y) comp = 3'b001;
else comp = 3'b111;

endmodule
```

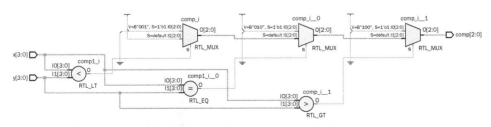

FIGURE 8.4 RTL schematic of four-bit comparator.

8.2.2 Comparators in VHDL

We next provide the VHDL description of the one-bit comparator in terms of the dataflow modeling in Listing 8.7. As in the Verilog description in Listing 8.5, we directly implement the circuit in Fig. 8.3. Hence, input bits to be compared are represented by binary variables x and y. Output values are represented by binary variables g, eq, and l.

The VHDL description of an *N*-bit comparator using dataflow modeling will be complex. Therefore, behavioral modeling will be more appropriate for this case. To do so, we will introduce relational operators and conditional statements in VHDL. Let's start with relational operators.

8.2.2.1 Relational Operators in VHDL

While constructing a VHDL description, we may need to compare two variables or arrays. As in Verilog, VHDL has specific operators for this purpose. We provide the operators to be used in this book and their explanation using two binary variables x and y in Table 8.5. As these operations are executed, their result will be either logic level 0 or 1 based on whether the given condition is satisfied or not.

Listing 8.7 VHDL Description of One-Bit Comparator

```
library ieee;
use ieee.std_logic_1164.all;

entity one_bit_comparator is
port (x : in std_logic;
      y : in std_logic;
      g : out std_logic;
      l : out std_logic;
     eq : out std_logic);
end one_bit_comparator;

architecture dataflow of one_bit_comparator is
begin
 g <= x and not y;
eq <= not(x xor y);
 l <= not x and y;
end dataflow;
```

Operation	Explanation
x=y	x is equal to y
x/=y	x is not equal to y
x>y	x is greater than y
x<y	x is less than y
x>=y	x is greater than or equal to y
x<=y	x is less than or equal to y

TABLE 8.5 Relational Operators in VHDL

8.2.2.2 *Conditional Statements in VHDL*

As in Verilog, VHDL allows adding conditional statements to a behavioral description using if keyword. Via this keyword, given statements can be executed if the condition is satisfied. The condition can be formed by a single signal, two or more signals combined with logical operators, or relational operators. The syntax of a conditional statement using the if keyword is as follows:

```
if (condition) then statements;
end if;
```

The if keyword also allows using elsif and else keywords. The syntax for their usage is as follows. The elsif keyword allows adding a new condition (in a sequential manner). The else keyword is executed if none of the above conditions are satisfied.

```
if (condition) then statements;

elsif (condition) then statements;

else statements;
end if;
```

An *N*-bit comparator can be constructed by the if keyword. We provide such a VHDL description only for behavioral modeling in Listing 8.8. Here, two arrays each being four-bits (x and y) are compared and the result is written to another array comp. If the first array is greater than the second one, then comp(2)=1. If the second array

Listing 8.8 VHDL Description of Four-Bit Comparator Using the if Keyword

```
library ieee;
use ieee.std_logic_1164.all;

entity N_bit_comparator is
port(x : in std_logic_vector (3 downto 0);
     y : in std_logic_vector (3 downto 0);
  comp : out std_logic_vector (2 downto 0));
end N_bit_comparator;

architecture behavioral of N_bit_comparator is
begin

process(x,y)
begin
comp <="000";
if      (x>y) then comp<="100";
elsif   (x=y) then comp<="010";
elsif   (x<y) then comp<="001";
else
comp<="111";
end if;
end process;

end behavioral;
```

Inputs		Outputs			
x0	x1	y0	y1	y2	y3
0	0	1	0	0	0
0	1	0	1	0	0
1	0	0	0	1	0
1	1	0	0	0	1

TABLE 8.6   Truth Table of Two-to-Four Decoder

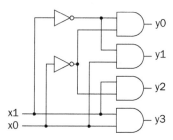

FIGURE 8.5   Circuit diagram of two-to-four decoder.

is greater than the first one, then comp(0)=1. Finally, if the two arrays are equal, then comp(1)=1. The RTL schematic of the VHDL description will be as in Fig. 8.4.

8.3   Decoders

Basic usage of a decoder is to decode its input and give specific output corresponding to it. In general, the decoder has N inputs and 2^N outputs to cover all input combinations. Let's focus on the two-to-four decoder with the truth table presented in Table 8.6. As can be seen in this table, there are two inputs and four (2^2) outputs. The output corresponding to a given input will be at logic level 1. For this combination, all other outputs will be at logic level 0. Hence, the input is decoded.

The decoder can be constructed by AND and NOT gates. The circuit diagram of the two-to-four decoder will be as in Fig. 8.5. As can be seen in this figure, the decoder is constructed by two NOT and four AND gates. If we consider $y0$, it gives logic level 1 only when $x0$ and $x1$ are at logic level 0. Therefore, zeroth input sets output $y0$. This input combination sets all other outputs to logic level 0.

8.3.1   Decoders in Verilog

We provide the Verilog description of the two-to-four decoder in Listing 8.9. Here, we implemented the circuit in Fig. 8.5. Therefore, input to the decoder is represented by the two-element vector x. The output of the decoder is represented by the four-element vector y.

The Verilog description of the three-to-eight decoder using dataflow and structural modeling will be complex. Behavioral modeling will be more appropriate for this case. Here, we can use the if keyword to construct conditional statements. However, Verilog

Listing 8.9 Verilog Description of Two-to-Four Decoder

```
module two_to_four_decoder(y,x);

// Port definitions
input [1:0] x;
// for structural and functional modeling
output [3:0] y;

// for structural modeling
wire o1,o2;

// Structural modeling
not g1(o1,x[0]);
not g2(o2,x[1]);
and g3(y[0],o1,o2);
and g4(y[1],x[0],o2);
and g5(y[2],o1,x[1]);
and g6(y[3],x[0],x[1]);

// Dataflow modeling
assign out[0] = ~x[0] & ~x[1];
assign out[1] = x[0] & ~x[1];
assign out[2] = ~x[0] & x[1];
assign out[3] = x[0] & x[1];

endmodule
```

also has another keyword which is more appropriate for the decoder structure. This keyword is case with the syntax as follows:

```
case(variable)
    value 1 : statement;
    value 2 : statement;
    value 3 : statement;
    value 4, value 5 : statement;
        ...
    default : statements;
endcase
```

As can be seen here, the variable to be used in the case statement is defined in parentheses just after the keyword. For each value of this variable, a statement is assigned. If we have more than one statement for a variable, then we should use block keywords (begin and end) to encapsulate them. Note that variable values need not be exhaustive. We can only define values of interest. Then, we can define a default value for the rest. Moreover, we can group variable values by adding a comma in between. This way, we can eliminate duplicates. We provided such an example on the fourth and fifth values above.

The three-to-eight decoder can be constructed by the case keyword. We provide such a Verilog description only for behavioral modeling in Listing 8.10. In this description, the input to the decoder is represented by the three-element vector x. The output of the decoder is represented by the eight-element vector y. We provide the RTL schematic of the three-to-eight decoder in Fig. 8.6. As can be seen in this figure, only a block memory element (ROM, to be explored in Sec. 9.5) is used in synthesizing the

---

Listing 8.10 Verilog Description of Three-to-Eight Decoder Using `case` Keyword

```verilog
module three_to_eight_decoder(y,x);

input [2:0] x;
output reg [7:0] y;

initial
y = 8'b0;

always @ (x)
case(x)
  3'b000 : y = 8'b00000001;
  3'b001 : y = 8'b00000010;
  3'b010 : y = 8'b00000100;
  3'b011 : y = 8'b00001000;
  3'b100 : y = 8'b00010000;
  3'b101 : y = 8'b00100000;
  3'b110 : y = 8'b01000000;
  3'b111 : y = 8'b10000000;
endcase

endmodule
```

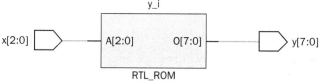

FIGURE 8.6 RTL schematic of three-to-eight decoder.

Verilog description. We will analyze how the decoder is implemented in an FPGA in Sec. 8.8.

8.3.2 Decoders in VHDL

We next provide the VHDL description of the two-to-four decoder in terms of dataflow modeling in Listing 8.11. As in the Verilog description in Listing 8.9, we directly implement the circuit in Fig. 8.5. Therefore, the input to the decoder is represented by a two-element array x. The output of the decoder is represented by a four-element array y.

As in Verilog, the VHDL description of the three-to-eight decoder using dataflow and structural modeling will be complex. Behavioral modeling will be more appropriate for this case. Here, we can use the `if` keyword to construct conditional statements. However, VHDL also has another keyword which is more appropriate for the decoder structure. This keyword is `case` with the syntax as follows:

```vhdl
case variable is
    when value 1 =>  statement;
    when value 2 =>  statement;
    when value 3 =>  statement;
    . . .
    when others =>  statement;
end case
```

Listing 8.11 VHDL Description of Two-to-Four Decoder

```vhdl
library ieee;
use ieee.std_logic_1164.all;

entity two_to_four_decoder is
port(x : in std_logic_vector (1 downto 0);
     y : out std_logic_vector (3 downto 0));
end two_to_four_decoder;

architecture dataflow of two_to_four_decoder is
begin
y(0) <= not x(0) and not x(1);
y(1) <= x(0) and not x(1);
y(2) <= not x(0) and x(1);
y(3) <= x(0) and x(1);
end dataflow;
```

Listing 8.12 VHDL Description of Three-to-Eight Decoder Using case Keyword

```vhdl
library ieee;
use ieee.std_logic_1164.all;

entity three_to_eight_decoder is
port(x : in std_logic_vector (2 downto 0);
     y : out std_logic_vector (7 downto 0));
end three_to_eight_decoder;

architecture behavioral of three_to_eight_decoder is
begin
process(x)
begin
case x is
    when "000" =>  y<="00000001";
    when "001" =>  y<="00000010";
    when "010" =>  y<="00000100";
    when "011" =>  y<="00001000";
    when "100" =>  y<="00010000";
    when "101" =>  y<="00100000";
    when "110" =>  y<="01000000";
    when others => y<="10000000";
end case;
end process;
end behavioral;
```

As can be seen here, the variable to be used in the case statement is defined just after the keyword. For each value of this variable, a statement is assigned. The reader can use the others keyword to define the default case.

The three-to-eight decoder can be constructed by the case keyword. We provide such a VHDL description only for behavioral modeling in Listing 8.12. In this description, the input to the decoder is represented by a three-element array x. The output of the decoder is represented by an eight-element array y. The RTL schematic of the VHDL description will be as in Fig. 8.6.

8.4   Encoders

The encoder works just as the opposite of the decoder. Its function is to encode a given input and provide encoded output. In general, an encoder has at most 2^N inputs and N outputs. Let's focus on the four-to-two encoder with the truth table presented in Table 8.7.

As can be seen in Table 8.7, the output of the encoder is the binary representation of the input. While constructing the truth table, we assumed that no two inputs will have logic level 1 at the same time. If such a case occurs, then the output of the decoder becomes unpredictable. To overcome this problem, we can add priority to inputs such that the output is the one with the higher precedence. Based on this form, the new truth table becomes as in Table 8.8. Here, don't care conditions are represented by " - " sign. Within the priority encoder, we still assume that all inputs will not be zero at the same time. To check whether such an input comes, we can add a valid signal, v, to the output. This will indicate that the obtained output is either valid or not.

Based on Table 8.8, we can construct the combinational circuit of the four-to-two priority encoder as follows:

$$y0 = x2 + x3$$
$$y1 = x3 + x1 \cdot \overline{x2}$$
$$v = x0 + x1 + x2 + x3$$

The above input–output relations lead to the circuit diagram of the four-to-two priority encoder as in Fig. 8.7.

Inputs				Outputs	
x0	x1	x2	x3	y0	y1
1	0	0	0	0	0
0	1	0	0	0	1
0	0	1	0	1	0
0	0	0	1	1	1

TABLE 8.7   Truth Table of the Four-to-Two Encoder

Inputs				Outputs		
x0	x1	x2	x3	y0	y1	v
0	0	0	0	-	-	0
1	0	0	0	0	0	1
-	1	0	0	0	1	1
-	-	1	0	1	0	1
-	-	-	1	1	1	1

TABLE 8.8   Truth Table of Four-to-Two Priority Encoder

8.4.1 Encoders in Verilog

We provide the Verilog description of the four-to-two priority encoder in Listing 8.13. Here, we directly implement the circuit in Fig. 8.7. Therefore, the input of the encoder is represented by a four-element vector x. The output of the encoder is represented by a two-element vector y and a binary variable v.

We next focus on an eight-to-three priority encoder. Unfortunately, the dataflow and structural models will be complex for this device. Therefore, we will provide only the behavioral model in Verilog. Here, we will again benefit from the case keyword. However, since we have don't care conditions in operation, we will use the casex keyword instead. We provide the Verilog description in Listing 8.14. In this description, the input to the encoder is represented by an eight-element vector x. The output of the encoder is represented by a three-element vector y. Within this description, we discarded the valid (v) output. Instead, we set the output to high impedance (z) for such cases. We provide the RTL schematic of the eight-to-three priority encoder in Fig. 8.8. As can be

---

Listing 8.13 Verilog Description of Four-to-Two Priority Encoder

```verilog
module four_to_two_encoder(y,v,x);

// Port definitions
input [3:0] x;
// for structural and functional modeling
output [1:0] y;
output v;

// for structural modeling
wire o1,o2;

// Structural modeling
or  g1(y[0],x[2],x[3]);
not g2(o1,x[2]);
and g3(o2,o1,x[1]);
or  g4(y[1],x[3],o2);
or  g5(v,x[3],x[2],x[1],x[0]);

// Dataflow modeling
assign y[0] = x[2] | x[3];
assign y[1] = x[3] | (x[1] & ~x[2]);
assign v = x[0] | x[1] | x[2] | x[3];

endmodule
```

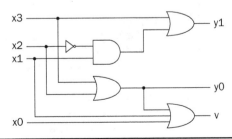

FIGURE 8.7 Circuit diagram of four-to-two priority encoder.

Listing 8.14 Verilog Description of Eight-to-Three Priority Encoder Using `casex` Keyword

---

```verilog
module eight_to_three_encoder(y,x);

input [7:0] x;
output reg [2:0] y;

initial
y = 3'bzzz;

always @ (x)
casex(x)
  8'b1xxxxxxx : y = 3'b111;
  8'b01xxxxxx : y = 3'b110;
  8'b001xxxxx : y = 3'b101;
  8'b0001xxxx : y = 3'b100;
  8'b00001xxx : y = 3'b011;
  8'b000001xx : y = 3'b010;
  8'b0000001x : y = 3'b001;
  8'b00000001 : y = 3'b000;
  default :     y = 3'bzzz;
endcase
endmodule
```

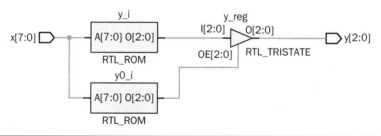

FIGURE 8.8 RTL schematic of an eight-to-three priority encoder.

---

seen in this figure, two block memory elements (ROM) are used in synthesizing the Verilog description. We will analyze how priority encoder is implemented in an FPGA in Sec. 8.8.

8.4.2 Encoders in VHDL

We next provide the VHDL description of the four-to-two priority encoder in terms of the dataflow modeling in Listing 8.15. As in the Verilog description in Listing 8.9, we directly implement the circuit in Fig. 8.7. Therefore, the input to the encoder is represented by a four-element array x. The output of the encoder is represented by a two-element array y and a binary variable v.

As in Verilog, the dataflow modeling of an eight-to-three encoder in VHDL will be complex. Therefore, we consider only the behavioral model of this device in VHDL. We next provide this description in Listing 8.16. Here, the input to the encoder is represented by an eight-element array x. The output of the encoder is represented by a three-element array y. The RTL schematic of the VHDL description will be as in Fig. 8.8.

Listing 8.15 VHDL Description of Four-to-Two Priority Encoder

```vhdl
library ieee;
use ieee.std_logic_1164.all;

entity four_to_two_encoder is
port(x : in std_logic_vector (3 downto 0);
     y : out std_logic_vector (1 downto 0)
     v : out std_logic);
end four_to_two_encoder;

architecture dataflow of four_to_two_encoder is
begin
y(0) <= x(2) or x(3);
y(1) <= x(3) or (x(1) and not x(2));
v <= x(0) or x(1) or x(2) or x(3);
end dataflow;
```

Listing 8.16 VHDL Description of Eight-to-Three Priority Encoder

```vhdl
library ieee;
use ieee.std_logic_1164.all;

entity eight_to_three_encoder is
port(x : in std_logic_vector (7 downto 0);
     y : out std_logic_vector (2 downto 0));
end eight_to_three_encoder;

architecture behavioral of eight_to_three_encoder is
begin
process(x)
begin
y<="000";

if x(1) = '1' then y<="001"; end if;
if x(2) = '1' then y<="010"; end if;
if x(3) = '1' then y<="011"; end if;
if x(4) = '1' then y<="100"; end if;
if x(5) = '1' then y<="101"; end if;
if x(6) = '1' then y<="110"; end if;
if x(7) = '1' then y<="111"; end if;

end process;
end behavioral;
```

8.5 Multiplexers

We have introduced the multiplexer in Sec. 2.2.3. Moreover, we have used it in the combinational circuit implementation in Chap. 7. For completeness, let's review its fundamental properties. The multiplexer is a combinational circuit that transfers data coming from several inputs to single output. Therefore, it can be used to select a specific input from a group of inputs and feed it to the output. To perform this task, the multiplexer has N select pins, 2^N input pins, and one output pin.

Select pins		Output
s1	**s0**	**y**
0	0	x0
0	1	x1
1	0	x2
1	1	x3

TABLE 8.9 Truth Table of Four-to-One Multiplexer

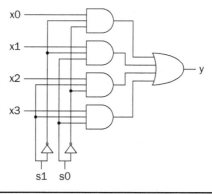

FIGURE 8.9 Circuit diagram of a four-to-one multiplexer.

Let's focus on a four-to-one multiplexer. This device has four inputs, two select pins, and one output with the truth table presented in Table 8.9. In this table, select pins are represented by binary variables $s0$ and $s1$. Inputs are labeled as $x0, \cdots, x3$. The output of the multiplexer is represented by binary variable y.

We can implement a four-to-one multiplexer as in Fig. 8.9. As can be seen in this figure, only one AND gate is enabled for each select input sequence. For instance, the first AND gate is enabled when $s1$ and $s0$ are at logic level 0. All other AND gates are disabled for this sequence. Hence, only input $x0$ appears at output y.

8.5.1 Multiplexers in Verilog

We provide the Verilog description of a four-to-one multiplexer in Listing 8.17. Here, we directly implemented the circuit in Fig. 8.9. Therefore, select values are represented by a two-element vector s; inputs are represented by a four-element vector x; and the output is represented by binary variable y.

For the eight-to-one multiplexer, the dataflow and structural representations in Verilog will be complex. On the other hand, the behavioral model in Verilog will be neat. We next provide such a description in Listing 8.18. As in a four-to-one multiplexer, select pins are represented by a three-element vector s; inputs are represented by an eight-element vector x; and the output is represented by a binary variable y in this description. We provide the RTL schematic of an eight-to-one multiplexer in Fig. 8.10. As can be seen in this figure, the multiplexer is used in synthesizing the Verilog description as it is.

Listing 8.17 Verilog Description of Four-to-One Multiplexer

```verilog
module four_to_one_multiplexer(y,x,s);

// Port definitions
input [1:0] s;
input [3:0] x;

// for structural and functional modeling
output y;

// for structural modeling
wire w1,w2,w3,w4,w5,w6,w7,w8;

// Structural modeling
or  g1(y,w1,w2,w3,w4);
and g2(w1,w5,w6,x[0]);
and g3(w2,sel[0],w7,x[1]);
and g4(w3,w8,sel[1],x[2]);
and g5(w4,s[0],s[1],x[3]);
not g6(w5,s[0]);
not g7(w6,s[1]);
not g8(w7,s[1]);
not g9(w8,s[0]);

// Dataflow modeling
assign y = (x[0] & ~s[1] & ~s[0]) | (x[1] & ~s[1] & s[0]) | (x[2] & s
    [1] & ~s[0]) | (x[3] & s[1] & s[0]);

endmodule
```

Listing 8.18 Verilog Description of Eight-to-One Multiplexer

```verilog
module eight_to_one_multiplexer(y,x,s);

input [2:0] s;
input [7:0] x;
output reg y;

always @ (s or x)
case(s)
  3'b000 : y = x[0];
  3'b001 : y = x[1];
  3'b010 : y = x[2];
  3'b011 : y = x[3];
  3'b100 : y = x[4];
  3'b101 : y = x[5];
  3'b110 : y = x[6];
  3'b111 : y = x[7];
endcase

endmodule
```

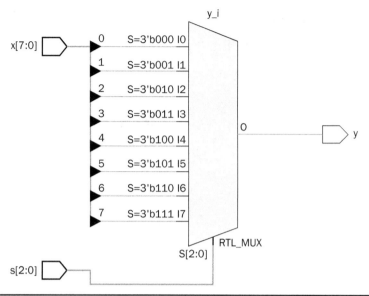

FIGURE 8.10 RTL schematic of eight-to-one multiplexer.

Listing 8.19 VHDL Description of Four-to-One Multiplexer

```
library ieee;
use ieee.std_logic_1164.all;

entity four_to_one_multiplexer is
port(s : in std_logic_vector (1 downto 0);
     x : in std_logic_vector (3 downto 0)
     y : out std_logic);
end four_to_one_multiplexer;

architecture dataflow of four_to_one_multiplexer is
begin
y = (x(0) and not s(1) and not s(0)) or (x(1) and not s(1) and s(0)) or
    (x(2) and s(1) and not s(0)) or (x(3) and s(1) and s(0));
end dataflow;
```

8.5.2 Multiplexers in VHDL

We next provide the VHDL description of a four-to-one multiplexer in Listing 8.19. As in the Verilog description, here we directly implemented the circuit in Fig. 8.9. Therefore, select pins are represented by a two-element array s; inputs are represented by a four-element array x; and the output is represented by a binary variable y.

As in Verilog, the dataflow model of an eight-to-one multiplexer in VHDL will be complex. Therefore, we consider only the behavioral model for this device. We next provide this description in Listing 8.20. As in a four-to-one multiplexer, select pins are represented by a three-element array s; inputs are represented by an eight-element array x; and the output is represented by a binary variable y in this description. The RTL schematic of the VHDL description will be as in Fig. 8.10.

Listing 8.20 VHDL Description of Eight-to-One Multiplexer

```
library ieee;
use ieee.std_logic_1164.all;

entity eight_to_one_multiplexer is
port(s : in std_logic_vector (2 downto 0);
     x : in std_logic_vector (7 downto 0);
     y : out std_logic);
end eight_to_one_multiplexer;

architecture behavioral of eight_to_one_multiplexer is
begin
process(x)
begin
case sel is
    when "000" =>  y<=x(0);
    when "001" =>  y<=x(1);
    when "010" =>  y<=x(2);
    when "011" =>  y<=x(3);
    when "100" =>  y<=x(4);
    when "101" =>  y<=x(5);
    when "110" =>  y<=x(6);
    when others => y<=x(7);
end case;
end process;
end behavioral;
```

8.6 Parity Generators and Checkers

While transferring or storing binary data, some bit values may change because of a physical effect or an unpredicted disturbance. To check whether such an undesired change has occurred or not, extra bits can be added to the data. This is called parity generation. The idea here is setting standard characteristics to data such that when a change occurs, it can be detected easily.

8.6.1 Parity Generators

One simple method in parity generation is adding an extra bit to set the total number of bits in a binary data block as even or odd. The idea here is as follows. If a bit value changes from logic level 1 to 0 (or vice versa) by an undesired effect, the total number of even (or odd) bits will not satisfy the initial condition. Therefore, the change can be detected easily. There are two options here. The first option is setting the total number of ones to be even. This is called even parity. The second option is setting the total number of ones to be odd. This is called odd parity.

Let's assume three-bit data. Furthermore, assume that even parity will be applied to it. We can form a truth table to generate the parity bit for each input data combination as in Table 8.10. Here, the three-bit data is represented by binary variables $b0$, $b1$, and $b2$. The generated even-parity bit is represented by binary variable pe.

Based on Table 8.10, the even-parity bit can be generated by the logic function $pe = b0 \oplus b1 \oplus b2$. Therefore, the even-parity generator can be composed of two XOR gates with two inputs. The corresponding circuit diagram will be as in Fig. 8.11.

Data bits			Parity bit
b0	**b1**	**b2**	**pe**
0	0	0	0
0	0	1	1
0	1	0	1
0	1	1	0
1	0	0	1
1	0	1	0
1	1	0	0
1	1	1	1

TABLE 8.10 Truth Table of Three-Bit Even-Parity Generator

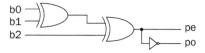

FIGURE 8.11 Circuit diagram of three-bit even-parity generator.

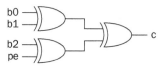

FIGURE 8.12 Circuit diagram of three-bit even-parity checker.

If odd parity is required, the only change needed will be inverting the generated parity bit. Therefore, for the above example, the odd-parity bit will be $po = \overline{pe}$. An N-bit parity generator can be constructed in the same way. Here, we will need $N - 1$ XOR gates. Besides, the architecture will be the same.

8.6.2 Parity Checkers

After adding a parity bit, we can construct a combinational circuit to check whether an undesired change has occurred in the data during transmission or storage. To do so, we can use the same circuitry as in the parity generator with an additional parity bit. This is called parity checker. Let's continue with the previous example having even parity for three bits of data. Parity checker circuitry can be constructed by logic function $c = b0 \oplus b1 \oplus b2 \oplus pe$. Hence, we will use four XOR gates each having two inputs. The circuit diagram for this setup will be as in Fig. 8.12. An N-bit parity checker can be constructed in the same way. Here, we will need N XOR gates. Besides, the architecture will be the same.

8.6.3 Parity Generators and Checkers in Verilog

We provide the Verilog description of a three-bit even-parity generator in Listing 8.21. Here, we directly implement the circuit in Fig. 8.11. Therefore, the input is represented by a three-bit vector b. The generated parity bit is represented by binary variable pe. We provide the RTL schematic of a three-bit even-parity generator in Fig. 8.13. As can be seen in this figure, two XOR gates are used in synthesizing the Verilog description.

Listing 8.21   Verilog Description of Three-Bit Even-Parity Generator

```verilog
module three_bit_even_parity_generator(pe,b);

// Port definitions
input [2:0] b;
output pe;

// for structural modeling
wire w1;

// Structural modeling
xor g1(w1,b[0],b[1]);
xor g2(pe,w1,b[2]);

// Dataflow modeling
assign pe = b[0] ^ b[1] ^ b[2];

endmodule
```

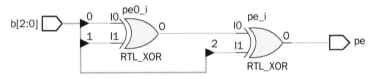

FIGURE 8.13   RTL schematic of three-bit even-parity generator.

Listing 8.22   Verilog Description of Three-Bit Even-Parity Checker

```verilog
module three_bit_even_parity_checker(c,pe,b);

// Port definitions
input [2:0] b;
input pe;
output c;

// for structural modeling
wire w1,w2;

// Structural modeling
xor g1(w1,b[0],b[1]);
xor g2(w2,pe,b[2]);
xor g3(c,w1,w2);

// Dataflow modeling
assign c = b[0] ^ b[1] ^ b[2] ^ pe;

endmodule
```

We next provide the Verilog description of a three-bit even-parity checker in Listing 8.22. Here, we directly implement the circuit in Fig. 8.12. Different from the three-bit parity generator, this description has pe bit as input. The output of the parity checker is c in the description. We provide the RTL schematic of a three-bit even-parity checker in Fig. 8.14. As can be seen in this figure, three XOR gates are used in synthesizing the Verilog description.

8.6.4 Parity Generators and Checkers in VHDL

We next provide the VHDL description of the three-bit even parity generator in Listing 8.23. As in Verilog, we directly implement the circuit in Fig. 8.11. Hence, the input is represented by the three-bit array b. The generated parity bit is represented by binary variable pe. The RTL schematic of the VHDL description will be as in Fig. 8.13.

We finally provide the VHDL description of a three-bit even-parity checker in Listing 8.24. As in Verilog, we directly implement the circuit in Fig. 8.12. Different from the three-bit parity generator, this description has the pe bit as an input. The output of the

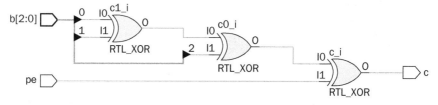

FIGURE 8.14 RTL schematic of three-bit even-parity checker.

Listing 8.23 VHDL Description of Three-Bit Even-Parity Generator

```vhdl
library ieee;
use ieee.std_logic_1164.all;

entity three_bit_even_parity_generator is
port(b : in std_logic_vector (2 downto 0);
    pe : out std_logic);
end three_bit_even_parity_generator;

architecture dataflow of three_bit_even_parity_generator is
begin
pe <= b(0) xor b(1) xor b(2);
end dataflow;
```

Listing 8.24 VHDL Description of Three-Bit Even-Parity Checker

```vhdl
library ieee;
use ieee.std_logic_1164.all;

entity three_bit_even_parity_checker is
port(b : in std_logic_vector (2 downto 0);
    pe : in std_logic;
     c : out std_logic);
end three_bit_even_parity_checker;

architecture dataflow of three_bit_even_parity_checker is
begin
c <= b(0) xor b(1) xor b(2) xor pe;
end dataflow;
```

parity checker is c in the description. The RTL schematic of the VHDL description will be as in Fig. 8.14.

8.7 Applications on Combinational Circuit Blocks

We can improve applications in previous chapters using combinational circuit blocks. Therefore, we will reconsider calculator, home alarm, and car park occupied slot counting systems in this section.

8.7.1 Improving the Calculator

We can improve the calculator introduced in Sec. 6.10 using the case keyword. The modified Verilog description for the calculator will be as in Listing 8.25. As can be seen in this description, the case keyword improved the readability of the description.

We represent this module as an IP block and provide a modified top module for the calculator in Listing 8.26. Here, the calculator IP is represented as calculator_0.

8.7.2 Improving the Home Alarm System

We can improve the home alarm system using a seven-segment display. When the system is active, the display will show character A. When it is closed, the display will show

---

Listing 8.25 Modified Calculator Using the case Keyword in Verilog

```verilog
module calculator(a,b,op,result);

parameter number_length=4;
input [number_length-1:0] a;
input [number_length-1:0] b;
input [2:0] op;
output reg [15:0] result;

always @ (*)
case(op)
  2'b00 : result <= a + b;
  2'b01 : result <= a - b;
  2'b10 : result <= a * b;
  2'b11 : result <= a / b;
endcase

endmodule
```

Listing 8.26 Improved Calculator Implemented on the Basys3 Board in Verilog

```verilog
module calculator_topmodule(sw,led);

input [9:0] sw; //sw[7:4],sw[3:0] numbers, sw[9:8] operation
output [15:0] led;

calculator_0 calc1(sw[7:4],sw[3:0],sw[9:8],led);

endmodule
```

Listing 8.27 Verilog Description of the Seven-Segment Display Decoder Module

```verilog
module decoder_7seg(in1,out1);

input [3:0] in1;
output reg [6:0] out1;

always @ (in1)
case (in1)
    4'b0000 : out1=7'b1000000; //0
    4'b0001 : out1=7'b1111001; //1
    4'b0010 : out1=7'b0100100; //2
    4'b0011 : out1=7'b0110000; //3
    4'b0100 : out1=7'b0011001; //4
    4'b0101 : out1=7'b0010010; //5
    4'b0110 : out1=7'b0000010; //6
    4'b0111 : out1=7'b1111000; //7
    4'b1000 : out1=7'b0000000; //8
    4'b1001 : out1=7'b0010000; //9
    4'b1010 : out1=7'b0001000; //A
    4'b1011 : out1=7'b0000011; //B
    4'b1100 : out1=7'b1000110; //C
    4'b1101 : out1=7'b0100001; //D
    4'b1110 : out1=7'b0000110; //E
    4'b1111 : out1=7'b0001110; //F
  endcase

endmodule
```

character O. To do so, we should add a seven-segment display decoder module to the system. This module converts the provided hexadecimal number to the corresponding seven-segment display pattern as introduced in Exercise 7.14. We provide the Verilog description of the seven-segment display decoder module in Listing 8.27. We should form an IP block for this module to be used in the application. The VHDL version of the seven-segment display decoder module is also available in Listing 8.28.

We provide the modified Verilog description for the application in Listing 8.29. Here, the home alarm system in Listing 7.5 is taken as an IP block. Therefore, we assume that the reader has converted it to an IP block and added it to the project.

8.7.3 Improving the Car Park Occupied Slot Counting System

We can improve the car park occupied slot counting system in two ways. First, we can extend the number of slots to be examined to nine. We provide the modified Verilog description for the car park occupied slot counting system in Listing 8.30. We should form an IP block for this part to be used in the project.

Second, we can display the number of occupied slots on the rightmost seven-segment display of the Basys3 board. To do so, we should add the seven-segment display decoder module in Listing 8.27. Based on these modifications, the Verilog description of the top module for car park occupied slot counting system will be as in Listing 8.31.

Listing 8.28 VHDL Description of the Seven-Segment Display Decoder Module

```vhdl
library ieee;
use ieee.std_logic_1164.all;

entity decoder_7seg is
port(in1 : in std_logic_vector (3 downto 0);
    out1 : out std_logic_vector (6 downto 0));
end decoder_7seg;

architecture behavioral of decoder_7seg is

begin
process (in1)
begin
case in1 is
    when "0000" =>  out1 <= "1000000";
    when "0001" =>  out1 <= "1111001";
    when "0010" =>  out1 <= "0100100";
    when "0011" =>  out1 <= "0110000";
    when "0100" =>  out1 <= "0011001";
    when "0101" =>  out1 <= "0010010";
    when "0110" =>  out1 <= "0000010";
    when "0111" =>  out1 <= "1111000";
    when "1000" =>  out1 <= "0000000";
    when "1001" =>  out1 <= "0010000";
    when "1010" =>  out1 <= "0001000";
    when "1011" =>  out1 <= "0000011";
    when "1100" =>  out1 <= "1000110";
    when "1101" =>  out1 <= "0100001";
    when "1110" =>  out1 <= "0000110";
    when "1111" =>  out1 <= "0001110";
end case;
end process;

end behavioral;
```

Listing 8.29 Improved Home Alarm System Implemented on the Basys3 Board in Verilog

```verilog
module home_alarm_topmodule(sw,led,seg,an);

input [4:0] sw;
output [0:0] led;
output [6:0] seg;
output [3:0] an;

wire [3:0] act;

//rightmost seven-segment display digit is selected
assign an = 4'b1110;

//set 0 and A patterns to be displayed based on sw[4]
assign act = {sw[4],1'b0,sw[4],1'b0};

decoder_7seg ss(act,seg);
home_alarm_0 HA(.a(led),.s(sw[3:0]),.m(sw[4]));

endmodule
```

Listing 8.30 Verilog Description of the Car Park Occupied Slot Counting System for Nine Cars

```verilog
module car_park(c,s);

input [8:0] s;
output reg [3:0] c;

always @(s)
c = s[8]+s[7]+s[6]+s[5]+s[4]+s[3]+s[2]+s[1]+s[0];
endmodule
```

Listing 8.31 Improved Car Park Occupied Slot Counting System Implemented on the Basys3 Board in Verilog

```verilog
module car_park_topmodule(led,sw,seg,an);

input [8:0] sw;
output [3:0] led;
output [6:0] seg;
output [3:0] an;

//use the righmost seven segment display digit.
assign an = 4'b1110;

car_park_0 park1(.c(led),.s(sw));
decoder_7seg ss(led,seg);

endmodule
```

8.8   FPGA Building Blocks Used in Combinational Circuit Blocks

We have provided the RTL schematic of combinational circuit blocks considered in previous sections. The reader can observe that different RTL building blocks are used in implementing the comparator, decoder, encoder, multiplexer, parity generators, and checkers. In fact, all these combinational circuit blocks are implemented by LUTs on an FPGA. To be more specific, four-bit comparator requires five LUTs. An eight-to-three decoder needs four LUTs. In a similar manner, a three-to-eight encoder needs four LUTs. An eight-to-one multiplexer needs two LUTs. Finally, a three-bit parity generator and checker requires one LUT for each. Hence, CLBs will be the main block to be used in this chapter. Besides, interconnect resources and input/output blocks are needed while implementing combinational circuit blocks considered in this chapter.

8.9   Summary

This chapter discussed the combinational circuit blocks extensively used in digital design. We specifically focused on adders, comparators, decoders, encoders, multiplexers, parity generators, and checkers. We provided Verilog and VHDL descriptions of each building block. We also introduced conditional statements and relational operators while constructing implementations. These will be extensively used in the following chapters. Therefore, the reader should practice using these.

8.10   Exercises

8.1 Use the full adder block in Sec. 8.1.2 to add two four-bit numbers.

    a. Implement this device in Verilog or VHDL.

    b. Compare this implementation with the one realized by the " + " operator introduced in Chap. 6.

8.2 Implement a four-bit full adder/subtractor. The user decides on operation type by a control input. When the control input is logic level 1, subtraction will be done. When the control input is logic level 0, addition is done. Implement this device in Verilog or VHDL.

8.3 Design an eight-bit comparator for unsigned numbers. Implement this device in Verilog or VHDL.

8.4 Repeat Exercise 8.3 for eight-bit signed numbers.

8.5 Design an eight-bit comparator for unsigned numbers. The output of the comparator will be the larger number. Implement this device in Verilog or VHDL.

8.6 Repeat Exercise 8.5 for eight-bit signed numbers.

8.7 Implement the two-to-four decoder in Verilog or VHDL using

    a. `case` keyword.

    b. `if` keyword.

8.8 How can we realize a two-input logic function $z = F(x, y)$ using a two-to-four decoder and four-input OR gate.

8.9 A combinational circuit is represented in a SOP form $F(x, y, z) = \sum(0, 2, 4)$. Implement this circuit using one decoder and one three-input OR gate.

8.10 Implement the four-to-two encoder in Verilog or VHDL using

    a. `case/casex` keyword.

    b. `if` keyword.

8.11 Represent logic function of the four-to-one multiplexer in Fig. 8.9

    a. in SOP form.

    b. in POS form.

8.12 Implement the four-to-one multiplexer in Verilog or VHDL using

    a. `case/casex` keyword.

    b. `if` keyword.

8.13 Find the SOP form of three-bit even

    a. parity generator.

    b. parity checker.

8.14 Use multiplexers and memory elements to realize a three-bit even

    a. parity generator.

    b. parity checker.

8.15 **(Arithmetic operations on signed numbers.)** Use conditional statements to apply arithmetic operations on fixed-point signed numbers introduced in Chap. 6. Implement these operations in Verilog or VHDL.

8.16 (Car park occupied slot counting system.) Redesign the car park occupied slot counting system in Sec. 7.6.3 using conditional statements and arithmetic operations in Verilog or VHDL. The new park has 16 slots.

8.17 (Fire alarm system.) Redesign the fire alarm system in Exercise 7.13 using conditional statements in Verilog or VHDL.

8.18 (Keypad decoder.) Redesign the keypad decoder system in Exercise 7.15 using conditional statements in Verilog or VHDL.

8.19 Repeat the Exercise 7.16 using conditional statements in Verilog or VHDL.

8.20 (Even/odd number detector.) Design a combinational circuit to detect whether a given N-bit number is even or odd. Implement the designed circuit using arithmetic operations and conditional statements in Verilog or VHDL.

8.21 (ASCII lowercase/uppercase converter.) Design a combinational circuit to detect whether a given ASCII code corresponds to a character in lowercase form. If this is the case, then the circuit converts the character to uppercase form. Implement the designed circuit using arithmetic operations and conditional statements in Verilog or VHDL.

8.22 (Joystick application.) Repeat Exercise 7.20 using conditional statements.

8.23 (Moving LEDs.) Write a complete Verilog or VHDL description for the following operation. Four switches on the Basys3 board (sw[0], sw[1], sw[14], sw[15]) will control the pattern of 16 LEDs (from led[0] to led[15]). Here
- led[7] and led[8] will turn on when all switches are in off condition (initial condition).
- led[0] and led[1] will turn on when only sw[0] is on.
- led[1] and led[2] will turn on when only sw[1] is on.
- led[13] and led[14] will turn on when only sw[14] is on.
- led[14] and led[15] will turn on when only sw[15] is on.
- led[7] and led[8] will turn on for all other combinations of these switches.

Also each pattern has a condition number which will be displayed on the leftmost seven-segment display digit as follows. The seven-segment display shows 0 for initial condition, 1 when sw[0] is on, 2 when sw[1] is on, 3 when sw[14] is on, 4 when sw[15] is on, and 0 for all other conditions.

8.24 (Car door alarm system.) In this application, we will design a car door alarm system. The system should allow checking all four doors and the trunk (back). We will use five buttons on the Basys3 board. In our application, btnL and btnR represent front doors. btnU and btnD represent back doors. btnC represents the trunk door. We will show whether a door is open or closed by the rightmost seven-segment display digit on the Basys3 board. Based on the label of segments in Fig. 7.25, segments F and B will show status of front doors. Segments E and C will show status of back doors. Segment D will show status of the trunk door. When a segment is on, it indicates that the corresponding door is open. When all the doors are closed, led[0] on the Basys3 board should turn on. Write a complete Verilog or VHDL description to realize this application.

8.25 (Displaying numbers.) Write a complete Verilog or VHDL description on the Basys3 board which will take four-bit input from switches (from sw0 to sw3) and show it on the seven-segment display as a decimal number. Conditions for displaying the number is given below:

- If `btnL` is pressed, the number will be displayed only on the first seven-segment display digit.
- If `btnD` is pressed, the number will be displayed only on the second seven-segment display digit.
- If `btnR` is pressed, the number will be displayed only on the third seven-segment display digit.
- If `btnU` is pressed, the number will be displayed only on the fourth seven-segment display digit.
- If `btnC` is pressed, the number will be displayed on all seven-segment display digits at the same time.
- If more than one button is pressed at the same time, the number should be displayed on the corresponding display digit according to the conditions given above.
- If none of the buttons are pressed, all display digits should be turned off.
- If the number to be displayed is greater than 9, character E should be displayed.

CHAPTER 9

Data Storage Elements

Data storage in a digital system can be made in two ways. First, the system can be designed as a sequential circuit, which will be introduced in the next chapter. In such a circuit, the output depends on past input or output besides current input values. Hence, the data should be stored within the system. This operation is generally performed by flip-flops. Second, the data can be stored in a memory block associated with the system. The memory block can also be constructed by flip-flops. Therefore, we will introduce data storage elements starting from latches as basic building block of flip-flops in this chapter. Then, we will introduce different flip-flop types. Flip-flops can be used to form registers as basic elements of memory blocks. Therefore, we will evaluate register formation next. Afterward, we will focus on read-only memory (ROM) and random access memory (RAM). In constructing ROM and RAM, we will extensively use IP blocks provided by Xilinx.

9.1 Latches

A latch is a basic data storage element that can store one bit of data. Next, we introduce SR and D latches.

9.1.1 SR Latch

An SR latch is the simplest data storage element composed of either two cross-coupled NAND or NOR gates. Let's look at an SR latch composed of two NOR gates with circuit diagram in Fig. 9.1. As can be seen in this figure, an SR latch has two inputs as set (s) and reset (r). It has two outputs as q and $\bar{q}$ which are inverse of each other.

We can represent input/output characteristics of an SR latch in tabular form in a characteristic table. The difference between this table and the truth table is that it can also represent previous and future output values. The characteristic table of an SR latch is presented in Table 9.1. As can be seen in this table, when inputs s and r have logic

FIGURE 9.1 Circuit diagram of SR latch.

179

Inputs		Output
s	**r**	**q**
0	0	q_{prev}
0	1	0
1	0	1
1	1	U

TABLE 9.1   Characteristic Table of SR Latch

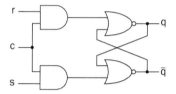

FIGURE 9.2   Circuit diagram of SR latch with control input.

Inputs			Output
s	**r**	**c**	**q**
–	–	0	q_{prev}
0	0	1	q_{prev}
0	1	1	0
1	0	1	1
1	1	1	U

TABLE 9.2   Characteristic Table of SR Latch with Control Input

levels 0 and 1, respectively, the output q will be at logic level 0. In other words, the SR latch is reset. When inputs s and r have logic levels 1 and 0, respectively, the output q will be at logic level 1. We can call this as setting the SR latch. When both s and r have logic level 0, the SR latch stays in its previous state q_{prev}. Hence, it stores the previous output value. When s and r are at logic level 1, we can call this input as both setting and resetting (SR) latch at the same time. Here, a contradiction occurs such that both q and $\bar{q}$ should be at logic level 1. However, $\bar{q}$ is the inverse of q. The output cannot be predicted due to race conditions in transistor level for such a condition. Hence, this input combination should be avoided while using the SR latch. The output at this stage is represented by the undefined symbol (U) in Table 9.1.

We can add a control input to the SR latch. Via this input, we can control when to operate the device. The circuit diagram of the SR latch with control input is shown in Fig. 9.2.

The characteristic table of the SR latch with control input is presented in Table 9.2. Here, the only difference from the SR latch is control input c. When this input is at logic level 0, the output of the SR latch will be kept in its previous value independent of inputs applied to it. Therefore, inputs are represented by the don't care symbol in Table 9.2 when

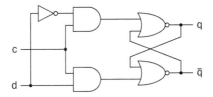

Figure 9.3   Circuit diagram of D latch.

Inputs		Output
d	**c**	**q**
–	0	q_{prev}
0	1	0
1	1	1

Table 9.3   Characteristic Table of D Latch with Control Input

Figure 9.4   Symbol of D latch with control input.

c is at logic level 0. The SR latch becomes active when c is set to logic level 1. Afterward, its output can be changed by s and r inputs.

9.1.2   D Latch

One way to avoid setting and resetting the SR latch at the same time is always feeding inverse inputs to s and r. We can achieve this by adding a NOT gate between them. We call the new structure a D (or data) latch since it saves one bit of data. The circuit diagram of a D latch is depicted in Fig. 9.3.

The characteristic table of the D latch (with control input) is presented in Table 9.3. As can be seen in this table, when the control input is at logic level 0, the D latch keeps its previous output value. We can save the data in the D latch by providing logic level 1 to its control input. Afterward, the bit value at the input d will be saved in the latch. Hence, when d has logic level 0, q will be at logic level 0. When d has logic level 1, q will be at logic level 1. Therefore, the D latch simply stores one bit of the data. The symbol of the D latch (with control input) is presented in Fig. 9.4.

9.1.3   Latches in Verilog

We can form the latch description in two different ways in Verilog. The first one is by using the circuit diagram of the latch and forming the corresponding structural or dataflow model. We will form such a description only for an SR latch. The second way of describing a latch is by using a behavioral model. This will be the form we will be using extensively in describing latches.

9.1.3.1   SR Latch

We provide the Verilog description of the SR latch in Listing 9.1. Here, we have structural and dataflow models of the latch based on the circuit diagram in Fig. 9.1. Besides,

---

Listing 9.1 Verilog Description of SR Latch

```verilog
module SR_latch(s,r,q,qn);

// Port definitions
input s,r;

// for structural and dataflow modeling
//output q,qn;

// for behavioral modeling
output reg q;
output reg qn;

// Structural modeling
//nor(q,r,qn);
//nor(qn,q,s);

// Dataflow modeling
//assign q = ~(r | qn);
//assign qn = ~(s | q);

// Behavioral modeling
always @ (s or r)
if (s) {q,qn} <= 2'b10;
else if (r) {q,qn} <= 2'b01;

endmodule
```

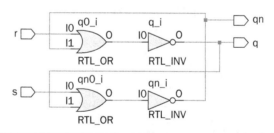

FIGURE 9.5 RTL schematic of SR latch described using dataflow model.

the behavioral model of the SR latch is available in the description. The reader should enable the model to be implemented while disabling other models. In all three models, inputs of the SR latch are represented by s and r. Outputs of the latch are denoted by q and qn. Please note the nonblocking assignment usage in behavioral modeling. As explained in Sec. 5.1.4, we will be using nonblocking assignments in the behavioral model of sequential circuits.

We provide the RTL schematic of the SR latch using the dataflow model in Fig. 9.5. As can be seen in this figure, the RTL schematic is the same as the circuit diagram in Fig. 9.1. Vivado synthesizes the SR latch description in the dataflow model as in Fig. 9.6. Here, three-input and two-input look-up tables (LUTs) are used in implementation. There is a feedback loop between the output and input of the three-input LUT which establishes

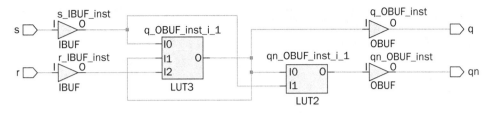

Figure 9.6 Synthesization result of SR latch described using dataflow model.

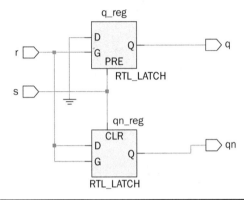

Figure 9.7 RTL schematic of SR latch described using behavioral model.

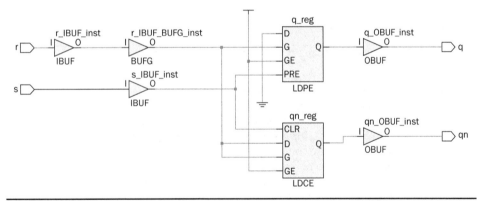

Figure 9.8 Synthesization result of SR latch described using behavioral model.

the data storage capability of the module. Remember that feedback loops from the output to input are general characteristics of latches.

We next provide the RTL schematic of the SR latch using the behavioral model in Fig. 9.7. As can be seen in this figure, the RTL schematic is composed of two latches. Therefore, the RTL schematic and circuit diagram in Fig. 9.1 is not the same. Furthermore, Vivado synthesizes the SR latch description in the behavioral model as in Fig. 9.8. As can be seen in this figure, two flip-flops are used in implementation. Therefore, dataflow and behavioral models of the same SR latch are implemented in different ways in Vivado. We will analyze the difference between these two implementations in detail in Sec. 9.8.

Listing 9.2 Verilog Description of SR Latch with Control Input

```verilog
module SR_latch_control(s,r,c,q,qn);

input s,r,c;
output reg q;
output reg qn;

always @ (s or r or c)
if (c & s) {q,qn} <= 2'b10;
else if (c & r) {q,qn} <= 2'b01;

endmodule
```

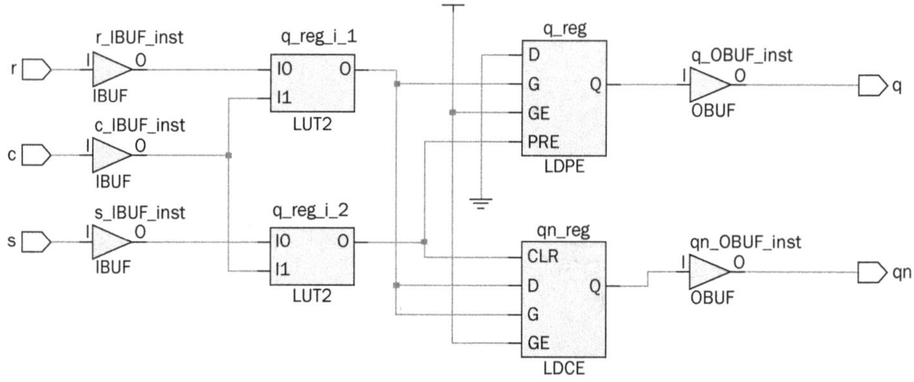

FIGURE 9.9   Synthesization result of SR latch with control input.

9.1.3.2 SR Latch with Control Input

We next provide the Verilog description of the SR latch with control input in Listing 9.2. Here, we have only the behavioral model of the circuit in Fig. 9.2. As in the SR latch description in Listing 9.1, inputs of the SR latch are represented as s and r with an extra control input c. When the control input c is at logic level 0, the SR latch does not respond to other inputs. Outputs of the latch are denoted by q and qn.

The synthesization result of the SR latch with control input is as in Fig. 9.9 which is almost the same as Fig. 9.8. The only difference is the control input. Therefore, the behavioral model of the SR latch with and without control input is implemented in a similar way in Vivado.

9.1.3.3 D Latch

We finally provide the Verilog description of the D latch using the behavioral model in Listing 9.3. In this description, inputs of the latch are represented as d and c. As in the SR latch with control input, when c is at logic level 0, the D latch does not respond to d input. Outputs of the latch are denoted by q and qn in the description.

We provide the RTL schematic of a D latch in Fig. 9.10. As can be seen in this figure, the RTL schematic consists of two D latches (one for each output). In fact, if we had only one output as q, then the RTL schematic would consist of one D latch. We will see in Sec. 9.8 why this is the case. The synthesization result of the D latch description will be

Listing 9.3 Verilog Description of D Latch

```verilog
module D_latch(d,c,q,qn);

input d,c;

output reg q;
output reg qn;

always @ (d or c)
if (c) {q,qn} <= {d,~d};

endmodule
```

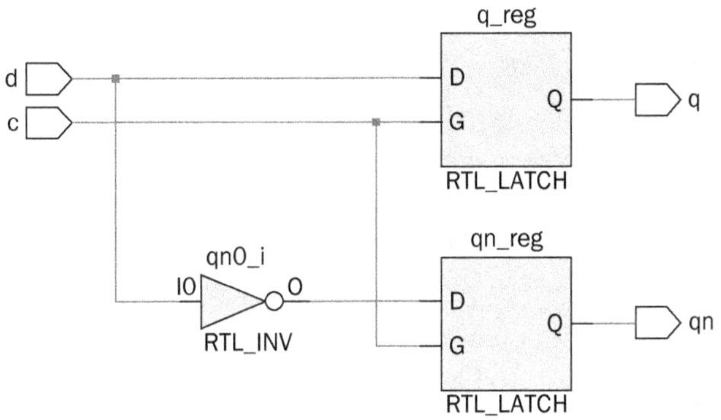

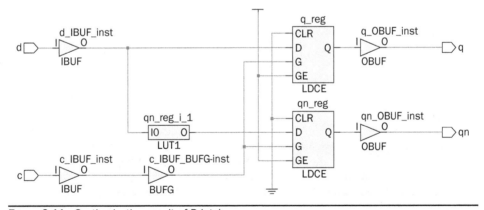

FIGURE 9.10 RTL schematic of D latch.

FIGURE 9.11 Synthesization result of D latch.

as in Fig. 9.11. As can be seen in this figure, one-input LUT and two flip-flops are used in implementation. We will analyze this implementation in detail in Sec. 9.8.

9.1.4 Latches in VHDL

As in Verilog, we can form the latch description in two different ways in VHDL. The first one is using circuit diagram of the latch and forming the corresponding dataflow

---

Listing 9.4 VHDL Description of SR Latch

```
library ieee;
use ieee.std_logic_1164.all;

entity SR_latch is
port(s : in std_logic;
     r : in std_logic;
     q : inout std_logic;
    qn : inout std_logic);
end SR_latch;

architecture dataflow of SR_latch is
begin
q <= r nor qn;
qn<= s nor q;
end dataflow;

architecture behavioral of SR_latch is
begin
process(s,r)
begin
if    ((s='0') and (r='1')) then q<='0'; qn<='1';
elsif ((s='1') and (r='0')) then q<='1'; qn<='0';
elsif ((s='1') and (r='1')) then q<='-'; qn<='-';
end if;
end process;
end behavioral;
```

---

model. We provide this description only for an SR latch. The second way of describing the latch is by using a behavioral model as it is easier to interpret. We do not provide the RTL schematic and synthesization results in this section since these will be almost the same as in Sec. 9.1.3. However, we suggest the reader to observe them in Vivado.

9.1.4.1   SR Latch

We provide the VHDL description of the SR latch in dataflow and behavioral models in Listing 9.4. Here, the dataflow model of the SR latch is based on the circuit diagram in Fig. 9.1. As in the corresponding Verilog description, inputs of the SR latch are represented as s and r. Outputs of the latch are denoted by q and qn. In the behavioral model, the undefined output when $s = 1$ and $r = 1$ is represented by don't care symbol in Listing 9.4.

9.1.4.2   SR Latch with Control Input

We provide the VHDL description of the SR latch with control input in Listing 9.5. As in the SR latch description in Listing 9.4, inputs of the latch are represented as s and r with extra control input c. When control input c is at logic level 0, the SR latch does not respond to other inputs. Outputs of the latch are denoted by q and qn. Again, the undefined output when $s = 1$ and $r = 1$ is represented by don't care symbol in Listing 9.5 as in the SR latch description.

Listing 9.5 VHDL Description of SR Latch with Control Input

```vhdl
library ieee;
use ieee.std_logic_1164.all;

entity SR_latch_control is
port(s : in std_logic;
     r : in std_logic;
     c : in std_logic;
     q : out std_logic :='0';
     qn : out std_logic :='1');
end SR_latch_control;

architecture behavioral of SR_latch_control is
begin
process(s,r,c)
begin
if (c='1') then
if    ((s='0') and (r='1')) then q<='0'; qn<='1';
elsif ((s='1') and (r='0')) then q<='1'; qn<='0';
elsif ((s='1') and (r='1')) then q<='-'; qn<='-';
end if;
end if;
end process;
end behavioral;
```

Listing 9.6 VHDL Description of D Latch

```vhdl
library ieee;
use ieee.std_logic_1164.all;

entity D_latch is
port(d : in std_logic;
     c : in std_logic;
     q : out std_logic :='0';
     qn : out std_logic :='1');
end D_latch;

architecture behavioral of D_latch is
begin
process(d,c)
begin
if (c='1') then q<=d; qn<=not d;
end if;

end process;
end behavioral;
```

9.1.4.3 D Latch

We finally provide the VHDL description of the D latch with control input in Listing 9.6. In this description, inputs of the latch are represented as d and c. As in the SR latch with control input, when c is at logic level 0, the D latch does not respond to d input. Outputs of the latch are denoted by q and qn in the description.

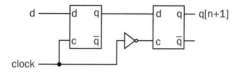

FIGURE 9.12   Constructing D flip-flop from two latches.

FIGURE 9.13   Symbol of D flip-flop.

9.2   Flip-Flops

A latch may change its output whenever its input changes. This may cause inconsistency in the operation of a sequential circuit. To overcome this problem, the clock signal introduced in Sec. 2.2.7 may be used. Therefore, the change at the output of a circuit may occur on either the rising or falling edge of the clock signal. To distinguish such devices from the latches introduced in the previous section, we will call them flip-flops. There are basically three flip-flop types: D, JK, and T.

9.2.1   D Flip-Flop

The D flip-flop can be constructed by connecting two D latches with control input as in Fig. 9.12. In this setup, let's call the two latches a leader and a follower, respectively. We can explain the working principles of the D flip-flop as follows. The control input of both the leader and follower latches are connected to the same clock signal. However, the follower latch receives the inverted clock signal. Therefore, when the clock signal reaches logic level 1 from 0 (rising edge of the clock), the leader latch is enabled and the follower latch is disabled. At this time, the output of the leader latch can be changed by its input (hence the input of the flip-flop). The output of the follower latch (hence the output of the flip-flop) does not change during this time interval since its control input is at logic level 0. When the clock signal reaches logic level 0 from 1 (falling edge of the clock), the control input of the leader latch will be at logic level 0. Hence, its output will be kept in its previous value. In other words, the output of the leader latch will reflect its input when the clock signal was at logic level 1. As can be seen in Fig. 9.12, the output of the leader latch is connected to the input of the follower latch. Since the control input of the follower latch is at logic level 1, its output is set to its input. Therefore, the output of the D flip-flop changes. This operation is specifically called edge-triggered since flip-flop changes its output during rising (or falling) edge of the clock signal.

The symbol of the D flip-flop is presented in Fig. 9.13. Here, the control input is specifically represented by a triangle to indicate that this device changes its output on the rising edge of the clock signal.

The characteristic table of the D flip-flop is presented in Table 9.4. Here, we represent the output of the flip-flop as $q[n + 1]$ to indicate the value at the next clock cycle. We implicitly assume the present clock cycle as n. Within the characteristic table, the clock operation is not explicitly shown.

Input	Output
d	q[n+1]
0	0
1	1

TABLE 9.4 Characteristic Table of D Flip-Flop

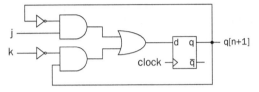

FIGURE 9.14 Circuit diagram of JK flip-flop.

Inputs		Output
j	k	q[n+1]
0	0	$q[n]$
0	1	0
1	0	1
1	1	$\bar{q}[n]$

TABLE 9.5 Characteristic Table of JK Flip-Flop

9.2.2 JK Flip-Flop

The D flip-flop provides a good option to save one bit of data. We can form a more general flip-flop structure using it. The new device will act similar to the SR latch while eliminating its undeterminate state. We call this device the JK flip-flop. The circuit diagram of the JK flip-flop constructed by a D flip-flop is shown in Fig. 9.14.

The characteristic table of the JK flip-flop is presented in Table 9.5. As can be seen in this table, the JK flip-flop acts similar to the SR latch. However, there is no undetermined output here. We can assume the j input as set, the k input as reset. As in the D flip-flop, the output at the next clock cycle is represented by $q[n+1]$. The output at the present clock cycle is represented by $q[n]$. When both j and k inputs are at logic level 1, the output of the JK flip-flop toggles.

9.2.3 T Flip-Flop

We can obtain a specific structure called a T (toggle) flip-flop by connecting input pins of a JK flip-flop. Although this new structure may seem redundant, it will be of great use in counters to be introduced in Sec. 10.4. The characteristic table of the T flip-flop is presented in Table 9.6. As can be seen in this table, the T flip-flop is, in fact, a limited version of a JK flip-flop such that it either gets input of logic level 0 or 1. When the input is at logic level 0, the output of the flip-flop does not change. When the input is at logic level 1, the output of the flip-flop toggles.

Input	Output
t	q[n+1]
0	$q[n]$
1	$\overline{q}[n]$

TABLE 9.6 Characteristic Table of T Flip-Flop

Listing 9.7 Verilog Description of D Flip-Flop

```
module D_flip_flop(d,clk,clr,q,qn);

input d,clk,clr;

output reg q;
output reg qn;

always @ (posedge clk, negedge clr)
if (clr == 0) begin
  q <= 0;
  qn <= 1;
  end
  else begin
  q <= d;
  qn <= ~d;
  end
endmodule
```

9.2.4 Flip-Flops in Verilog

Flip-flops introduced in the previous section can be described in Verilog. Behavioral modeling is the most suitable form to describe a flip-flop since it operates on clock cycles. Let's start with the D flip-flop.

9.2.4.1 D Flip-Flop

The D flip-flop can be described by using behavioral modeling as in Listing 9.7. In this description, inputs of the flip-flop are d (data), clk (clock), and clr (clear). The data input is for a bit value to be saved in the flip-flop. The clock input is for the clock-based operation. The clear input resets the flip-flop output independent of its input. Outputs of the flip-flop are denoted by q and qn in the description. The flip-flop is reset when a negative edge of the clear signal comes. This is achieved by the Verilog keyword negedge. The flip-flop operates whenever a positive edge of the clock signal comes. Again, this is achieved by the Verilog keyword posedge. As a result, the sensitivity list in behavioral modeling becomes posedge clk, negedge clr.

We provide the RTL schematic of a D flip-flop in Fig. 9.15. As can be seen in this figure, the RTL schematic consists of two D flip-flops (one for each output). In fact, if we had only one output as q, then the RTL schematic would consist of one D flip-flop. We will see in Sec. 9.8 why this is the case. The synthesization result of the D flip-flop description will be as in Fig. 9.16. As can be seen in this figure, two one-input LUTs and D flip-flops are used in implementation. We will analyze this implementation in detail in Sec. 9.8.

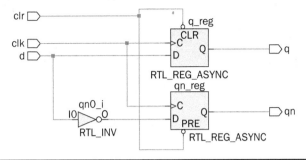

Figure 9.15 RTL schematic of D flip-flop.

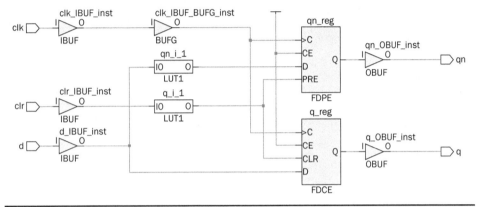

Figure 9.16 Synthesization result of D flip-flop.

9.2.4.2 JK Flip-Flop

We next provide the Verilog description of a JK flip-flop in Listing 9.8. In this description, inputs are represented as j (set), k (reset), clr (clear), and clk (clock). Outputs of the flip-flop are denoted by q and qn in the description. The working principles of a JK flip-flop are similar to those of a D flip-flop. The only difference is that the JK flip-flop has two inputs to set and reset output.

Vivado synthesizes the JK flip-flop description as in Fig. 9.17. As can be seen in this figure, four LUTs and two D flip-flops are used in implementation. This is in line with the circuit diagram of the JK flip-flop constructed from the D flip-flop in Fig. 9.14. In other words, the JK flip-flop is implemented by D flip-flops in Vivado. We will analyze this implementation in detail in Sec. 9.8.

9.2.4.3 T Flip-Flop

We finally provide the Verilog description of a T flip-flop in Listing 9.9. In this description, inputs are represented as t (toggle), clr (clear), and clk (clock). Outputs of the flip-flop are denoted by q and qn in the description. As explained in Sec. 9.2.3, a T flip-flop is a special type of a JK flip-flop. Therefore, the reader can deduce the working principles of a T flip-flop by referring to a JK flip-flop.

Vivado synthesizes the T flip-flop description in Verilog as in Fig. 9.18. Similar to the synthesis result of the JK flip-flop in Fig. 9.17, three LUTs and two D flip-flops are used in implementation. This is expected since the T flip-flop is a special case of a JK flip-flop.

Listing 9.8 Verilog Description of JK Flip-Flop

```verilog
module JK_flip_flop(j,k,clr,clk,q,qn);

input j,k,clr,clk;
output reg q;
output reg qn;

always @ (posedge clk, negedge clr)
if (clr == 0)
begin
q <= 0;
qn<= 1;
end
else
case({j,k})
  2'b01 : begin q<=1'b0; qn<=1'b1;end
  2'b10 : begin q<=1'b1; qn<=1'b0;end
  2'b11 : begin q<=qn; qn<=q;end
endcase

endmodule
```

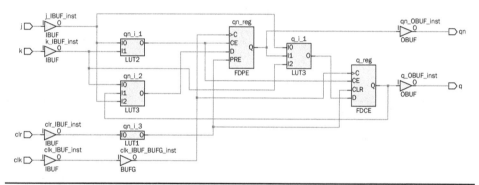

FIGURE 9.17 Synthesization result of JK flip-flop.

Listing 9.9 Verilog Description of T Flip-Flop

```verilog
module T_flip_flop(t,clk,clr,q,qn);

input t,clk,clr;
output reg q;
output reg qn;

always @ (posedge clk, negedge clr)
if (clr == 0) begin
  q <= 0;
  qn <= 1;
  end
  else begin
  q <= q ^ t;
  qn<= ~(q ^ t);
end
endmodule
```

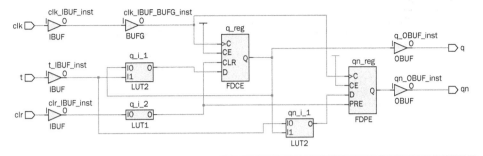

FIGURE 9.18 Synthesization result of T flip-flop in behavioral model.

Listing 9.10 VHDL Description of D Flip-Flop

```vhdl
library ieee;
use ieee.std_logic_1164.all;

entity D_flip_flop is
port (d : in std_logic;
    clk : in std_logic;
    clr : in std_logic;
      q : inout std_logic;
     qn : out std_logic);
end D_flip_flop;

architecture behavioral of D_flip_flop is
begin
process(clk)
begin
if clr='1' then q<='0';
elsif rising_edge(clk) then q<=d;
end if;
end process;
qn <= not q;
end behavioral;
```

9.2.5 Flip-Flops in VHDL

We next provide the VHDL description of D, JK, and T flip-flops. As in the previous section, we consider behavioral modeling here. We did not provide the RTL schematic and synthesization results in this section since these will be almost the same as in Sec. 9.2.4. However, the reader can observe them in Vivado if needed.

9.2.5.1 D Flip-Flop

We provide the VHDL description of a D flip-flop in Listing 9.10. As in the corresponding Verilog description, the inputs of the flip-flop are d (data), clk (clock), and clr (clear). The data input is for the bit value to be saved in flip-flop. The clock input is for clock-based operation. The clear input resets the flip-flop output independent of its input. The outputs of the flip-flop are denoted by q and qn in the description. The sensitivity list of the process in behavioral modeling contains only the clock signal. The flip-flop is reset when a clear signal comes and the clock is at logic level 1. The flip-flop operates whenever the rising edge of the clock signal comes. This is achieved by the VHDL keyword

---

Listing 9.11   VHDL Description of JK Flip-Flop

```vhdl
library ieee;
use ieee.std_logic_1164.all;

entity JK_flip_flop is
port(j : in std_logic;
     k : in std_logic;
   clk : in std_logic;
   clr : in std_logic;
     q : inout std_logic :='0';
    qn : inout std_logic :='1');
end JK_flip_flop;

architecture behavioral of JK_flip_flop is
signal jk : std_logic_vector (1 downto 0);

begin
process(clk)
begin

jk <= j&k;

if clr='1' then q<='0';
elsif rising_edge(clk) then
 case jk is
    when "01" => q<='0';
    when "10" => q<='1';
    when "11" => q<=not q;
    when others => null;
 end case;
end if;
end process;
qn <= not q;

end behavioral;
```

---

rising_edge. If the falling edge of the clock was required as the triggering signal, then the corresponding VHDL keyword would be falling_edge.

9.2.5.2   *JK Flip-Flop*

We next provide the VHDL description of a JK flip-flop in Listing 9.11. In this description, inputs are represented as j (set), k (reset), clr (clear), and clk (clock). Outputs of the flip-flop are denoted by q and qn in the description. The working principles of the JK flip-flop are similar to those of a D flip-flop. The only difference is that the JK flip-flop has two inputs to set and reset output.

9.2.5.3   *T Flip-Flop*

We finally provide the VHDL description of a T flip-flop in Listing 9.12. In this description, inputs are represented as t (toggle), clr (clear), and clk (clock). Outputs of the flip-flop are denoted by q and qn in the description. As explained in Sec. 9.2.3, the T flip-flop is a special type of a JK flip-flop. Therefore, the reader can deduce the working principles of the T flip-flop by referring to the JK flip-flop.

Listing 9.12 VHDL Description of T Flip-Flop

```
library ieee;
use ieee.std_logic_1164.all;

entity T_flip_flop is
port(t  :  in std_logic;
    clk :  in std_logic;
    clr :  in std_logic;
      q :  inout std_logic := '0';
     qn :  out std_logic  := '1');
end T_flip_flop;

architecture behavioral of T_flip_flop is
begin
process(clk)
begin
if clr='1' then q<='0';
elsif rising_edge(clk) then q<=t xor q;
end if;
end process;
qn <= not q;
end behavioral;
```

9.3 Register

A register is an N-bit data storage element constructed by N flip-flops. In forming a register, flip-flops are connected in parallel in such a way that data can be processed all at once. We provide the block diagram of a four-bit register constructed by four D flip-flops in Fig. 9.19. As can be seen in this figure, flip-flops share the same clock. Besides, the input to each flip-flop is independent of the other. Hence, four bits can be stored to the register in a parallel manner. In the same way, the output of each flip-flop is independent of the other. Therefore, stored N-bit data can be observed in a parallel manner. The symbol of a four-bit register is provided in Fig. 9.20.

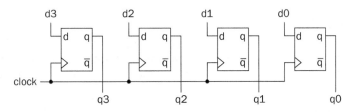

FIGURE 9.19 Block diagram of four-bit register.

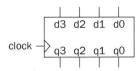

FIGURE 9.20 Symbol of four-bit register.

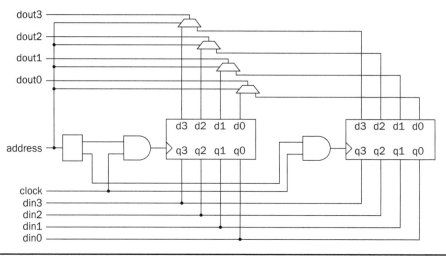

FIGURE 9.21   Circuit diagram of 2 × 4 bit memory.

9.4   Memory

The memory is a data storage element constructed by registers. Within memory, a specific register should be reached. This is achieved by its address. More generally, the wires holding the address data are called address bus. We should be able to write or read the data from a specific register. The wires used for this operation are called data bus. We provide a sample memory implementation by using two four-bit registers in Fig. 9.21. As can be seen in this figure, the data input to two separate registers are done in parallel. An input register is selected by a one-to-two encoder in such a way that the selected register gets the clock signal. The other register not receive the clock. Hence, it will be disabled. The data output from registers are selected by multiplexers. Both data input and output locations are selected by the address bit. Although this is a simple setup, it shows how the memory works.

9.5   Read-Only Memory

The stored data may be taken as static during operation of a digital system. In other words, the data in a specific memory location should not be altered within the system. Such a location is called read-only memory (ROM). We can represent ROM both in Verilog and VHDL.

9.5.1   ROM in Verilog

We provide the Verilog description of a 4×8 bit ROM in Listing 9.13. One can think of this module as composed of four registers each holding eight bits. The input of the module is address. The output of the module is data. The ROM content can be loaded either in a binary or a hexadecimal form. To use the binary form, the command $readmemb should be used. Entries of the ROM are saved in the text file ROM_entries_bin.txt for this case. To use the hexadecimal form, the command $readmemh should be used. Here, entries of the ROM are saved in the text file ROM_entries_hex.txt for this case.

Listing 9.13 Verilog Description of 4 × 8 bit ROM Module

```verilog
module Eight_bit_ROM(data,address);

input [1:0] address;
output [7:0] data;

reg [7:0] ROM [3:0];

assign data=ROM[address];

//load binary ROM content from ROM_entries_bin.txt file
initial $readmemb("H:/Xilinx_projects/project_ROM/ROM_entries_bin.txt",
    ROM);
//load hexadecimal ROM content from ROM_entries_hex.txt file
//initial $readmemh("H:/Xilinx_projects/project_ROM/ROM_entries_hex
    .txt", ROM);

endmodule
```

9.5.2 ROM in VHDL

We provide the VHDL description of a 4 × 8 bit ROM in Listing 9.14. This module has the same naming convention as the corresponding Verilog description. The ROM content is loaded from the text file ROM_entries_hex.txt similar to the application in Sec. 5.4. The only difference is using the for keyword and to_integer implicit function. The for keyword is used to form a loop. The to_integer function converts a given value to an integer form. Besides, the file reading operation is the same.

9.5.3 ROM Formation Using IP Blocks

Xilinx offers IP blocks for memory construction, with two options: distributed and block memory formation. Distributed memory is composed of LUTs. In fact, the ROM descriptions in the previous section are good examples of distributed memory formation. Block memory uses the FPGA parts dedicated for this operation as explained in Chap. 2.

Let's start with distributed ROM generation using IP. Here, we will explain the concept using the Verilog description. The same idea applies to the VHDL description as well. Assume that a Vivado project is opened as explained in Chap. 4. We can add the distributed ROM by selecting it under IP catalog following Memories & Storage Elements → RAMs & ROMs → Distributed Memory Generator. Then the customized IP window appears. In this window, the user can configure the memory element at hand. Since we plan to generate the distributed ROM, we should apply the following steps. First, we should set the depth and data width of the memory block in the "memory config" tab. Assume that we need a 16-element ROM, each element with eight bits. Hence, the depth will be 16 and data width will be eight. Next, we should select the memory type. Here, we will select the ROM. We can set input and output port properties in the "port config" tab. Finally, we can add an initialization file from the "RS & initialization" tab. We can add the text file ROM_entries_hex.txt here with little modification. The IP accepts files in coe format which is easy to construct [30]. The modified file (to be added) will be ROM_entries_hex.coe. As we add the modified IP block to the project,

Listing 9.14 VHDL Description of 4 × 8 bit ROM Module

```vhdl
library ieee;
use ieee.std_logic_1164.all;
use ieee.numeric_std.all;
use std.textio.all;
use ieee.std_logic_textio.all;

entity Eight_bit_ROM is
port(clk : in std_logic;
 address : in std_logic_vector (1 downto 0);
    data : out std_logic_vector (7 downto 0) );
end Eight_bit_ROM;

architecture behavioral of Eight_bit_ROM is

type ROM_type is array (0 to 3) of std_logic_vector (7 downto 0);

shared variable ROM : ROM_type;

begin

process

variable rdline : line;
file file_input : text open read_mode is "H:/Xilinx_projects/project_19
    /ROM_entries_hex.txt";

begin
for i in ROM_type'range loop
  readline(file_input, rdline);
  hread(rdline, ROM(i));
  wait for 5 ns;
end loop;
wait;
end process;

process(clk) begin
if rising_edge(clk) then
 data<= ROM(to_integer(unsigned(address)));
end if;

end process;
end behavioral;
```

we can form a top module as in Listing 9.15. Afterward, we can reach a specific ROM content by providing its address. For more information on the distributed ROM, please see [30].

We can also use the block memory IP to construct a ROM module. As in the distributed ROM formation example, we will only handle the Verilog description here. Assume that a Vivado project is opened as explained in Chap. 4. We can add a block ROM by selecting it under the IP catalog following Memories & Storage Elements → RAMs & ROMs & BRAMs → Block Memory Generator. Then, customized IP window appears.

Listing 9.15 Verilog Description of Distributed ROM Using IP

```
'timescale 1ns / 1ps

module distributed_ROM(data,address);

input [3:0] address;
output [7:0] data;

dist_mem_gen_0 ROM(.a(address),.spo(data));

endmodule
```

Listing 9.16 Verilog Description of Block ROM Using IP

```
'timescale 1ns / 1ps

module block_ROM(data,address,clk);

input [3:0] address;
output [7:0] data;
input clk;

blk_mem_gen_0 ROM(.clka(clk),.addra(address),.douta(data));

endmodule
```

In this window, the user can configure the memory element at hand. Since we plan to generate a block ROM, we should apply the following steps. First, we should set the interface type as "Native" and the memory type as "Single Port ROM" from the "Basic" tab. Then, we should switch to the "Port A Options" tab and set the "Port A Width" and "Port A Depth." Assume that we need a 16-element ROM, each element with eight bits. Hence, the width will be eight and depth will be 16. Finally, we can add an initialization file from the "Other Options" tab. We can add the file ROM_entries_hex.coe here. As we add the modified IP block to the project, we can form a top module as in Listing 9.16. Afterward, we can reach a specific ROM content by providing its address. For more information on block ROM, please see [31].

9.6   Random Access Memory

The stored data may be taken as dynamic during operation of a digital system. In other words, the data in a specific memory location can be altered within the system. Such a location is called random access memory (RAM). We can represent the RAM both in Verilog and VHDL by modifying ROM descriptions in Sec. 9.5. The only difference will be adding a data write option to descriptions. Instead, we will directly use IP blocks introduced in the previous section to construct the RAM.

Let's start with the distributed RAM generation using IP. We will follow the steps in forming the distributed ROM in the previous section. Different from there, we should select the memory type as "Single Port RAM." In the "Port Config" tab, we can also set output options as "registered." We can add the initial RAM content by including the file RAM_entries_hex.coe. As we add the modified IP block to the project, we

---

Listing 9.17 Verilog Description of Distributed RAM Using IP

```verilog
'timescale 1ns / 1ps

module distributed_RAM(data_out,clk,we,addr_out);

reg [3:0] address=4'b0000;
reg [7:0] data_in;
output [7:0] data_out;
input clk, we;
output reg [3:0] addr_out;

dist_mem_gen_0 RAM(.a(address),.d(data_in),.clk(clk),.we(we),.qspo
    (data_out));

always @(posedge clk)
begin
address<=address+4'b0001;
if (we==1'b1)
data_in<={address,address};

addr_out <=address;

end
endmodule
```

---

can form a top module as in Listing 9.17. The top module writes numbers to specific memory locations when the write enable value is at logic level 1. When this value goes to logic level 0, the user can read a specific memory location. For more information on the distributed RAM, please see [30].

We can modify the distributed RAM application by using the block RAM. Here, we will follow the steps in forming the block ROM in the previous section. Different from there, we should set the memory type as "Single Port RAM" from the "Basic" tab. Then, we should switch to "Port A Options" tab and set the Memory Size as "Write Width" to eight bits, "Read Width" to eight bits, and "Write Depth" to 16 bits. "Read Depth" will be set automatically based on this value. As we add the modified IP block to the project, we can form a top module as in Listing 9.18. Similar to Listing 9.17, the top module writes numbers to specific memory locations when the write enable value is at logic level 1. When this value goes to logic level 0, the user can read a specific memory location. For more information on block RAM, please see [31].

9.7 Application on Data Storage Elements

We can improve the calculator by adding memory to it. We provide the top module for the improved calculator in Listing 9.19. Here, the calculator IP is represented as calculator_0. To keep the result of an operation in memory, the user should press btnC button on the Basys3 board. If the user wants to add a number to the one in memory, he or she should press btnL button on the Basys3 board. If subtraction is required, then the user should press btnR button on the Basys3 board. If the user wants to turn back to normal operation (without using the value in memory) then he or she should press btnD button on the Basys3 board.

Listing 9.18 Verilog Description of Block RAM Using IP

```verilog
'timescale 1ns / 1ps

module block_RAM(data_out,clk,we,addr_out,en);

reg [3:0] address=4'b0000;
reg [7:0] data_in;
output [7:0] data_out;
input clk, we, en;
output reg [3:0] addr_out;

blk_mem_gen_0 RAM(.clka(clk),.ena(en),.wea(we),.addra(address),.dina
    (data_in),.douta(data_out));

always @(posedge clk)
begin
address<=address+4'b0001;
if (we==1'b1)
data_in<={address,address};

addr_out <=address;
end

endmodule
```

Since buttons are used in all operations, we should eliminate their malfunction known as "debouncing." This problem occurs when physical properties of the button result in more than one button press effect when it is actually pressed once. There are two ways to eliminate debouncing. One is using the physical resistor and capacitor circuitry [32]. Although this is a good solution, we should avoid adding discrete circuit elements at this step. Therefore, the second solution is adding a delay element to the button press port. We provide the Verilog module performing this operation in Listing 9.20.

In Listing 9.20, the inputs to the debounce module are btn (representing button press) and clk (representing clock signal). The output of the module is btn_clr which indicates the button press signal without any (possible) debouncing effect. The module works as follows. The delay parameter is set as 650000. Assume that we feed the Basys3 clock with a frequency 100 MHz that corresponds to 10-ns clock period. Hence, the delay parameter corresponds to 6.5-ms time duration. The module provides clean button press output if it stays unchanged in this time interval.

9.8   FPGA Building Blocks Used in Data Storage Elements

Data storage elements require different FPGA building blocks compared to the ones used in previous chapters. Let's start with the FPGA building blocks used in latch implementation. As indicated in Sec. 9.1.3, while implementing the SR latch the model used affects the FPGA building blocks used. To be more specific, the dataflow model of the SR latch in Listing 9.1 needs two LUTs used as logic elements. Here, the data storage is performed by a feedback loop as in Fig. 9.6. On the other hand, the behavioral model of the SR latch requires one LUT and two D latches. Therefore, dataflow and behavioral model implementations require different FPGA building blocks. Moreover, elements used in

---

Listing 9.19 Improved Calculator Implemented on the Basys3 Board in Verilog

```verilog
module calculator_topmodule(clk,sw,btnL,btnC,btnR,btnD,led);

input clk;
input [9:0] sw;
input btnL,btnC,btnR,btnD;
output [15:0] led;

reg [7:0] number1;
reg [7:0] number2;
reg [1:0] op;
reg [7:0] mem;
wire [15:0] result;

wire btnCclr,btnLclr,btnRclr,btnDclr;
reg btnCclr_prev,btnLclr_prev,btnRclr_prev,btnDclr_prev;

debounce_0 dbc(clk,btnC,btnCclr);
debounce_0 dbl(clk,btnL,btnLclr);
debounce_0 dbr(clk,btnR,btnRclr);
debounce_0 dbd(clk,btnD,btnDclr);

calculator_0 calc(number1,number2,op,result);

always @ (posedge clk) begin
btnCclr_prev <= btnCclr;
btnLclr_prev <= btnLclr;
btnRclr_prev <= btnRclr;
btnDclr_prev <= btnDclr;

if (btnCclr_prev == 0 && btnCclr == 1) mem <= result[7:0];
else if (btnLclr_prev == 0 && btnLclr == 1) begin
    number1 <= result[7:0];
    number2 <= mem;
    op <= 2'b00;
    end
else if (btnRclr_prev == 0 && btnRclr == 1) begin
    number1 <= result[7:0];
    number2 <= mem;
    op <= 2'b01;
    end
else if (btnDclr_prev == 0 && btnDclr == 1) begin
    number1 <= {4'b0000,sw[7:4]};
    number2 <= {4'b0000,sw[3:0]};
    op <= sw[9:8];
    end
end

assign led = result;

endmodule
```

Listing 9.20 Verilog Description of Debounce Module

```verilog
module debounce(clk,btn,btn_clr);

input clk;
input btn;
output reg btn_clr;

parameter delay = 650000; //6.5ms delay
integer count=0;

reg xnew=0;

always @(posedge clk)
if (btn != xnew)
  begin
  xnew <= btn;
  count <= 0;
  end
else if (count == delay) btn_clr <= xnew;
else count <= count + 1;

endmodule
```

implementing behavioral model of the SR latch are formed of D latches. This may seem contradictory since we need D latches to construct the SR latch. However, the reader should remember that there are only D latches in the Artix-7 XC7A35T FPGA. Therefore, this is the main latch structure to be used in Vivado. We can confirm this by looking at Figs. 9.10 and 9.11.

Next, let's focus on the flip-flop implementation details. Again, here the main building block used in the FPGA implementation is the D flip-flop independent of flip-flop type considered. This is also because of the fact that there are only D flip-flops in the Artix-7 XC7A35T FPGA. Therefore, these are the main building blocks in operation. Let's focus on the D, JK, and T flip-flop implementation details. The D flip-flop requires two LUTs used as logic elements, two slices, and one LUT flip-flop pairs in implementation. The JK flip-flop, on the other hand, requires four LUTs (two being used as logic elements), two slices, and one LUT flip-flop pairs in implementation. Finally, the T flip-flop requires three LUTs used as logic elements, two slices, and two LUT flip-flop pairs in implementation.

Since a register is composed of flip-flops, it is implemented in a similar way. The distributed ROM and RAM will also be based on flip-flop and LUTs. However, as the name implies the block ROM and RAM is specifically based on the block RAM in the FPGA as explained in Chap. 2. The reader can check this property while implementing these elements in Secs. 9.5 and 9.6. There, the block RAM is used to construct memory elements.

We can summarize the fundamental results while implementing data storage elements in the FPGA as follows. Since D latches and flip-flops reside in CLBs in the FPGA, basically they are used in implementation. The distributed ROM and RAM is also constructed in the same way. The block ROM and RAM will be based on specific FPGA blocks for implementation. Besides, interconnect resources and input/output blocks are also needed while implementing data storage elements, as considered in this chapter.

We should warn the reader about one important implementation detail of latches and flip-flops. The provided Verilog and VHDL descriptions work without any problem in the simulation level. However, they may not work as expected (or the corresponding bitstream cannot be generated) when implemented on the Basys3 or Arty board. The reason for this shortcoming is as follows. Vivado specifically asks for any sensitivity list entry labeled by `posedge` or `nededge` to be a clock signal. If this is not satisfied, then a bitstream cannot be generated. To overcome this problem, an edge detector circuit should be used in the description. We provide such an edge detector for Verilog in Listing 10.33.

9.9 Summary

Data storage is a necessary property for most digital systems. A latch can be taken as the basic data storage element to be used for this purpose. However, its usage in an actual FPGA implementation is not desired since a latch lacks a synchronization signal. On the other hand, flip-flops can be constructed by using latches. Therefore, exploring the latch structure was necessary. We will be using flip-flops extensively in constructing sequential circuits. The specific type to be used in implementation will be the D flip-flop because of its availability in the Artix-7 XC7A35T FPGA. Therefore, the reader should understand its working principles. D flip-flops lead to registers and they lead to memory blocks. If the block data is to be saved in an FPGA, these should be used in implementation.

9.10 Exercises

9.1 Construct the SR latch in Sec. 9.1.1 using NAND gates.

9.2 Describe the SR latch with control input in Verilog using
   a. structural modeling.
   b. dataflow modeling.

9.3 Describe the D latch with control input in Verilog using
   a. structural modeling.
   b. dataflow modeling.

9.4 Describe the SR latch with control input in VHDL using dataflow modeling.

9.5 Describe the D latch with control input in VHDL using dataflow modeling.

9.6 Obtain the RTL schematic of SR and D latches in Sec. 9.1.4. Compare the obtained results with the ones in Sec. 9.1.3.

9.7 How would the FPGA building block usage change if only the q output of the D flip-flop is required?

9.8 Use a button and a LED on the Basys3 (or Arty) board such that when the button is pressed once, the LED turns on. When it is pressed twice, the LED turns off. Use a suitable flip-flop description for this operation in Verilog or VHDL.

CHAPTER 10

Sequential Circuits

$\mathbf{F}$lip-flops introduced in the previous chapter allow us to design sequential circuits. The common characteristic of these circuits is that they have memory. Hence, their behavior depend not only on the current input but also on the past input and output. Flip-flops serve as memory elements for this purpose. In this chapter, we will extensively use the D flip-flop since it is available in the Artix-7 XC7A35T FPGA. To understand sequential circuits, we will start with their analysis. This will be different from combinational circuit analysis due to memory elements in the sequential circuit. Therefore, we will introduce new methods specific for this purpose. Then, we will explore the timing concept in sequential circuits. Afterward, we will explain working principles of two sequential circuit families used extensively. These are shift registers and counters. As in combinational circuits, we will review the basic design methodology for sequential circuits by adding extra tools. Finally, we will focus on how sequential circuits can be implemented on the field-programmable gate array (FPGA).

10.1 Sequential Circuit Analysis

We can analyze characteristics of a sequential circuit in three different ways using state equation, state table, and state diagram. This section is on these concepts. Let's first start with defining what a state is.

10.1.1 Definition of State

A flip-flop can store one bit of data as either logic level 0 or 1. Therefore, we can say that it can be in one of two states. If a sequential circuit has N flip-flops, then it can store N bits of data having one of 2^N combinations. In other words, the sequential circuit can be in one of 2^N states. Since there are finite number of states the sequential circuit can be in, it is also called a finite state machine. Throughout the book, we will use both names interchangeably.

10.1.2 State and Output Equations

A sequential circuit changes its state by an input signal and/or clock fed to it. Hence, we can characterize the sequential circuit using its state transitions described by state equations. The aim here is representing the next state using the present state and input values. To represent the output of a sequential circuit, we can use two different models as Mealy and Moore. In Mealy model, the output is a function of both present state and input. In Moore model, the output is a function of the present state only. For more information on Mealy and Moore models, please see [26,33].

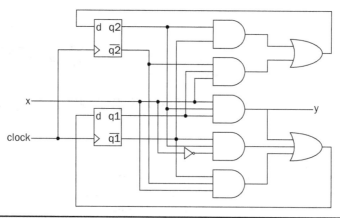

FIGURE 10.1 Circuit diagram of the sample sequential circuit.

Let's take the sequential circuit in Fig. 10.1 as an example and form its state and output equations. As can be seen in this figure, the sequential circuit contains two D flip-flops and logic gates. Let's call the first and second flip-flops as $q1$ and $q2$, respectively. Based on these, possible state values in the circuit will be as $\{q2q1\} \in \{00,01,10,11\}$.

By analyzing the circuit diagram in Fig. 10.1, we can form state and output equations of the corresponding sequential circuit. The output of a D flip-flop can be taken as its present state. This can be represented as $q[n]$ where n indicates the present clock cycle. Therefore, we will have present state values as $q1[n]$ and $q2[n]$ in the sequential circuit. The input of a D flip-flop can be taken as its next state since it will be fed to the output by the next clock cycle. Hence, $q1[n+1]$ and $q2[n+1]$ will be taken as next state values where $n+1$ indicates the next clock cycle. These definitions lead to state and output equations. Here, we will take next states and the output separately as if they are simple combinational circuits. Using techniques introduced in Chap. 7, we can form state and output equations for the sequential circuit in Fig. 10.1 as follows:

$$q2[n+1] = \overline{q2}[n] \cdot q1[n] \cdot x + q2[n] \cdot \overline{q1}[n]$$
$$q1[n+1] = \overline{q2}[n] \cdot \overline{q1}[n] \cdot x + q2[n] \cdot \overline{q1}[n] \cdot \bar{x} + q2[n] \cdot q1[n] \cdot x$$
$$y = q2[n] \cdot q1[n] \cdot x$$

10.1.3   State Table

The state (characteristic) table of a sequential circuit is similar to the truth table of a combinational circuit. However, the state table holds all input and present state combinations at its first section. The second section of the state table holds both output and next state values.

We can form the state table of the sequential circuit in Fig. 10.1 by using its state and output equations. Using these, the state table can be constructed as presented in Table 10.1. This table summarizes characteristics of the sequential circuit. By looking at it, we can know what the next state and output will be based on the present state and input values.

Present State		Input	Next State		Output
q2[n]	q1[n]	x	q2[n + 1]	q1[n + 1]	y
0	0	0	0	0	0
0	0	1	0	1	0
0	1	0	0	0	0
0	1	1	1	0	0
1	0	0	1	1	0
1	0	1	1	0	0
1	1	0	0	0	0
1	1	1	0	1	1

TABLE 10.1 State Table of the Example Sequential Circuit

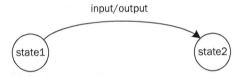

FIGURE 10.2 Part of a generic state diagram for Mealy model.

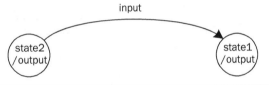

FIGURE 10.3 Part of a generic state diagram for Moore model.

10.1.4 State Diagram

Although the state table characterizes a sequential circuit, it may not be descriptive enough. Therefore, the third method to describe the sequential circuit is using a state diagram composed of circles and directed arcs. Each circle represents a state. A directed arc represents the transition between states. The directed arc also holds the required input value for transition to occur. However, transition timings are not explicitly shown in the state diagram.

If the sequential circuit is of the Mealy type, the directed arc holds what the corresponding output will be after the state transition. Let's provide part of a generic state diagram (for the Mealy model) in Fig. 10.2. As can be seen in this figure, the directed arc holds information on what the input value should be for transition to the next state to occur. The directed arc also holds information on the output value after this transition.

The state diagram based on the Moore model requires outputs to be defined along with states. Therefore, directed arcs will have only input values. Next, we provide part of a generic state diagram for the Moore model in Fig. 10.3. As can be seen in this figure, the directed arc only contains the input value required for transition. The circle representing the state also holds the corresponding output value.

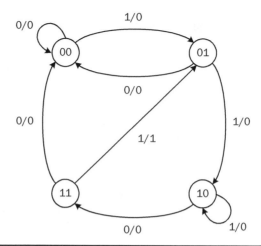

Figure 10.4   State diagram of the example sequential circuit.

Let's turn back to the sequential circuit characterized by its state table in Table 10.1. The output equation of the circuit clearly indicates that this is a Mealy model. Besides, there are four states based on two flip-flops in the circuit. Hence, there will be four circles in the state diagram. Since there is one input and output in the sequential circuit, its state diagram will be as presented in Fig. 10.4.

The state diagram in Fig. 10.4 can be read as follows. There are four states labeled as 00, 01, 10, and 11. Directed arcs have labels such as 1/0. Here, the number before the slash represents input. The number after the slash represents the output. As an example, the directed arc between states 00 and 01 is labeled as 1/0. This indicates that when the system is at state 00 and an input with logic level 1 comes, the system goes to state 01 while producing the output 0. In a similar manner, when the system is at state 00 and an input with logic level 0 comes, the system stays at the same state while producing output 0.

We should mention what the initial state of the sequential circuit should be. We implicitly assumed that the circuit under consideration starts its operation with state 00. In other words, both flip-flops were reset when the first input comes. This setup can be taken as default unless a specific state is taken as the initial state.

We are in a position to judge what the sequential circuit in Fig. 10.4 does. Here, the most helpful representation is its state diagram. Based on it, we can decide that the sequential circuit gives output of logic level 1 only when a sequence of inputs with pattern 1101 comes. Hence, this device is a sequence detector. Such devices are helpful in detecting specific patterns in a sequence.

10.1.5   State Representation in Verilog

We can represent the sequence detector in Fig. 10.1 in Verilog. The first method in describing it is using state and output equations. We provide Verilog description of the sequence detector using these in Listing 10.1.

Instead of representing the sequence detector as presented in Listing 10.1, we can take the advantage of the behavioral modeling in Verilog. The aim here is having a more descriptive representation of the device. Moreover, Verilog allows us to represent states in parametric form. This makes the description more readable. Let's apply this idea to

Listing 10.1 Verilog Description of the Sequence Detector

```
module sequence_detector(y,q1,q2,x,clk,clr);

input x,clk,clr;
output reg y;
output reg q1=0;
output reg q2=0;

always @ (posedge clk)
if (clr==1)
begin
   q1 <= 1'b0;
   q2 <= 1'b0;
end
else
begin
   q1 <= (x&~q2&~q1) + (~x&q2&~q1) + (x&q2&q1);
   q2 <= (x&~q2&q1) + (q2&~q1);
   y  <= q2&q1&x;
end
endmodule
```

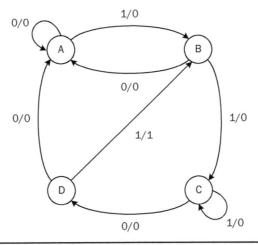

FIGURE 10.5 State diagram of the sequence detector using parametric form.

the sequence detector by representing state values {00, 01, 10, 11} in the device as {A, B, C, D}, respectively. Based on this representation, we will have the new state diagram as shown in Fig. 10.5.

Based on the state diagram in Fig. 10.5, we can reconstruct the Verilog description of the sequence detector. Here, we will represent states as A, B, C, and D. Besides, we will have the actual behavioral description such that state transitions are done by case statements. The final Verilog description of the sequence detector will be as presented in Listing 10.2. This description allows us to analyze working principles of the sequential

Listing 10.2 Verilog Description of the Sequence Detector in Behavioral Form

```verilog
module sequence_detector(y,x,clk,clr);

input x,clk,clr;
output reg y;

reg [1:0] state=2'b00;
parameter A=2'b00, B=2'b01, C=2'b10, D=2'b11;

always @ (posedge clk)
begin
if (clr == 1) state <= A;
else
begin
if (x == 0)
  case (state)
    A : state <= A;
    B : state <= A;
    C : state <= D;
    D : state <= A;
  endcase
else
  case (state)
    A : state <= B;
    B : state <= C;
    C : state <= C;
    D : state <= B;
  endcase
if ((x == 1) && (state == D)) y <= 1;
else y <= 0;
end
end
endmodule
```

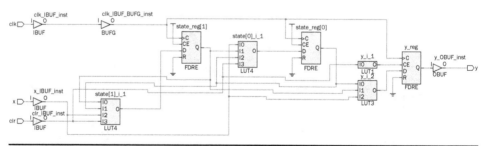

Figure 10.6 Synthesization result of the sequence detector in behavioral model.

circuit easily. Therefore, we will represent sequential circuits this way whenever possible from this point on.

Vivado synthesizes the sequence detector description in Listing 10.2 as presented in Fig. 10.6. Here, four LUTs and three D flip-flops are used in implementation. We will analyze this implementation in detail in Sec. 10.7.

10.1.6 State Representation in VHDL

The sequence detector in Fig. 10.1 can also be described in VHDL. As in the Verilog description in Listing 10.1, the first method is using state and output equations in describing the sequence detector. We provide the VHDL description of the sequence detector formed this way in Listing 10.3.

The second method in describing the sequence detector is using the power of behavioral modeling. VHDL provides an extra advantage compared to Verilog such that states in the device can be represented as a new data type by the VHDL keyword type. The usage of this keyword will be as type state_type is (A,B,C,D). This usage defines a new data type called state_type which can take four values as A, B, C, D. If a signal with the name state is to be defined by type state_type, this can be done by signal state : state_type. We provide the behavioral model of the sequence detector described this way in Listing 10.4. Compared to the description in Listing 10.3, this new form is more readable and explains working principles of the sequence detector clearly. Hence, we will use such a behavioral description whenever possible from this point on. The synthesization result of the sequence detector description in Listing 10.4 will be similar to the one in Fig. 10.6. Therefore, we did not provide it here.

Listing 10.3 VHDL Description of the Sequence Detector

```vhdl
library ieee;
use ieee.std_logic_1164.all;

entity sequence_detector is
port(x : in std_logic;
   clk : in std_logic;
   clr : in std_logic;
    q1 : inout std_logic := '0';
    q2 : inout std_logic := '0';
     y : out std_logic := '0');

end sequence_detector;

architecture behavioral of sequence_detector is

begin
process(clk)
begin

if clr='1' then q1<='0'; q2<='0'; y<='0';
elsif rising_edge(clk) then
    q1<= (x and not q2 and not q1) or (not x and q2 and not q1) or (x
        and q2 and q1);
    q2<= (x and not q2 and q1) or (q2 and not q1);
    y <= q2 and q1 and x;
end if;
end process;
end behavioral;
```

Listing 10.4 VHDL Description of the Sequence Detector in Behavioral Model

```vhdl
library ieee;
use ieee.std_logic_1164.all;

entity sequence_detector is
port(x : in std_logic;
   clk : in std_logic;
   clr : in std_logic;
     y : out std_logic := '0');
end sequence_detector;

architecture behavioral of sequence_detector is
type state_type is (A,B,C,D);
signal state : state_type := A;

begin
process(clk)
begin

if clr='1' then state<=A; y<='0';
elsif rising_edge(clk) then

if x='0' then
   case state is
   when A => state<=A;
   when B => state<=A;
   when C => state<=D;
   when others => state<=A;
   end case;
elsif x='1' then
   case state is
   when A => state<=B;
   when B => state<=C;
   when C => state<=C;
   when others => state<=B;
   end case;
end if;
if (x='1') and (state=D) then y<='1';
else y<='0';
end if;
end if;
end process;
end behavioral;
```

10.2   Timing in Sequential Circuits

Sequential circuits can operate in two different modes in terms of timing. These are synchronous and asynchronous operations. Let's start with the former one.

10.2.1   Synchronous Operation

What we mean by synchronous operation is as follows. All transitions within the sequential circuit are done in clock cycles. In other words, circuit elements share a common clock such that every operation is synchronized with it. The reason of using such a synchronization signal is as follows. When there are flip-flops in the circuit, we may need present state values in obtaining next state values. However, these operations should be done in order. Otherwise, the next state value may be used erroneously instead of the present state value. Hence, synchronization is necessary within the circuit. The sequence detector introduced in Sec. 10.1 is a good example of the synchronous sequential circuit. As can be seen in Fig. 10.1, there are two D flip-flops in the device sharing the same clock signal. The synchronization in the circuit is accomplished this way.

One method to perform synchronous operation in HDL is putting all state transition operations in the same block which is evoked by a change in clock signal. Let's focus on this operation in Verilog first. In Listing 10.2, the description under `always @ (posedge clk)` is responsible for state transitions and output formation. The `posedge` keyword indicates that the `always` block is executed whenever a rising edge of clock comes. Since all state transitions are performed in the `always` block, these operations are synchronized by the rising edge of clock. The same operation can be achieved by the falling edge of clock. Then, the keyword for this operation would be `negedge`.

The synchronization in the VHDL description can be performed by using the `process` block triggered by clock signal. In the VHDL description of the sequence detector given in Listing 10.4, the synchronization is done by putting all state transitions under `process(clk)`. Different from Verilog, VHDL does not allow adding a complex constraint to trigger the `process` block. Hence, it is triggered first by a change in clock signal. Then, state transitions are performed by the required transition type within block. For the sequence detector, this was the rising edge of the clock described by the condition `rising_edge(clk)` within the `if` condition. To perform the same operation in the falling edge of clock signal, the `falling_edge(clk)` condition should have been used.

10.2.2 Asynchronous Operation

There are also asynchronous sequential circuits. In these, there is no common clock shared by all sequential circuit elements. Although asynchronous operations may be beneficiary for some applications, such circuits are not easy to construct and analyze.

We can analyze how asynchronous operation can be achieved in HDL using a basic example. Let's start with the Verilog description in Listing 10.5. Here, there are two `always` blocks. The first one is triggered by the positive edge of the clock signal. The second block is triggered by the negative edge of the binary variable q in the first block. In other words, the execution of the second block depends on the first block, not on the clock signal. This is a simple example of asynchronous operation in Verilog.

The asynchronous operation in Listing 10.5 can also be performed in VHDL. The corresponding description will be in Listing 10.6. Here, there are two `process` blocks the first being triggered by clock signal. Within the first `process` block, a signal q changes its state in each rising edge of clock. This change triggers the second `process` block. Hence, the second block is not triggered by clock signal. Therefore, the overall operation within the device becomes asynchronous.

10.3 Shift Register as a Sequential Circuit

There are sequential circuit families extensively used in digital systems. One such family is the shift register which will be introduced in this section. The register introduced in

Listing 10.5 Asynchronous Operation Example in Verilog

```verilog
module asynchronous_operation(y,clk);

input clk;
output reg y;

reg q;

initial
begin
q =1'b0;
y =1'b0;
end

always @(posedge clk)
q <= ~q;

always @(negedge q)
y <= ~y;
endmodule
```

Listing 10.6 Asynchronous Operation Example in VHDL

```vhdl
library ieee;
use ieee.std_logic_1164.all;

entity asynchronous_operation is
port(clk : in std_logic;
        y : inout std_logic :='0');
end asynchronous_operation;

architecture behavioral of asynchronous_operation is

signal q : STD_LOGIC :='0';

begin
process(clk)
begin
if rising_edge(clk) then
 q<= not q;
end if;
end process;

process(q)
begin
if falling_edge(q) then
    y<= not y;
end if;
end process;

end behavioral;
```

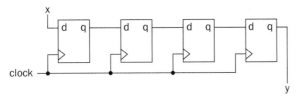

Sec. 9.3 can be modified such that bit locations can be altered in a sequential manner. The family of devices performing this operation is called shift register. There are four shift register types: serial in/serial out, parallel in/serial out, parallel in/parallel out, and serial in/parallel out.

In the serial in/serial out shift register, data is fed to the device in a serial manner. The output is also received in serial manner. This operation is especially useful when a sequence of bits is to be shifted to the left or right. The block diagram of the four-bit serial in/serial out shift register is as presented in Fig. 10.7. As can be seen in this figure, the shift register is constructed by four D flip-flops connected as a chain. Hence, the output of one flip-flop is connected to the input of the next flip-flop. New data bit is fed to the device through its x pin. At each clock cycle, bits are shifted to right between flip-flops. Last data bit is fed to output from y pin.

In the parallel in/serial out shift register, data is fed to the device in a parallel manner. Hence, data is fed all at once. Besides, shifting operation is the same as in serial in/serial out shift register.

Parallel in/parallel out and serial in/parallel out shift registers work similarly. In both devices, data is received in parallel manner. The only difference between these devices is how input is fed to the device. In the parallel in/parallel out shift register, data is fed all at once. In the serial in/parallel out shift register, data is fed bit by bit. Besides, shifting operation in these devices is the same as in the serial in/serial out shift register.

We can summarize working principles of four shift register types as follows. Shifting operation in all these devices is the same. The only difference between them is how the input and output is received. Therefore, let's consider N-bit serial in/serial out shift register to explain the overall operation. To construct the shift register, we should use N D flip-flops. Here, each bit in the sequence to be shifted is saved in a flip-flop named q_i. In this setup, let q_0 and q_{N-1} represent the least and most significant bits, respectively. This shift register can be explained best using its state and output equations as follows:

$$y = q_0[n]$$
$$q_0[n+1] = q_1[n]$$
$$\cdots$$
$$q_{N-2}[n+1] = q_{N-1}[n]$$
$$q_{N-1}[n+1] = x$$

As can be seen in above state equations, at every clock bits are shifted to the right flip-flop. The output equation indicates this is a Moore machine since the output depends only on the present state value.

State and output equations given above can be modified such that a left shift operation can be performed. Modified equations are given below. As can be seen in these equations, the mechanism of shifting operation is the same. Only the connection between the flip-flops, input, and output pins is altered.

$$
\begin{aligned}
y &= q_{N-1}[n] \\
q_{N-1}[n+1] &= q_{N-2}[n] \\
&\cdots \\
q_1[n+1] &= q_0[n] \\
q_0[n+1] &= x
\end{aligned}
$$

10.3.1 Shift Registers in Verilog

Verilog has predefined operators for shifting data in a vector. The shift right operator is ">>". The shift left operator is "<<". Let's assume that a vector Q is to be shifted to left by one bit. The Verilog description for this operation will be Q << 1.

Using predefined shifting operators in Verilog, we can describe shift registers. Let's focus on four-bit serial in/parallel out shift register which shifts data to right. We provide the Verilog description of this device in Listing 10.7. We deliberately handled the serial in/parallel out shift register to show how shifting operation is done in every clock cycle.

Vivado synthesizes the four-bit serial in/parallel out shift register description as presented in Fig. 10.8. Here, only four D flip-flops are used in implementation. This is in line with the block diagram of the shift register in Fig. 10.7.

10.3.2 Shift Registers in VHDL

The easiest way to construct a shift register in VHDL is using the array assignment operator. Through it, we can copy and replace portion of the array to be shifted. We provide the VHDL description of the serial in/parallel out shift register in Listing 10.8. As can be seen in this description, shifting is performed by array operators. The synthesization result of the shift register description will be similar to the one in Fig. 10.8. Therefore, we did not provide it here.

Listing 10.7 Verilog Description of Four-Bit Serial In/Parallel Out Shift Register

```
module SIPO_shift_register(x,clk,q);

input x,clk;
output reg [0:3] q;

initial q=4'b0000;

always @ (posedge clk)
begin
q <= q >> 1;
q[0]<=x;
end

endmodule
```

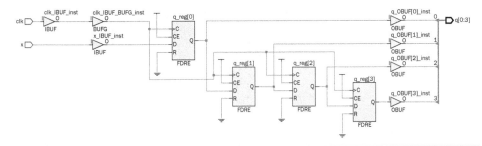

Figure 10.8   Synthesization result of four-bit serial in/parallel out shift register.

Listing 10.8   VHDL Description of Four-Bit Serial In/Parallel Out Shift Register

```vhdl
library ieee;
use ieee.std_logic_1164.all;

entity SIPO_shift_register is
port(x : in std_logic;
   clk : in std_logic;
      q : inout std_logic_vector (0 to 3) := "0000");
end SIPO_shift_register;

architecture behavioral of SIPO_shift_register is
begin
process(clk)
begin
if rising_edge(clk) then
  q(1 to 3)<= q(0 to 2);
  q(0)<= x;
end if;
end process;

end behavioral;
```

10.3.3   Multiplication and Division Using Shift Registers

We can use shift registers to multiply or divide a binary number by integer powers of two. Assume that we keep a binary number in shift register. As we shift all its bits to the left, while feeding input of logic level 0, the result will be the multiplication of original number by two. We can shift the result again to obtain multiplication by four. This operation can be repeated many times to obtain the multiplication of original number by a power of two. Here, the reader should be aware of overflow possibility such that the most significant bit may be lost during operation. Therefore, this bit should be handled specifically during shifting. If shifting is done to the right, then division of the original number by the power of two will be obtained.

Let's consider a simple example on binary multiplication and division operations by powers of two in HDL. We can start with the Verilog description in Listing 10.9. Here, we use an eight-bit parallel in/parallel out shift register for multiplication and division operations. Within the description, p2 represents the power of two for multiplication or division operation. Variable md can be set to logic level 1 for the multiplication operation. It can be set to logic level 0 for the division operation. If an overflow occurs, it is saved in ovr.

Listing 10.9 Verilog Description of the Eight-bit Parallel In/Parallel Out Shift Register for Multiplication and Division Operations

```verilog
module PIPO_shift_register(number,p2,md,clk,result,ovr);

input [7:0] number;
input [1:0] p2;
input md,clk;

output reg [7:0] result;
output reg ovr;

initial ovr =1'b0;

always @ (posedge clk)
begin
if (md==1)
{ovr,result} <= number << p2;
else
result <= number >> p2;
end

endmodule
```

Listing 10.10 VHDL Description of the Eight-bit Parallel In/Parallel Out Shift Register for Multiplication and Division Operations

```vhdl
library ieee;
use ieee.std_logic_1164.all;

entity PIPO_shift_register is
generic (p2 : integer := 2);
port(number : in std_logic_vector (7 downto 0);
        md : in std_logic;
       clk : in std_logic;
    result : inout std_logic_vector (7 downto 0);
       ovr : out std_logic := '0');
end PIPO_shift_register;

architecture behavioral of PIPO_shift_register is

begin
process(clk)
begin
result<=number;
if rising_edge(clk) then
  if md='1' then
    ovr <=result(7-p2+1);
    result(7 downto p2) <= result((7-p2) downto 0);
    result((p2-1) downto 0)<= (others =>'0');
  else
    result((7-p2) downto 0) <= result(7 downto p2);
    result(7 downto (7-p2))<= (others =>'0');
 end if;
end if;
end process;

end behavioral;
```

We provide the VHDL description of the binary multiplication and division example in Listing 10.10. As in the Verilog description, we use eight-bit parallel in/parallel out shift register for multiplication and division operations. Signal names here are same as in the corresponding Verilog description. While setting bits to logic level 0, we used `others =>'0'`. This description is very useful when the total number of bits to be processed is not known in advance in VHDL.

In Listing 10.10, we used VHDL keyword `generic` to pass a specific information into an entity. More specifically, we used it to define constant p2 to be used throughout the shift register architecture. We will use `generic` in the following chapters for such purposes as well.

10.4 Counter as a Sequential Circuit

The counter is another sequential circuit family used in digital systems. As the name implies, the first usage area of this circuit is counting number of input occurrences. The second usage area of a counter is in time-based operations. Here, a number of clock pulses are counted. If the period of the clock is known, then the total time passed during counting operation can be calculated. The third usage area of the counter is in frequency division operation. Here, the frequency of the input clock signal is divided by powers of two.

Working principles of a counter are as follows. Whenever an input signal comes, the counter circuit changes its state. If we assign successive numbers to states in the circuit, then the device visits each number successively. Here, the total number of states indicate the capacity of the counter. Based on the number assignment to states, upward or downward counting can be done.

The counter can best be explained by its state diagram. Let's pick a two-bit (four state) up counter as example. States of this circuit will be 00, 01, 10, and 11. Hence, the circuit will count upwards. If the count value reaches state 11, then the next state will be 00. To indicate that the count reached the final value and restarted counting, we can set the output as logic level 1 at this transition. The corresponding state diagram for the overall operation will be as presented in Fig. 10.9.

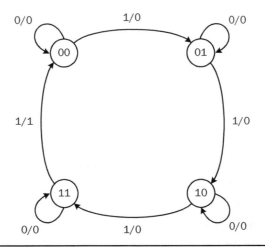

FIGURE 10.9 State diagram of two-bit up counter.

Present State		Input	Next State		Output
q2[n]	q1[n]	x	q2[n+1]	q1[n+1]	y
0	0	0	0	0	0
0	0	1	0	1	0
0	1	0	0	1	0
0	1	1	1	0	0
1	0	0	1	0	0
1	0	1	1	1	0
1	1	0	1	1	0
1	1	1	0	0	1

TABLE 10.2 State Table of a Two-Bit Synchronous Up Counter

A counter can be realized in two different ways as a synchronous or an asynchronous sequential circuit. Next, we explain each realization in detail.

10.4.1 Synchronous Counter

In asynchronous counter, all flip-flops within the sequential circuit are clocked with the same clock signal. We can implement the two-bit synchronous up counter as presented in Fig. 10.9. Since there are four states in the circuit, we will need two flip-flops in implementation. We can form the state table for the counter as presented in Table 10.2. Here, the input to the counter is represented by the binary variable x. The output of the counter is denoted by y.

We can form state and output equations by referring to Table 10.2 as follows:

$$q2[n+1] = q2[n] \cdot \overline{q1}[n] + \overline{q2}[n] \cdot q1[n] \cdot x + q2[n] \cdot \overline{x}$$
$$q1[n+1] = q1[n] \oplus x$$
$$y = q1[n] \cdot q2[n] \cdot x$$

Based on these state and output equations, the final circuit for two-bit synchronous up counter will be as presented in Fig. 10.10.

10.4.2 Asynchronous Counter

There is another way of implementing the two-bit up counter. To do so, we should analyze the state table in Table 10.2 more closely. As can be seen in this table, $q1$ toggles its state whenever rising edge of clock comes and input x equals to logic level 1. $q2$ toggles its state whenever falling edge of $q1$ comes and input x equals to logic level 1. This leads to asynchronous (ripple) counter in which clock signal is fed only to the first flip-flop. The second flip-flop changes its state based on the output of the first flip-flop.

We provide the circuit diagram of the two-bit asynchronous up counter in Fig. 10.11. Here, we use two D flip-flops. As can be seen in this figure, no extra combinational circuit is needed.

10.4.3 Counters in Verilog

Counters can be described in Verilog using arithmetic operations. Let's start with the two-bit synchronous up counter in Fig. 10.10. We can describe this circuit as presented in

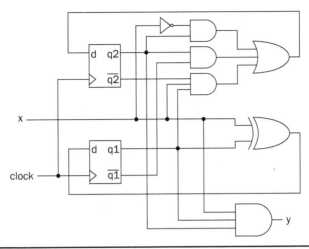

FIGURE 10.10 Circuit diagram of two-bit synchronous up counter.

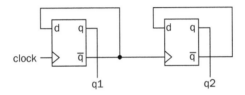

FIGURE 10.11 Circuit diagram of a two-bit asynchronous up counter.

Listing 10.11 Verilog Description of Two-Bit Synchronous Up Counter

```
module two_bit_sync_counter(x,clk,q,y);

input x,clk;
output reg [1:0] q;
output reg y;

initial
begin
q=2'b00;
y=1'b0;
end

always @ (negedge clk)
if (x == 1) {y,q} <= q + 1'b1;

endmodule
```

Listing 10.11. As can be seen in this description, counting operation is done by arithmetic addition by one at every clock cycle when input x is at logic level 1.

Vivado synthesizes the two-bit synchronous up counter in Listing 10.11 as presented in Fig. 10.12. Here, three LUTs and D flip-flops are used in implementation.

We can generalize the two-bit synchronous counter to N bits. Moreover, we can add up or down counting, and clearing the count value functionality. We provide the N-bit

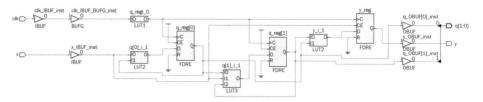

FIGURE 10.12 Synthesization result of two-bit synchronous up counter.

Listing 10.12 Verilog Description of N-bit Synchronous Up/Down Counter

```verilog
module N_bit_sync_counter(x,ud,clk,clr,q,y);

parameter N = 4;

input x,ud,clk,clr;

output reg [N-1:0] q;
output reg y;

always @ (negedge clk,posedge clr)
if (clr==1)
  begin
  q<=0;
  y<=0;
  end
else
if (x == 1)
  if (ud==1) {y,q} <= q + 1'b1;
  else
  {y,q} <= q - 1'b1;
endmodule
```

counter having all these properties in Listing 10.12. Here, ud decides the count direction. If this variable is set to logic level 1, then up counting is performed. Otherwise, down counting is done. Variable clr can be used to clear the count value.

Next, we consider the two-bit asynchronous up counter in Fig. 10.11. We provide the Verilog description of this circuit in Listing 10.13. As can be seen in this description, two always blocks are used to perform the asynchronous operation.

Vivado synthesizes the two-bit asynchronous up counter in Listing 10.13 as presented in Fig. 10.13. Here, four LUTs and three D flip-flops are used in implementation. This implementation clearly shows asynchronous operation if the reader follows clock signal connections.

10.4.4   Counters in VHDL

Counters can also be described in VHDL. Let's reconsider the two-bit synchronous up counter in Fig. 10.10. We can describe this circuit as presented in Listing 10.14. As can be seen here, counting operation is done by arithmetic addition by one. The synthesization result of the two-bit synchronous up-counter description will be similar to the one in Fig. 10.12. Therefore, we did not provide it here.

Listing 10.13   Verilog Description of Two-Bit Asynchronous Up Counter

```verilog
module two_bit_async_counter(x,clk,q,y);

input x,clk;
output reg [1:0] q;
output reg y;

initial
begin
q=2'b00;
y=1'b0;
end

always @ (negedge clk)
if (x == 1) q[0] <= q[0] + 1'b1;

always @ (negedge q[0])
if (x == 1) {y,q[1]} <= q[1] + 1'b1;

endmodule
```

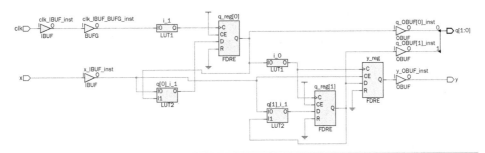

FIGURE 10.13   Synthesization result of two-bit asynchronous up counter.

Listing 10.14   VHDL Description of Two-Bit Synchronous Up Counter

```vhdl
library ieee;
use ieee.std_logic_1164.all;
use ieee.numeric_std.all;

entity two_bit_sync_counter is
port(x : in std_logic;
   clk : in std_logic;
     q : inout signed (1 downto 0) := "00";
     y : out std_logic := '0');
end two_bit_sync_counter;

architecture behavioral of two_bit_sync_counter is
begin
process(clk)
begin
if falling_edge(clk) then
  if x='1' then
    q <=q+1;
    if (q="11") then y<='1';
```

```
       else y <='0';
       end if;
     end if;
 end if;
 end process;


 end behavioral;
```

Listing 10.15 VHDL Description of *N*-bit Synchronous Up/Down Counter

```
library ieee;
use ieee.std_logic_1164.all;
use ieee.numeric_std.all;

entity N_bit_sync_counter is
generic (N : integer :=3);
port(x : in std_logic;
    ud : in std_logic;
   clk : in std_logic;
   clr : in std_logic;
     q : inout signed ((N-1) downto 0);
     y : out std_logic := '0');
end N_bit_sync_counter;

architecture behavioral of N_bit_sync_counter is

begin
process(clk,clr)
begin
if rising_edge(clr) then
q<= (others=>'0');
y <= '0';
end if;

if falling_edge(clk) then
  if ((x='1') and (ud='1')) then
    q <= q + 1;
    if (q="11") then y <= '1';
    else y <= '0';
    end if;
   elsif ((x='1') and (ud='0')) then
    q <= q - 1;
    if (q="00") then y <= '1';
    else y <= '0';
    end if;
  end if;
end if;

end process;
end behavioral;
```

As in the previous section, we can generalize the counter to have up/down and clear properties. We provide the VHDL description for this setup in Listing 10.15. Here, variable names are the same as the ones used in corresponding Verilog description. Hence, the reader can associate both descriptions.

Listing 10.16 VHDL Description of Two-Bit Asynchronous Up Counter

```vhdl
library ieee;
use ieee.std_logic_1164.all;
use ieee.numeric_std.all;

entity two_bit_async_counter is
port(x : in std_logic;
   clk : in std_logic;
      q : inout signed (1 downto 0) := "00";
      y : out std_logic := '0');
end two_bit_async_counter;

architecture behavioral of two_bit_async_counter is

begin
process(clk)
begin
if falling_edge(clk) then
  if x='1' then
    q(0) <= not q(0);
  end if;
    if (q="11") then y <= '1';
    else y <= '0';
    end if;
end if;
end process;

process(q(0))
begin
if falling_edge(q(0)) then
  if x='1' then
    q(1) <= not q(1);
  end if;
end if;
end process;

end behavioral;
```

Finally, we handle the two-bit asynchronous up counter in Fig. 10.11. We provide the VHDL description of this circuit in Listing 10.16. As can be seen in this description, two `process` blocks are used to perform asynchronous operation. The synthesization result of the two-bit asynchronous up-counter description will be similar to the one in Fig. 10.13. Therefore, we did not provide it here.

10.4.5 Frequency Division Using Counters

The clock frequency of a digital system may not be suitable for operation. Hence, we may need to change it. Module performing this is called frequency divider. Counters can be used for this purpose. What we have to do is feeding the clock signal as input and obtaining new clock signal with the frequency divided by powers of two from the output of counter flip-flops. We provide such a synchronous frequency divider in Fig. 10.14. Here, we use T (toggle) flip-flops introduced in Sec. 9.2.3.

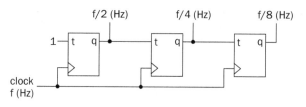

Figure 10.14 Block diagram of a synchronous frequency divider.

Listing 10.17 Verilog Description of Synchronous Frequency Divider

```verilog
module sync_frequency_divider(clk,Fout);

input clk;
output reg [2:0] Fout;

initial
Fout=3'b000;

always @ (negedge clk)
Fout <= Fout + 1'b1;
endmodule
```

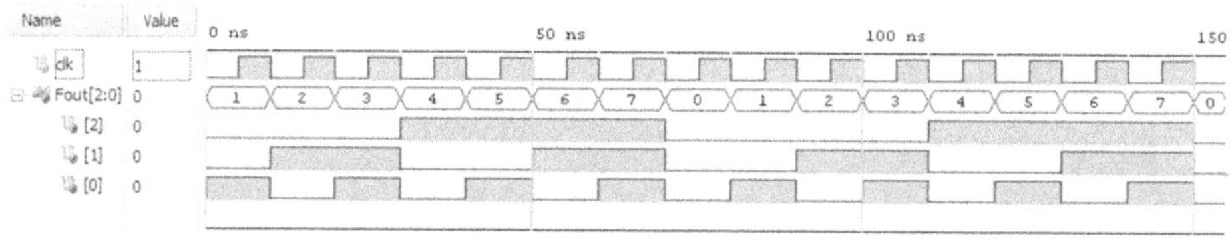

Figure 10.15 Frequency division results of synchronous frequency divider.

Next, we consider HDL description of this frequency divider circuit. We provide the corresponding Verilog description in Listing 10.17. Here, the synchronous counting is performed. Divided frequency values are taken from count digits. We also provide the frequency division result of the clock signal obtained from Vivado in Fig. 10.15. As can be seen in this figure, at each output the digit frequency of an input clock is divided by two, four, and eight. We provide the VHDL description of the synchronous frequency divider working on the same principle in Listing 10.18.

10.5 Sequential Circuit Design

We have introduced combinational circuit design steps in Sec. 7.5. These apply to sequential circuit design as well. However, the designer has to plan state representations and transitions besides usual input/output relationship in designing a sequential circuit. In fact, the main design criterion is deciding which states to be used. We can benefit from either the state diagram or state table for this purpose. The easiest way is using the state diagram. Through it, the designer can plan state transitions and corresponding input/output pairs visually. This leads to state and output equations.

The reader can either implement the state and output equations using HDL or, implementation can be done in HDL by taking benefit of behavioral modeling. We strongly

Listing 10.18 VHDL Description of Synchronous Frequency Divider

```vhdl
library ieee;
use ieee.std_logic_1164.all;
use ieee.numeric_std.all;

entity sync_frequency_divider is
port(clk : in std_logic;
    Fout : inout signed (2 downto 0) := "000");
end sync_frequency_divider;

architecture behavioral of sync_frequency_divider is

begin
process(clk)
begin
if falling_edge(clk) then
  Fout <= Fout+1;
end if;
end process;

end behavioral;
```

suggest the latter method in implementation since it is easy to describe working principles of the sequential circuit this way. Again let's emphasize that ideas mentioned here do not reflect a complete design methodology. The reader should take these just as introductory steps. Designing a sequential circuit should be mastered by consulting related literature.

10.6 Applications on Sequential Circuits

We can use sequential circuits to further improve applications introduced in previous chapters. Therefore, we will reconsider home alarm, digital safe, and car park occupied slot counting systems. We will also introduce two new applications on sequential circuits as vending machine and digital clock in this section.

10.6.1 Improving the Home Alarm System

Using sequential circuits can improve the home alarm system. To do so, we can add password, buzzer, and LED blink modules to the system. The modified system works as follows. Once the alarm is activated by pressing btnC on the Basys3 board, the rightmost seven-segment display digit shows character A. This indicates that the system is active. If one of the windows is opened, then the alarm LED turns on to indicate that alarm has turned on. Hence, the buzzer starts working. If the user enters the correct password, then the alarm is deactivated. Hence, the buzzer stops. If the door is opened, then the user has 20 seconds to enter the password. If the correct password is entered within this time slot, then the alarm turns off. Otherwise, the alarm LED turns on and buzzer starts working. Counting of 20 seconds is displayed on the two leftmost seven-segment display digits of the Basys3 board.

We provide the modified Verilog description for the home alarm module in Listing 10.19. This module has seven inputs. These are clk (main clock signal), pass (eight-bit password), act (activation signal), door (door input), win1, win2, win3 (window inputs). Door and window inputs are at logic level 1 when they are open.

Otherwise they are at logic level 0. The module has five outputs. These are `blinkled` (warning LED during alarm countdown), `alarmled` (shows the alarm status), `seg`, `an` (seven-segment display ports), and `buzzer` (the buzzer output).

Listing 10.19 Verilog Description of the Modified Home Alarm Module via Sequential Circuits

```verilog
module home_alarm2(clk,pass,act,door,win1,win2,win3,blinkled,alarmled,seg,
    an,buzzer);

input clk;
input [7:0] pass;
input act, door, win1, win2, win3;
output reg blinkled=0, alarmled=0;
output [6:0] seg;
output [3:0] an;
output reg buzzer=1;

localparam AOFF=2'b00, AON=2'b01, PASSCHECK=2'b10, SOUND=2'b11;
reg [1:0] state=AOFF;

integer passcounter=0;
localparam secondtime=100000000; //1 second
reg [7:0] seconds=0;
parameter password=8'h55;
reg [3:0]active=4'b0000;

wire [3:0] thos,huns,tens,ones;

binarytoBCD_0 bin({4'b0000,seconds},thos,huns,tens,ones);
sevenseg_driver_0 seg7(clk,1'b0,tens,ones,4'b0000,active,seg,an);

always @ (posedge clk)
case(state)
AOFF:
        if (act == 1) state <= AON;
AON:
        if (door != 0) state <= PASSCHECK;
        else if ((win1 | win2 | win3) != 0) state <= SOUND;
PASSCHECK:
        if (passcounter == secondtime)
        begin
        seconds <= seconds + 1'b1;
        blinkled <= ~blinkled;
        passcounter <= 0;
        end
        else
        if (pass == password) begin
        blinkled <= 0;
        seconds <= 0;
        state <= AOFF;
        end
        else if (seconds == 8'b0001_0100)
        begin
        blinkled <= 0;
        seconds <= 0;
        state <= SOUND;
        end
        else passcounter <= passcounter + 1;
```

```
SOUND:
        if (pass == password)
        begin
        alarmled <= 0;
        state <= AOFF;
        buzzer <= 1;
        end
        else begin
        alarmled <= 1;
        buzzer <= 0;
        end
endcase

always @ (posedge clk)
case (state)
AOFF:
        active <= 4'b0000;
AON:
        active <= 4'b1010;
endcase

endmodule
```

Working principles of the modified home alarm module (as a state machine) are as follows. The state machine has four states as AOFF, AON, PASSCHECK, and SOUND. When act goes to logic level 1, state changes from AOFF to AON. There are two options here. If one of the windows are opened, state machine directly goes to SOUND state which sounds the alarm. If the door is opened, state machine goes to PASSCHECK state. Here, the machine waits for 20 seconds for the user to enter password. The password is initially set to 55 in the hexadecimal form. If the entered password is correct, then the machine goes to AOFF state. Otherwise, the state machine goes to SOUND state which sounds the alarm. IP modules binarytoBCD_0 and sevenseg_driver_0 should be added to the project.

Let's explain the sevenseg_driver module first. The main purpose of this module is to drive the seven-segment display on the Basys3 board. This display is a "common anode" type four-digit display. There are seven signals, named as seg, to drive four digits commonly, and four-digit enable signal, named an, to enable each digit. Since these are all common anode signals, they should be set to logic level 0 when they are active. Since four digits have common seg signals, an signal should be periodically changed at a rate faster than the human eye can catch. In every step, seven-bits seg data will be fed to the selected digit. We provide the Verilog description of the module in Listing 10.20. The module has six inputs as clk (main clock), clr (active-high reset), and four four-bit digit inputs in1, in2, in3, in4. Outputs of the module are seg and an. The VHDL version of the seven-segment display driver module is available in Listing 10.21.

Listing 10.20 Seven-Segment Display Driver Module in Verilog for Four Digits on the Basys3 Board

```
module sevenseg_driver(clk,clr,in1,in2,in3,in4,seg,an);

input clk, clr;
input [3:0] in1, in2, in3, in4;
output reg [6:0] seg;
output reg [3:0] an;
```

```verilog
wire [6:0] seg1, seg2, seg3, seg4;
reg [12:0] segclk;

localparam LEFT = 2'b00, MIDLEFT = 2'b01, MIDRIGHT = 2'b10, RIGHT = 2'b11;
reg [1:0] state=LEFT;

decoder_7seg disp1(in1,seg1);
decoder_7seg disp2(in2,seg2);
decoder_7seg disp3(in3,seg3);
decoder_7seg disp4(in4,seg4);

always @ (posedge clk)
segclk <= segclk + 1'b1;

always @(posedge segclk[12] or posedge clr)
begin
if (clr == 1)
begin
  seg <= 7'b0000000;
  an <= 7'b0000;
  state <= LEFT;
end

else
begin
case(state)
LEFT:
        begin
        seg <= seg1;
        an <= 4'b0111;
        state <= MIDLEFT;
        end
MIDLEFT:
        begin
        seg <= seg2;
        an <= 4'b1011;
        state <= MIDRIGHT;
        end
MIDRIGHT:
        begin
        seg <= seg3;
        an <= 4'b1101;
        state <= RIGHT;
        end
RIGHT:
        begin
        seg <= seg4;
        an <= 4'b1110;
        state <= LEFT;
        end
    endcase
end
end
endmodule
```

Listing 10.21 Seven-Segment Display Driver Module in VHDL for Four Digits on the Basys3 Board

```vhdl
library ieee;
use ieee.std_logic_1164.all;
use ieee.numeric_std.all;

entity sevenseg_driver is
port(clk : in std_logic;
     clr : in std_logic;
     in1 : in std_logic_vector (3 downto 0);
     in2 : in std_logic_vector (3 downto 0);
     in3 : in std_logic_vector (3 downto 0);
     in4 : in std_logic_vector (3 downto 0);
     seg : out std_logic_vector (6 downto 0);
      an : out std_logic_vector (3 downto 0));
end sevenseg_driver;

architecture behavioral of sevenseg_driver is

type state_type is (LEFT, MIDLEFT, MIDRIGHT, RIGHT);
signal state  : state_type := LEFT;
signal seg1   : std_logic_vector (6 downto 0);
signal seg2   : std_logic_vector (6 downto 0);
signal seg3   : std_logic_vector (6 downto 0);
signal seg4   : std_logic_vector (6 downto 0);
signal segclk : std_logic_vector (12 downto 0) := (others => '0');

component decoder_7seg
port(in1 : in std_logic_vector (3 downto 0);
    out1 : out std_logic_vector (6 downto 0));
end component;

begin
process (clk)
begin
if rising_edge(clk) then
segclk <= std_logic_vector( unsigned(segclk) + 1 );
end if;
end process;


process (segclk(12), clr)
begin
if clr = '1' then
    seg <= "0000000";
    an <= "0000";
    state <= LEFT;
elsif rising_edge(segclk(12)) then

case state is
when LEFT =>
        seg <= seg1;
        an <= "0111";
        state <= MIDLEFT;
when MIDLEFT =>
        seg <= seg2;
        an <= "1011";
        state <= MIDRIGHT;
```

```
when MIDRIGHT =>
        seg <= seg3;
        an <= "1101";
        state <= RIGHT;
when RIGHT =>
        seg <= seg4;
        an <= "1110";
        state <= LEFT;
end case;
end if;
end process;

disp1 : decoder_7seg port map(in1 => in1,out1 => seg1);
disp2 : decoder_7seg port map(in1 => in2,out1 => seg2);
disp3 : decoder_7seg port map(in1 => in3,out1 => seg3);
disp4 : decoder_7seg port map(in1 => in4,out1 => seg4);

end behavioral;
```

Listing 10.22 Binary to BCD Converter Module in Verilog

```
module binarytoBCD(binary,thos,huns,tens,ones);

input [11:0] binary;
output reg [3:0] thos, huns, tens, ones;

reg [11:0] bcd_data=0;

always @ (binary)
begin
bcd_data = binary;
thos = bcd_data / 1000;
bcd_data = bcd_data % 1000;
huns = bcd_data / 100;
bcd_data = bcd_data % 100;
tens = bcd_data / 10;
ones = bcd_data % 10;
end

endmodule
```

The working principle of the seven-segment display driver module is as follows. First, we need to divide input clock by 2^{12} to drive segments. If clr goes to logic level 1 at anytime, both seg and an go to logic level 0. Then, all digits and segments are turned on and 8888 is seen on the display. Otherwise, a state machine starts. We have four states to indicate the position of digits from left to right. Note that four-bit inputs have been converted into seven-bit seven-segment codes with the module decoder_7seg in Listing 8.27. The state machine starts in LEFT state where the decoded pattern is loaded to seg and the first digit from left is selected by loading 0111 to an. Next, state turns to MIDLEFT. The same operation is done to drive the second digit from left. This continues in a loop for all four digits in the display.

Let's explain the binarytoBCD module next. This module converts a binary number to the corresponding binary coded decimal (BCD) form. For example, when we have an eight-bit binary number 11111111, we cannot show it directly on the seven-segment

Listing 10.23 Binary to BCD Converter Module in VHDL

```vhdl
library ieee;
use ieee.std_logic_1164.all;
use ieee.numeric_std.all;

entity binarytoBCD is
port(binary : in std_logic_vector (11 downto 0);
        thos : out std_logic_vector (3 downto 0);
        huns : out std_logic_vector (3 downto 0);
        tens : out std_logic_vector (3 downto 0);
        ones : out std_logic_vector (3 downto 0));
end binarytoBCD;

architecture behavioral of binarytoBCD is

begin
process(binary)

variable bcd_data : unsigned (11 downto 0);
variable thos_data: unsigned (11 downto 0);
variable huns_data: unsigned (11 downto 0);
variable tens_data: unsigned (11 downto 0);
variable ones_data: unsigned (11 downto 0);

begin
bcd_data := unsigned(binary);
thos_data := bcd_data / 1000;
bcd_data := bcd_data mod 1000;
huns_data := bcd_data / 100;
bcd_data := bcd_data mod 100;
tens_data := bcd_data / 10;
bcd_data := bcd_data mod 10;
ones_data := bcd_data;

thos <= std_logic_vector(thos_data(3 downto 0));
huns <= std_logic_vector(huns_data(3 downto 0));
tens <= std_logic_vector(tens_data(3 downto 0));
ones <= std_logic_vector(ones_data(3 downto 0));

end process;

end behavioral;
```

display. Therefore, we need to obtain every digit as a four-bit binary decimal code. For example, the corresponding decimal number is 255. Hence, we should have 0010 for decimal two in hundreds digit, 0101 for decimal five in tens digit, and 0101 for five in ones digit. We provide the Verilog description for binary to BCD converter in Listing 10.22. The module has one input as binary representing the binary number to be converted. Outputs of the module are thos, huns, tens, and ones. Here, we had to use blocking assignments in behavioral model to keep digit values. The VHDL version of the binary to BCD converter is available in Listing 10.23. We used variable definitions within the description to keep digit values. Moreover, we had to use the std_logic_vector function which converts its input to standard logic vector form.

Listing 10.24   Modified Home Alarm System Implemented on the Basys3 Board in Verilog

```verilog
module home_alarm_topmodule(clk,sw,btnC,btnU,btnD,btnR,btnL,led,seg,an,
    JA);

input clk;
input [7:0] sw; // password
//btnC: activation, btnU:door, rest windows
input btnC, btnU, btnD, btnR, btnL;
output [1:0] led;
output [6:0] seg;
output [3:0] an;
output [0:0] JA; //buzzer

wire btnCclr,btnUclr,btnDclr,btnRclr,btnLclr;

debounce_0 dbnC(clk,btnC,btnCclr);
debounce_0 dbnU(clk,btnU,btnUclr);
debounce_0 dbnD(clk,btnD,btnDclr);
debounce_0 dbnR(clk,btnR,btnRclr);
debounce_0 dbnL(clk,btnL,btnLclr);

home_alarm2_0 ha1(clk,sw,btnCclr,btnUclr,btnDclr,btnRclr,btnLclr,led
    [1],led[0],seg,an,JA);

endmodule
```

We provide the top module to implement the modified home alarm system on the Basys3 board in Listing 10.24. Here, we use buttons on the board to imitate the door and windows in the home alarm module. Hence, btnU represents the door, btnR, btnD, and btnL stand for windows. Debounce modules are employed within the top module to get the clear button output. Seven-segment display ports seg and an are connected to relevant ports on the Basys3 board. We connected a passive piezo buzzer module to JA port on Basys3 to sound the alarm. This buzzer has three ports as VCC, GND, and I/O. The first two ports are connected to V_{CC} and GND ports of JA on the Basys3 board. I/O is connected to JA[0]. When I/O goes to logic level 0, the buzzer sounds. We use the first eight switches to enter the password.

10.6.2   Improving the Digital Safe System

We can improve the digital safe system by using sequential circuits. Here, the user will have chance to enter his or her password instead of a fixed initial value. We provide the modified Verilog description for the digital safe module in Listing 10.25. This module has five inputs. These are clk (main clock signal), passinput (16-bit password), pass_set (input to change password), pass_reg (input to save new password), and pass_lock (to lock safe again after the password change). The output of the module is a two-bit vector safestate. This output indicates the state of lock, such that 00 shows locked; 01 indicates open; 10 represents enter new password; and 11 shows new password set.

Working principles of the modified digital safe module (as a state machine) are as follows. The state machine has two states: ENTERPASS and SETPASS. In ENTERPASS state, the machine checks whether the input matches the password. If this is the case, safestate changes to 01 which shows that lock is open. Besides, if pass_set is at

Listing 10.25 Verilog Description of the Modified Digital Safe Module via Sequential Circuits

```verilog
module digital_safe2(clk,passinput,pass_set,pass_reg,pass_lock,
    safestate);

input clk;
input [15:0] passinput;
input pass_set,pass_reg,pass_lock;
output reg [1:0] safestate;
//00:locked(c), 01:open(o), 10:enterpass, 11:pass changed(s)

localparam ENTERPASS=1'b0,SETPASS=1'b1;
reg [1:0] state=ENTERPASS;
reg [15:0] pass=16'h1234;

always @ (posedge clk)

case(state)
ENTERPASS:
        if (passinput == pass && pass_set == 1'b1)
        begin
        state <= SETPASS;
        safestate <= 2'b10;
        end
        else if (passinput == pass)
        safestate <= 2'b01;
        else safestate <= 2'b00;
SETPASS:
        if (pass_reg == 1'b1)
        begin
        pass <= passinput;
        safestate <= 2'b11; end
        else if (pass_lock == 1'b1)
        state <= ENTERPASS;
endcase

endmodule
```

logic level 1, then state of the machine goes to SETPASS where the new password is entered. After the user determines a new password, pass_reg should go to logic level 1 to save it. Then, pass_lock should go to logic level 1 to lock the safe again.

We can further improve the digital safe system to be implemented on the Basys3 board. Here, we can show state of the lock and the new password on the seven-segment display. To do so, we should add the seven-segment display module as an IP block. Inputs of the digital safe module will be connected to buttons and switches on the Basys3 board. Hence, we should also add the debounce module as an IP block. We provide the top module for this application in Listing 10.26.

In Listing 10.26, pass_set, pass_reg, and pass_lock inputs are assigned to btnU, btnD, and btnC of the Basys3 board respectively. Sixteen switches are used as passinput. The master clock of the board is connected to clk signal. The output safestate of the digital safe module is kept in a vector with the same name to control the seven-segment display on the board. Hence, when safestate is at 00 all four seven-segment display digits will show character C which stands for "Close". When safestate is at

---

Listing 10.26 Modified Digital Safe System Implemented on the Basys3 Board in Verilog

```verilog
module digital_safe_topmodule(clk,sw,btnC,btnU,btnD,led,an,seg);

input clk;
input [15:0] sw;
input btnC,btnU,btnD;
output [1:0] led;
output [3:0] an;
output [6:0] seg;

wire btnCclr,btnDclr,btnUclr;

debounce_0 dbc(clk,btnC,btnCclr);
debounce_0 dbu(clk,btnU,btnUclr);
debounce_0 dbd(clk,btnD,btnDclr);

reg [3:0] disp1=4'b0;
reg [3:0] disp2=4'b0;
reg [3:0] disp3=4'b0;
reg [3:0] disp4=4'b0;

sevenseg_driver_0 seg7(clk,1'b0,disp1,disp2,disp3,disp4,seg,an);

wire [1:0] safestate;

digital_safe2_0 ds(.clk(clk),.passinput(sw),.pass_set(btnUclr),.
    pass_reg(btnDclr),.pass_lock(btnCclr),.safestate(safestate));

always @ (posedge clk)
case(safestate)
2'b00 : {disp1,disp2,disp3,disp4} <= {4{4'b1100}}; //C
2'b01 : {disp1,disp2,disp3,disp4} <= {4{4'b0000}}; //O
2'b10 : {disp1,disp2,disp3,disp4} <= sw;
2'b11 : {disp1,disp2,disp3,disp4} <= {4{4'b0101}}; //S
endcase

assign led = safestate;

endmodule
```

01, all display digits will show the character O which stands for "Open". When safestate is at 11, all digits will show the character S which stands for "Set". In the 01 state (referring to the password change), digits show the password while the user changes it.

10.6.3 Improving the Car Park Occupied Slot Counting System

We can improve the car park occupied slot counting system using sequential circuits. Hence, we will use the seven-segment display to show total occupied slots. Since we can use more than one seven-segment display digit now, we extend the car park system to count for 16 slots. We provide the modified Verilog module in Listing 10.27.

We provide the top module to implement the modified car park system on the Basys3 board in Listing 10.28. As in previous applications, we used the Basys3 LEDs, switches, and seven-segment display in this top module. Besides, we added a proximity sensor

Listing 10.27 Verilog Description of the Modified Car Park Module via Sequential Circuits

```verilog
module car_park3(c,s);

input [15:0] s;
output reg [4:0] c;

always @ (s)
c = s[15]+s[14]+s[13]+s[12]+s[11]+s[10]+s[9]+s[8]+s[7]+s[6]+s[5]+s[4]+s
    [3]+s[2]+s[1]+s[0];

endmodule
```

Listing 10.28 Modified Car Park System Implemented on the Basys3 Board in Verilog

```verilog
module car_park_topmodule(clk,led,sw,JC,seg,an);

input clk;
input [15:0] sw;
input [3:3] JC;
output [4:0] led;
output [6:0] seg;
output [3:0] an;

wire [3:0] tens, ones;
wire [4:0] cars;

car_park3_0 cp(.c(cars),.s({sw[15:1],~JC[3]}));

binarytoBCD_0 bcd(.binary({7'b0,cars}),.thos(),.huns(),.tens(tens),.
    ones(ones));

sevenseg_driver_0 seg1(.clk(clk),.clr(1'b0),.in1(4'b0),.in2(4'b0),.in3(
    tens),.in4(ones),.seg(seg),.an(an));

assign led = cars;

endmodule
```

(working as a switch to one of the car park clot) to the system as well. The proximity sensor is connected to JC[3] port of the Basys3 board. It works in active-low form such that when no obstacle is detected, the sensor gives logic level 1, otherwise it gives logic level 0 as output. We included seven-segment display driver and binary to BCD IP blocks in the top module.

10.6.4 Vending Machine

We can construct a prototype vending machine using sequential circuits. Let's briefly explain how it works. The machine has two money inputs for 25 cents and 1 dollar (100 cents). Here, the reader can assume that the actual machine has one input for coins. However, 25 cents and 1 dollar are differentiated by a mechanism. Hence, we see two inputs. The vending machine is capable of offering four different products. Each product tray can keep up to 15 items. The user can select a product by its corresponding button. After the selection, the user should press the buy button to finalize the operation. The vending machine gives signal when a product goes out of stock. Then, the maintenance team can fill the corresponding tray and update the stock number by a button.

We provide the Verilog description of the vending machine module in Listing 10.29. This module has six inputs. These are clk (main clock signal), coin1 (25 cents input), coin2 (1 dollar input), select (selection input, every bit representing a different product), buy (buy the selected product), and load (load empty tray). The vending machine module has three outputs. These are money (total deposited money), products (triggers the tray of the corresponding product after a successful trade), and outofstock (to indicate a product has gone out of stock).

Listing 10.29 Verilog Description of the Vending Machine

```verilog
module vending_machine(clk,coin1,coin2,select,buy,load,money,products,
    outofstock);

input clk;
input coin1; //25 cents
input coin2; //1 dollar (100 cents)
input [3:0] select;
input buy;
input [3:0] load;
output reg [11:0] money=0;
output reg [3:0] products=0;
output reg [3:0] outofstock=0;

reg coin1_prev,coin2_prev;
reg buy_prev;

reg [3:0] stock1=4'b1010;
reg [3:0] stock2=4'b1010;
reg [3:0] stock3=4'b1010;
reg [3:0] stock4=4'b1010;

always @ (posedge clk)
begin
coin1_prev <= coin1;
coin2_prev <= coin2;
buy_prev <= buy;

if (coin1_prev == 1'b0 && coin1 == 1'b1) money <= money + 12'd25;
else if (coin2_prev == 1'b0 && coin2 == 1'b1) money <= money + 12'd100;
else if (buy_prev == 1'b0 && buy == 1'b1)

case (select)
4'b0001:
        if (money >= 12'd25 && stock1 > 0)
        begin
        products[0] <= 1'b1;
        stock1 <= stock1 - 1'b1;
        money <= money - 12'd25;
        end
4'b0010:
        if (money >= 12'd75 && stock2 > 0)
        begin
```

```
                  products[1] <= 1'b1;
                  stock2 <= stock2 - 1'b1;
                  money <= money - 12'd75;
                  end
4'b0100:
                  if (money >= 12'd150 && stock3 > 0)
                  begin
                  products[2] <= 1'b1;
                  stock3 <= stock3 - 1'b1;
                  money <= money - 12'd150;
                  end
4'b1000:
                  if (money >= 12'd200 && stock4 > 0)
                  begin
                  products[3] <= 1'b1;
                  stock4 <= stock4 - 1'b1;
                  money <= money - 12'd200;
                  end
endcase

else if (buy_prev == 1'b1 && buy == 1'b0)
        begin
        products[0] <= 1'b0;
        products[1] <= 1'b0;
        products[2] <= 1'b0;
        products[3] <= 1'b0;
        end

else begin
            if (stock1 == 4'b0) outofstock[0] <= 1'b1;
            else outofstock[0] <= 1'b0;
            if (stock2 == 4'b0) outofstock[1] <= 1'b1;
            else outofstock[1] <= 1'b0;
            if (stock3 == 4'b0) outofstock[2] <= 1'b1;
            else outofstock[2] <= 1'b0;
            if (stock4 == 4'b0) outofstock[3] <= 1'b1;
            else outofstock[3] <= 1'b0;

case (load)
            4'b0001: stock1 <= 4'b1111;
            4'b0010: stock2 <= 4'b1111;
            4'b0100: stock3 <= 4'b1111;
            4'b1000: stock4 <= 4'b1111;
endcase
end
end

endmodule
```

Working principles of the vending machine (as a state machine) are as follows. In every rising edge of clk, the machine looks for a rising edge on coin1, coin2, and buy. If coin1 goes to logic level 1, the machine adds 25 cents as a credit. If coin2 goes to logic level 1, the machine adds 1 dollar (100 cents) as a credit. If the user presses buy button, the machine first checks which product is selected. Then, it checks whether the total credit is enough and there is at least one product in stock. If all the conditions are satisfied, then the vending machine withdraws price of the product from total credit;

decreases stock of the product by one; and sets the relevant bit of the product output to logic level 1. When buy goes to logic level 0, products vector is also reset to logic level 0. At the end of each transaction, the vending machine checks whether any product has gone out of stock. Again, we should remind that each product is represented by a separate bit in input, output, and register vectors in the module. For example, if outof-stock is 0010, this means that the second product is out of stock. Or, if the maintenance team loads the tray of the fourth product, then load should be set to 1000.

We provide the top module to implement the vending machine on the Basys3 board in Listing 10.30. As in previous applications, we used the Basys3 LEDs, switches, and seven-segment display in this top module. Besides, we included the seven-segment display, binary to BCD, and debounce IP blocks in the top module.

10.6.5   Digital Clock

We can construct a digital clock using counters introduced in Sec. 10.4. Our clock displays hour and minute digits with 10^{-8}-second accuracy. The user can adjust the time by buttons.

We provide the Verilog description of the digital clock module in Listing 10.31. This module has five inputs. These are clk (main clock signal), en (active high enable signal), rst (resets all outputs when in logic level 1), hrup and minup (adjust hour and minute values). The digital clock module has six outputs each with four bits. These are s1 and s2 (for second digits), m1 and m2 (for minute digits), h1 and h2 (for hour digits).

Listing 10.30   Vending Machine Implemented on the Basys3 Board in Verilog

```
module vending_machine_topmodule(clk,btnC,btnR,btnL,sw,led,seg,an);

input clk;
input btnC,btnR,btnL;
input [7:0] sw;
output [7:0] led;
output [6:0] seg;
output [3:0] an;

wire [11:0] money;
wire btnCclk,btnLclr,btnRclr;

debounce_0 dbnC(clk,btnC,btnCclr);
debounce_0 dbnR(clk,btnR,btnRclr);
debounce_0 dbnL(clk,btnL,btnLclr);

wire [3:0] thos,huns,tens,ones;
binarytoBCD_0 bcd1(money,thos,huns,tens,ones);

sevenseg_driver_0 seg1(.clk(clk),.clr(1'b0),.in1(thos),.in2(huns),.in3(
    tens),.in4(ones),.seg(seg),.an(an));

vending_machine_0 vm(.clk(clk),.coin1(btnRclr),.coin2(btnLclr),.select(
    sw[3:0]),.buy(btnCclr),.load(sw[7:4]),.money(money),.products(led
    [3:0]),.outofstock(led[7:4]));
endmodule
```

Listing 10.31 Verilog Description of the Digital Clock

```verilog
module digital_clock(clk,en,rst,hrup,minup,s1,s2,m1,m2,h1,h2);

input clk;
input en,rst;
input hrup,minup;
output [3:0] s1,s2,m1,m2,h1,h2;

//time display
// h2 h1 : m2 m1

reg [5:0] hour=0,min=0,sec=0;

integer clkc=0;
localparam onesec=100000000; //1 second

always @ (posedge clk)
begin
//reset clock
if (rst == 1'b1)
  {hour,min,sec} <= 0;

//set clock
else if (minup == 1'b1)
  if (min == 6'd59)
  min<=0;
  else
  min <= min+1'd1;

else if (hrup == 1'b1)
  if (hour == 6'd23)
  hour<=0;
  else
  hour <= hour+1'd1;

// count
else if (en == 1'b1)
  if (clkc==onesec)
  begin
  clkc<=0;
  if (sec == 6'd59)
    begin
    sec<=0;
    if (min == 6'd59)
       begin
       min<=0;
       if(hour == 6'd23)
       hour<=0;
       else
       hour<=hour+1'd1;
       end
     else
     min<=min+1'd1;
     end
   else
   sec<=sec+1'd1;
  end
  else
```

```
    clkc<=clkc+1;

end

binarytoBCD_0 secs(.binary(sec),.thos(),.huns(),.tens(s2),.ones(s1));
binarytoBCD_0 mins(.binary(min),.thos(),.huns(),.tens(m2),.ones(m1));
binarytoBCD_0 hours(.binary(hour),.thos(),.huns(),.tens(h2),.ones(h1));

endmodule
```

Working principles of the digital clock (as a state machine) are as follows. There is an integer counter in the module. There is also a parameter onesecond representing one second when 100-MHz clock is used. In every rising edge of the clock, rst is checked. If it is at logic level 1, all output values are reset. Else the machine checks minup and hrup to increment minute or hour digits by one to adjust the clock. If rst, minup, and hrup are at logic level 0 and en is at logic level 1 then the clock starts operating. Here it waits for the counter to count up to onesecond to increment second digits. Afterward, minute and hour digits are incremented as in an actual digital clock operation.

We provide the top module to implement the digital clock on the Basys3 board in Listing 10.32. As in previous applications, we used the Basys3 LEDs, switches, and seven-segment display in this top module. Here, sw[0] enables the clock when it goes to logic level 1; btnC resets the clock; btnU increases the hour digit and btnR increases the minute digit. Besides, we included the seven-segment display, binary to BCD, and debounce IP blocks within the top module.

10.7 FPGA Building Blocks Used in Sequential Circuits

The FPGA building blocks used in this chapter are closely related to the ones considered in Sec. 9.8. Therefore, it is not necessary to reconsider them here. However, we strongly suggest the reader to observe Vivado synthesis result of sequential circuits evaluated in this chapter. This may allow understanding sequential circuit concepts better.

We should mention one important FPGA building block usage at this step. If sensitivity list of an always block in behavioral description depends on positive or negative edge of a clock signal (such as posedge clk or negedge clk), then any variable represented by the reg keyword will automatically have a D flip-flop. Hence, this value can be kept between clock cycles.

10.8 Summary

Sequential circuits allow constructing digital systems with memory. This opens up a new perspective which cannot be performed by combinational circuits. Therefore, we explored sequential circuits in detail in this chapter starting from basic definitions. Then, we analyzed timing in sequential circuits. Here, we can either use synchronous or asynchronous operations. We provided HDL examples for both. Afterward, we handled shift registers and counters as two popular sequential circuit families. Then, we briefly introduced sequential circuit design methodology. We suggest the reader to master how sequential circuits can be designed using related literature. We believe the overall handling of sequential circuits in this chapter will help the reader understand advanced concepts introduced in following chapters.

Listing 10.32 Digital Clock Implemented on the Basys3 Board in Verilog

```verilog
module digital_clock_topmodule(clk,sw,btnC,btnU,btnR,seg,an,led);

input clk;
input [0:0] sw;
input btnC,btnU,btnR;
output [6:0] seg;
output [3:0] an;
output [7:0] led;

wire [3:0] s1,s2,m1,m2,h1,h2;
reg hrup,minup;

wire btnCclr,btnUclr,btnRclr;
reg btnCclr_prev, btnUclr_prev, btnRclr_prev;

debounce_0 dbC(clk,btnC,btnCclr);
debounce_0 dbU(clk,btnU,btnUclr);
debounce_0 dbR(clk,btnR,btnRclr);

sevenseg_driver_0 seg7(clk,1'b0,h2,h1,m2,m1,seg,an);
digital_clock_0 clock1(clk,sw[0],btnCclr,hrup,minup,s1,s2,m1,m2,h1,h2);

always @ (posedge clk)
begin
btnUclr_prev <= btnUclr;
btnRclr_prev <= btnRclr;
if (btnUclr_prev == 1'b0 && btnUclr == 1'b1) hrup <= 1'b1; else hrup <=
    1'b0;
if (btnRclr_prev == 1'b0 && btnRclr == 1'b1) minup <= 1'b1; else minup
    <= 1'b0;
end

assign led[7:0] = {s2,s1};

endmodule
```

10.9 Exercises

10.1 A sequential circuit is represented by the state diagram in Fig. 10.16. The input to the circuit is x. The output of the circuit is y. Implement this sequential circuit in Verilog or VHDL using case statements.

10.2 Redo Exercise 9.8 such that the LED turns on after every five button presses. It turns off in the second button press after turned on.

10.3 Obtain the state diagram of the two-bit down counter in Sec. 10.4.

10.4 (**Barrel shifter.**) Design a barrel shifter in Verilog or VHDL.

10.5 (**Asynchronous frequency divider.**) Design an asynchronous frequency divider in Verilog or VHDL.

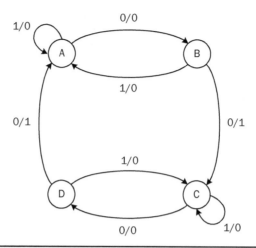

FIGURE 10.16   State diagram for Exercise **10.1**.

10.6 (Frequency divider.) Design a frequency divider module in Verilog or VHDL such that the user can select what the division ratio should be by selecting pins of the device. The device should feed the output as frequency of the input clock divided by one (no division), two, four, and eight.

10.7 (Frequency divider.) Design a frequency divider module in Verilog or VHDL such that the user can select what the division ratio should be by selecting pins of the device. The device should feed the output as frequency of the input clock divided by 1 (no division), 6, 10, and 12.

10.8 (Up-down counter.) Design an up-down counter in Verilog or VHDL such that the digital system counts up to the desired number. Then, it counts downward to zero. When the count value reaches zero, the output should be logic level 1.

10.9 (Frequency divider.) Design a frequency divider in Verilog or VHDL to generate a clock with one hertz frequency using the master clock of the Basys3/Arty board. Connect the generated clock to a LED on the board to observe how it turns on and off.

10.10 (Edge detector in Verilog.) An edge detector circuit is used to detect the rising edge of a signal with reference to the associated clock. Analyze the Verilog description of the edge detector in Listing 10.33.

10.11 (Edge detector in VHDL.) Design an edge detector in VHDL using the analysis in Exercise 10.10.

10.12 (Blink the LED.) Design a digital system to blink led[0] on the Basys3 board every second using a frequency divider and counter. In other words, the rightmost LED on the board will turn on one second and turn off one second periodically.
   a. Construct a synchronous frequency divider module in Verilog. Input to the module will be the clock of Basys3 board which has a frequency 100 MHz. The output of the module will be another clock with frequency 100/8 MHz.
   b. Write a complete top module in Verilog or VHDL to turn on and off led[0] every second. To do so, construct a 24-bit counter. Note that $2^{24}=16777216$.

Listing 10.33 Verilog Description of the Edge Detector

```verilog
module edge_detector(in,rst,clk,out);

input in,rst,clk;
output out;

reg [4:0]shift_reg;

always @ (posedge clk or posedge rst)
begin
  if(rst == 1) shift_reg <= 0;
  else shift_reg[4:0] <= {shift_reg[3:0],in};
end

assign out = ~shift_reg[4] & shift_reg[3];

endmodule
```

Therefore, the counter can count from 0 to 16777215 in decimal. Use the Verilog or VHDL description for frequency division in the previous part of the exercise.

10.13 **(DNA sequence detector.)** DNA is a helical structure of two conjugate strands. One strand in a string of about 3 billion organic molecules is named as nucleotides. There are four known nucleotides: Adenine (A), Thymine (T), Guanine (G), and Cytosine (C). The string of nucleotides tells us much about the organism. We can develop a state machine to detect a specific nucleotide sequence. Assume that we would like to detect the exact sequence composed of nucleotides ATTCGC. Form a Verilog or VHDL description to implement this detector. Here, it is assumed that nucleotides are read and provided to our system as four code values as 00, 01, 10, and 11 representing A, T, C, and G.

10.14 **(DNA sequence detector with empty slots.)** We can extend the sequence detector in Exercise 10.13 such that extra three nucleotides are allowed between A and T in the sequence ATTCGC. Modify the Verilog or VHDL description of your DNA sequence detector to handle this case.

10.15 **(Detecting the first logic level 1 in a binary sequence.)** Assume that we are fed with a binary sequence with values of logic 1 and 0. Design and implement a digital system in Verilog or VHDL for the following operations:

  a. When the first logic level 1 in the sequence is detected, the output of the system becomes logic level 1.

  b. The location of the detected logic level 1 is fed as another output of the digital system.

10.16 **(Snake game.)** We can design a snake game using seven-segment display and buttons on the Basys3 board. The aim of the game is extending the snake while not crossing over itself. We can form the snake on segments of the rightmost seven-segment display digit. Buttons btnU, btnD, btnL, and btnR can be used to enter the next direction of the snake. Here, we will take eight directions as up, down, left, right, up left, up right, down left, and down right. btnC acts as a reset button for the game.

a. Form the basic game in Verilog or VHDL such that snake is initially represented by a segment (let's take E) in the rightmost seven-segment display digit of the Basys3 board. As the user extends the snake fully, led[0] on the board will turn on. If the snake crosses itself during the game, the snake turns back to its initial state.

b. Add a timing module such that the snake extension should be done within limited time.

c. Is it possible to extend this game on four seven-segment display digits?

10.17 (Pulse width modulation.) The aim of the pulse width modulation (PWM) is forming a digital signal with constant period but with varying on and off time values within a period. Form a Verilog or VHDL description to observe basic working principles of the PWM.

10.18 (Digital clock.) How can we implement the digital clock application in Sec. 10.6 using asynchronous operations in Verilog or VHDL?

CHAPTER **11**

Embedding a
Soft-Core
Microcontroller

Most of the time, digital systems implemented on field-programmable gate array (FPGA) and microcontroller platforms seem like rivals. In fact, this is not the case. Both platforms have their advantages and disadvantages as mentioned in Chap. 2. Fortunately, the FPGA design allows including a microcontroller (in soft-core form) as an IP block. This opens up a way to benefit from advantages of both the FPGA and microcontroller platforms at once. Therefore, this chapter focuses on how a soft-core microcontroller can be implemented on the FPGA platform. To explain this process, we will start with introducing building blocks of a generic microcontroller. While doing this, we will reference combinational and sequential circuit blocks introduced in previous chapters. Then, we will introduce two Xilinx microcontroller IP cores named PicoBlaze and MicroBlaze. Finally, we will explore properties of both microcontrollers in detail as well as their usage in simple projects.

Before going further, let's clarify one point. This chapter is not on microcontroller programming which requires a study of its own. Hence, we will direct the reader to related references for this purpose [32]. Instead, the aim here is explaining basics of a microcontroller architecture from an hardware description language (HDL) point of view. This will be a very valuable insight for both FPGA and microcontroller users. The former will understand how a microcontroller can be constructed by HDL description. The latter will have a chance to observe what is going on inside a microcontroller in lowest (possible) level. Hence, the FPGA and microcontroller users can benefit from topics explored in this chapter.

11.1 Building Blocks of a Generic Microcontroller

There are several microcontroller families developed by different vendors. Although these have different properties, they share similar building blocks. In this section, we overview these building blocks by taking a generic microcontroller as benchmark. This will give an insight in the microcontroller development, discussed in the following sections.

11.1.1   Central Processing Unit

The central processing unit (CPU) is a sequential circuit in its basic sense. It is responsible for executing commands given to it in the form of instructions. Therefore, the CPU is the fundamental block responsible for working of a microcontroller.

Each CPU family has a specific instruction set of its own. The user should form a code block using these instructions to program the microcontroller. Generally, this is called assembly language programming. Recent microcontrollers also allow C, C++, or similar languages for programming. Each program is executed sequentially by the CPU. If allowed, interrupts are also served such that an asynchronous operation is performed.

Data or instructions to be processed by the CPU should be taken from other modules such as memory and peripheral units. Two set of wires are needed for this operation. The first set is address bus which holds the location of the data (or instruction) to be processed. The actual data (or instruction) is carried by the data bus which is the second set of wires. The size of the data bus helps decide the type of microcontroller, such as either eight bit or 16 bit.

Most CPUs have registers on them. These are data storage elements as explained in Sec. 9.3. The bit size of these registers should also be in line with the type of microcontroller. Hence, an eight-bit microcontroller will have registers formed also of eight bits. The first group of registers is reserved for the operation of the CPU. One such register is the program counter (PC) which holds the address of the next instruction to be executed. Depending on the type of the CPU, there may also be a specific register (status register) holding the status of the CPU after an instruction is executed. This is also called a flag. Some flags can be used to modify the working state of the CPU, such as allowing an operation to be executed. The second group of registers is provided to the usage of the programmer.

11.1.2   Arithmetic Logic Unit

An arithmetic logic unit (ALU) is a subpart of the CPU responsible for executing arithmetic and logic operations. Therefore, the ALU is basically composed of combinational circuits. Depending on the microcontroller type, the ALU can perform addition, subtraction, multiplication, and division operations. Most microcontrollers will only have addition and subtraction operations since they are easy to implement. These operations are done on fixed-point numbers. If floating-point operations need to be done, the microcontroller should have a floating-point unit. Logic operations to be performed in the ALU may be AND, OR, NOT, and XOR. Besides, the ALU may have a comparator unit.

11.1.3   Memory

A microcontroller needs memory for two reasons. First, instructions to be executed by the CPU should be kept somewhere. Program memory (ROM) is the place for this operation. Second, some instructions to be executed will work on data which need to be stored in memory. Data memory (RAM) will be the block for this operation. Therefore, a microcontroller should have ROM and RAM blocks. Depending on the microcontroller type, the size of these blocks will differ.

11.1.4   Oscillator/Clock

Being a sequential device, the microcontroller needs a clock signal to operate. A clock is generated either by an external source or internal oscillator in the microcontroller.

Initial microcontrollers had just one clock source to operate. Recent microcontrollers have more than one clock source such that each can be used by a different block in the microcontroller. This allows enabling and disabling different blocks based on the application at hand. As a result, power savings can be achieved by using only needed blocks.

11.1.5 General Purpose Input/Output

A microcontroller interacts with the outside world through its input and output pins. These are called general purpose input and output (GPIO). Most of the times, GPIO pins operate with digital data. These pins may be of use for other applications as well, such as the digital communication and analog-to-digital conversion. Therefore, the reader can use these pins for a wide range of applications.

11.1.6 Other Blocks

A microcontroller may also have other blocks generally called peripherals. The digital communication, analog-to-digital converter (ADC), and digital-to-analog converter (DAC) can be counted as such blocks. A microcontroller implemented on the FPGA has freedom on such blocks since any digital device can be implemented alongside it. In other words, the user is free to add any peripheral device to the microcontroller implemented on the FPGA.

11.2 Xilinx PicoBlaze Microcontroller

PicoBlaze is an eight-bit soft-core microcontroller developed by Xilinx. The specific core to be used in this book is called KCPSM6, which is suitable for the Xilinx Artix-7 FPGAs. PicoBlaze documentation and files can be downloaded from [34] as file `KCPSM6_Release9_30Sept14.zip`. We will next analyze functional blocks of PicoBlaze. Then, we will explore how it can be used in connection with Verilog and VHDL descriptions.

11.2.1 Functional Blocks of PicoBlaze

Functional block diagram of PicoBlaze (provided by Xilinx) is presented in Fig. 11.1. We will explain PicoBlaze using it based on definitions in Sec. 11.1. However, we will not cover all blocks and operations (such as stack) in this figure to simplify the explanation. More information on these issues can be found in [35, 36].

The CPU is not specifically shown in Fig. 11.1. However, we can consider the instruction decoder, PC, stack, registers, and flags as parts of the CPU. The instruction decoder is responsible for fetching and preparing the instruction to be executed. The PC holds the address of the next instruction to be executed. The PC is automatically incremented to the next instruction location when the present instruction is executed. PicoBlaze has a 10-bit PC that supports 1024 instruction address. PicoBlaze has 16 registers each holding eight bits. These are named as s0 to sF in Fig. 11.1. There are also three flags in PicoBlaze. The first one, called IE, enables and disables interrupts. The second and third flags are called Z and C, respectively. These are set when a zero or carry occurs after the ALU operation.

PicoBlaze has its own assembly language formed of instructions represented by 18 bits. The assembly language and its usage are explained in detail in [35, 36]. Instructions to be executed are saved in program memory called instruction PROM. The program written in assembly language should be embedded to PROM via a specific

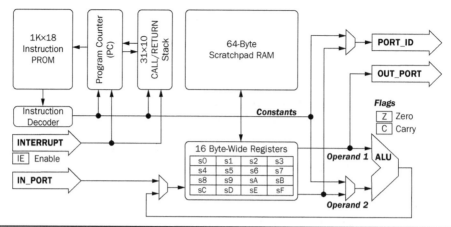

Figure 11.1 Functional block diagram of the PicoBlaze microcontroller.

procedure explained in Sec. 11.2.4. PicoBlaze has the data memory called Scratchpad RAM which can save up to 64 bytes of data.

The ALU in PicoBlaze can perform arithmetic and logic operations required by the instruction to be executed. Arithmetic operations that can be performed on PicoBlaze are addition and subtraction. Logic and compare operations that can be performed on the microcontroller are bitwise logic operations, arithmetic compare and bitwise test operations, and comprehensive shift and rotate operations.

The GPIO in PicoBlaze is indicated by IN_PORT, OUT_PORT, and PORT_ID in Fig. 11.1. PicoBlaze supports up to 256 input and 256 output pins or a combination of both. The interrupt is indicated as another input called INTERRUPT in Fig. 11.1.

The oscillator/clock module is not specifically shown in Fig. 11.1. However, the user should add it so that the microcontroller can operate. Moreover, the user is free to add any extra peripheral blocks to PicoBlaze through the HDL description.

11.2.2 PicoBlaze in Verilog

Xilinx provides the Verilog description of the PicoBlaze microcontroller in KCPSM6_ Release9_30Sept14.zip. The microcontroller module in Verilog is named as kcpsm6.v within this archive. The reader can use this module within his or her design by providing necessary peripheral devices including the clock signal. We provide the shortened version of kcpsm6.v in Listing 11.1. Here, all input and output connections of the microcontroller are defined.

To program the microcontroller, the reader should prepare a file (with extension psm) consisting of assembly language instructions. These should be embedded to PROM. To do so, the reader should run the assembler kcpsm6.exe which can be found in KCPSM6_ Release9_30Sept14.zip. The assembler will convert the file containing assembly language instructions to a Verilog file to be added to the project. At this point, please remember that the Verilog file ROM_form.v should be in the same folder with the assembler kcpsm6.exe. As all files are added and settings are done, the Verilog description can be implemented. We will provide such an example in Sec. 11.2.4.

11.2.3 PicoBlaze in VHDL

Xilinx also provides the VHDL description of the PicoBlaze microcontroller in KCPSM6_ Release9_30Sept14.zip. The microcontroller module in VHDL is named as

Listing 11.1 Verilog Description of the PicoBlaze Microcontroller in Shortened Form

```
module kcpsm6 (address,instruction,bram_enable,in_port,out_port,
    port_id,write_strobe,k_write_strobe,read_strobe,interrupt,
    interrupt_ack,sleep,reset,clk);

parameter [7:0] hwbuild = 8'h00 ;
parameter [11:0] interrupt_vector = 12'h3FF ;
parameter integer scratch_pad_memory_size = 64 ;

output [11:0] address;
input [17:0] instruction;
output bram_enable;
input [7:0] in_port;
output [7:0] out_port;
output [7:0] port_id;
output write_strobe;
output k_write_strobe;
output read_strobe;
input interrupt;
output interrupt_ack;
input sleep;
input reset;
input clk;

// Architecture of kcpsm6

endmodule
```

kcpsm6.vhd within the archive. The reader can use this module within his or her design by providing necessary peripheral devices including the clock signal. We provide the shortened version of kcpsm6.vhd in Listing 11.2. As in the Verilog description, all input and output connections of the microcontroller are defined.

Programming the microcontroller via its VHDL description is the same as in Verilog. The only difference here is that the VHDL file ROM_form.vhd should be in the same folder with the assembler kcpsm6.exe. Besides, the project should be implemented as in the previous section.

11.2.4 PicoBlaze Application on the Basys3 Board

To show how an actual Verilog project can be established using PicoBlaze, we direct the reader to Phil Tracton's GitHub repository in [37]. There are several projects in this address. However, we will only use digital I/O example in [38]. The idea of this project is controlling first eight LEDs of the Basys3 board by corresponding eight switches. The reader can benefit from this project such that he or she can observe how PicoBlaze can be used with GPIO. As we were writing this book, VHDL version of the project was not available. However, the reader can benefit from IP block operations to use the Verilog description in a VHDL project.

11.3 Xilinx MicroBlaze Microcontroller

MicroBlaze is a 32-bit soft-core microcontroller developed by Xilinx. It has a fairly complex architecture compared to PicoBlaze. Therefore, we will not handle its functional

Listing 11.2 VHDL Description of the PicoBlaze Microcontroller in Shortened Form

```vhdl
library ieee;
use ieee.std_logic_1164.all;
use ieee.std_logic_unsigned.all;

entity kcpsm6 is
generic( hwbuild : std_logic_vector (7 downto 0) := x"00";
interrupt_vector : std_logic_vector (11 downto 0) := x"3FF";
scratch_pad_memory_size : integer := 64);

port(address : out std_logic_vector (11 downto 0);
 instruction : in std_logic_vector (17 downto 0);
 bram_enable : out std_logic;
     in_port : in std_logic_vector (7 downto 0);
    out_port : out std_logic_vector (7 downto 0);
     port_id : out std_logic_vector (7 downto 0);
write_strobe : out std_logic;
k_write_strobe: out std_logic;
  read_strobe: out std_logic;
interrupt : in std_logic;
interrupt_ack: out std_logic;
    sleep : in std_logic;
    reset : in std_logic;
      clk : in std_logic);

end kcpsm6;

-- Architecture of kcpsm6
```

blocks in detail here. Instead, we refer the reader to related references [39, 40]. Fortunately, Xilinx offers MicroBlaze IP cores in Vivado which we will introduce next.

11.3.1 MicroBlaze as an IP Block in Vivado

Vivado has two MicroBlaze IP cores named MicroBlaze and MicroBlaze microcontroller system (MCS). Xilinx offers a detailed comparison of both cores on its website [41]. As can be seen there, MicroBlaze MCS is a lite version of MicroBlaze. However, it is easier to use. Therefore, we will focus on it here.

MicroBlaze MCS core can be reached from the IP catalog under the list "Embedded processing" and "processor." The reader can select MicroBlaze MCS for usage by pressing on it twice. Afterward, the customize IP window opens up as in Fig. 11.2. As can be seen in this figure, the user can modify almost all microcontroller properties directly in this window. Afterward, the same steps for adding an IP block to a project (as explained in Sec. 4.7) can be followed to add the MicroBlaze MCS.

At this stage, let's assume that we will be using MicroBlaze MCS to turn on and off LEDs by the corresponding switches as in Sec. 11.2.4. Therefore, we should set the clock, memory, gpi, and gpo properties from the MicroBlaze MCS customize IP window. Leave the clock frequency at 100.0 MHz and set memory size to 16 KB in the MCS tab. Then, switch to the GPO tab. Make sure General Purpose Output 1 is selected and set the number of bits to 16 since we will use all LEDs on the board. Likewise, in the GPI tab, select General Purpose Input 1 and set the number of bits to 16 since we will use all switches

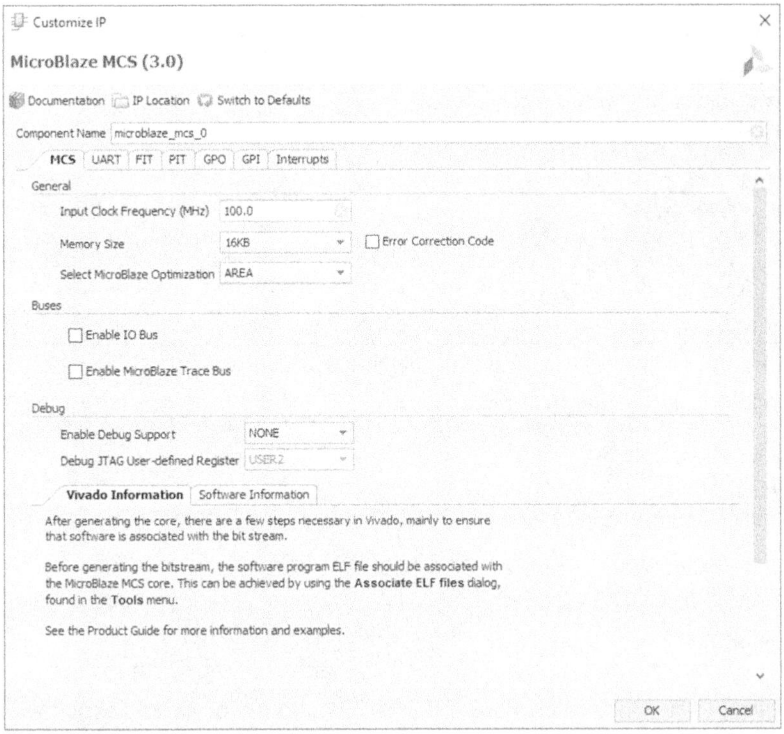

FIGURE **11.2** MicroBlaze MCS in IP catalog.

Listing 11.3 MicroBlaze MCS Instantiation Template in Verilog

```
microblaze_mcs_0 your_instance_name (
  .Clk(Clk),                      // input wire Clk
  .Reset(Reset),                  // input wire Reset
  .GPIO1_tri_i(GPIO1_tri_i),      // input wire [15 : 0] GPIO1_tri_i
  .GPIO1_tri_o(GPIO1_tri_o)       // output wire [15 : 0] GPIO1_tri_o
);
```

on the board. As all these settings are done, press OK. Generate Output Products window will appear. Here, click on Generate to proceed. You will be informed by a window saying "Out-of-context module run was launched for generating output products." Just click OK. Vivado generates the instantiation template for Verilog (in simplified form) as in Listing 11.3. The VHDL version of this template is presented in Listing 11.4. Next, we will use these instantiation templates in an application.

11.3.2 MicroBlaze MCS Application on the Basys3 Board

We will turn on/off LEDs via switches on the Basys3 board in this application as described in Sec. 11.2.4. Here, we will provide only the Verilog description since most steps will be similar in VHDL. Hence, we expect the reader to transfer this design to VHDL if needed. Let's first create a project named Basys3_Microblaze. Do not add any sources to the project at startup. As the project is created, add a new Verilog source

Listing 11.4 MicroBlaze MCS Instantiation Template in VHDL

```vhdl
component microblaze_mcs_0
port(Clk : in std_logic;
    Reset : in std_logic;
GPIO1_tri_i : in std_logic_vector (15 downto 0);
GPIO1_tri_o : out std_logic_vector (15 downto 0));
end component;

your_instance_name : microblaze_mcs_0 port map (Clk => Clk, Reset =>
    Reset, GPIO1_tri_i => GPIO1_tri_i, GPIO1_tri_o => GPIO1_tri_o);
```

Listing 11.5 Initial Verilog Description of the Top Module for MicroBlaze Application on Basys3

```verilog
module Basys3_mcs(clk,btnC,sw,led);

input clk, btnC;
input [15:0] sw;
output [15:0] led;

microblaze_mcs_0 microcontroller(.Clk(clk),.Reset(btnC),.GPIO1_tri_i
    (sw),.GPIO1_tri_o(led));

endmodule
```

file and include the MicroBlaze MCS IP block to the project as explained in Sec. 11.3.1. After adding and modifying the instantiation template, top module of the project will be as in Listing 11.5.

The top module in Listing 11.5 has a clk input as the master clock of the system. The center button (btnC) will be used as the reset button of the MicroBlaze microcontroller. Last, all switches and LEDs are associated to the Basys3 board items via the XDC file to be added to the project. Therefore, add it to the project as explained in Sec. 4.6.1. Do not forget to enable lines corresponding to the clock, 16 LEDs, and switches.

Follow the steps in Sec. 4.6.2 to generate the bitstream of the project. At this step, implementation in the Vivado side is complete. Once the bitstream is generated, from the Vivado main screen go to File → Export → Export Hardware. There, click OK by leaving "include bitstream" option unchecked. To program the MicroBlaze microcontroller in C language, we will use Xilinx software development kit (SDK) which comes with the Vivado WebPACK. If this module has not been installed yet, go to Help → Add design tools from Vivado selections and add it.

We will proceed with Xilinx SDK and embed a C code for the project. Here, we will benefit from Duckworth's tutorial titled *MicroBlaze MCS Tutorial, v2* and available at his website [42]. Launch Xilinx SDK by pressing File → Launch SDK item. A welcome screen appears asking for the "exported location" and "Workspace." Leave both options as "Local to Project." A project explorer window should appear as in Fig. 11.3. As can be seen in this figure, Vivado project properties are transferred to the SDK as well.

The next step in the SDK is programming the microcontroller in C language. Therefore, we will generate a new project under Xilinx SDK by pressing File → New Application Project. Let's name this project Basys3_mcs. Then, the window should look like as in Fig. 11.4. We will press "Next" and select a project template. Here, we will select the predefined "Hello World" project and modify it to fit our needs.

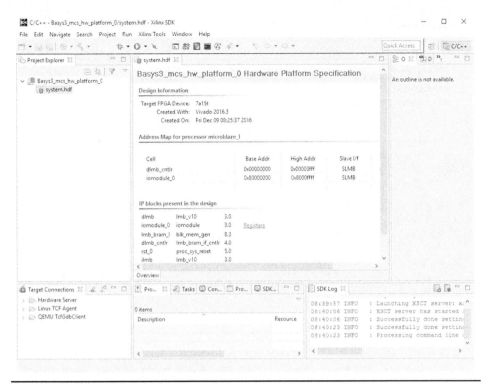

FIGURE 11.3 Xilinx SDK project explorer window.

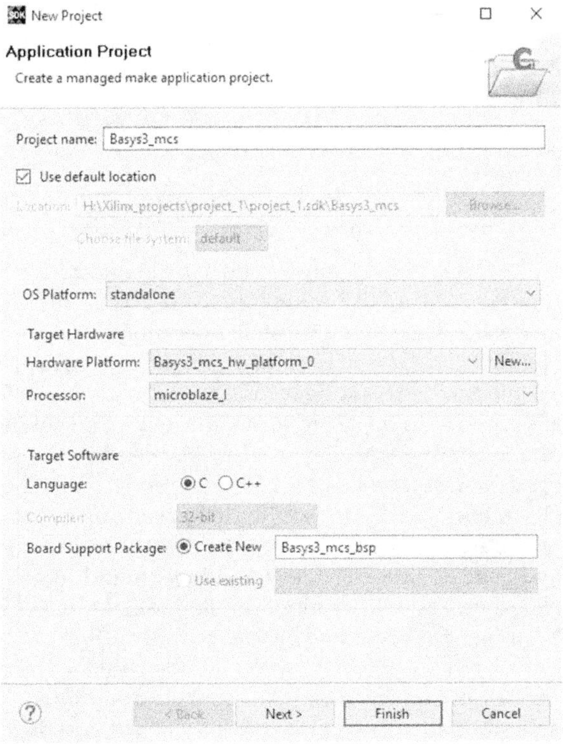

FIGURE 11.4 Xilinx SDK new project window.

Listing 11.6 C Code for the Xilinx SDK Project

```c
#include <stdio.h>
#include "platform.h"
#include "xiomodule.h"

int main()
{
//initialize the microcontroller
init_platform();

//define variables
u32 data;
XIOModule gpi, gpo;

//initialize microcontroller input pins
data = XIOModule_Initialize(&gpi, XPAR_IOMODULE_0_DEVICE_ID);
data = XIOModule_Start(&gpi);

//initialize microcontroller output pins
data = XIOModule_Initialize(&gpo, XPAR_IOMODULE_0_DEVICE_ID);
data = XIOModule_Start(&gpo);

//Check switches and modify LEDs accordingly in an infinite loop

while(1)
{
//read switch values
data= XIOModule_DiscreteRead(&gpi,1);
//modify LED values
XIOModule_DiscreteWrite(&gpo,1, data);
}
cleanup_platform();
return 0;
}
```

To modify the "Hello World" project, go to `helloworld.c` source file located under the directory `Basys3_mcs` → `src`. Replace the source code as in Listing 11.6. Here, predefined C functions are used for almost all operations.

As the modified C code is saved, the Xilinx SDK generates an executable and linkable formatted file `Basys3_msc.elf`. We can use this file in Vivado to associate it to the Verilog description. To do so, go to Vivado again and select Tools → Associate ELF files. Ignore the upcoming warning window by pressing Continue. A pop-up window will appear titled as Associate ELF File. Here, we should add the generated elf file under the Xilinx SDK to our Vivado project. To do so, click on browse under Design sources and add the generated elf file under `H:/Xilinx_Projects/project_1/project_1.sdk/Basys3_msc/Debug`. The add window should look like as in Fig. 11.5. Final view of the Associate ELF File window should look like as in Fig. 11.6.

We should update the generated bitstream after adding elf file to the project. Click on `Generate Bitstream` to regenerate your bit file with the embedded C code in it. Once generation finishes, open hardware manager and program your FPGA as explained in Sec. 4.6.2. As the program is run on the Basys3 board, a switch should turn on and off the corresponding LED.

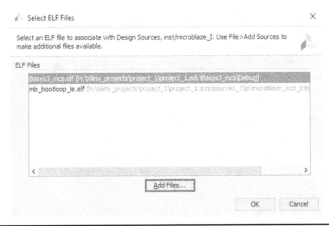

FIGURE 11.5 Adding the elf file to the Vivado project.

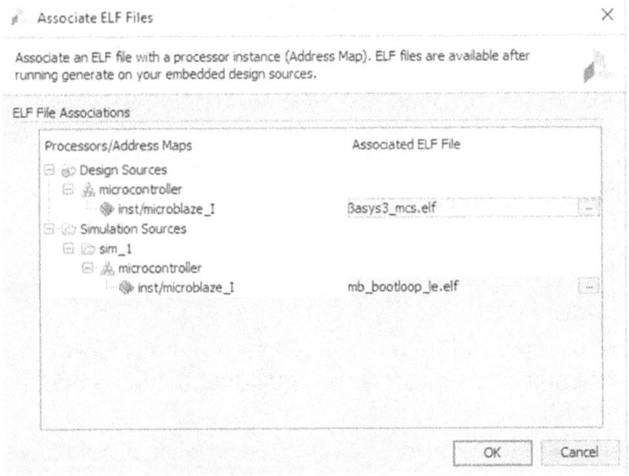

FIGURE 11.6 Final view of the adding ELF file window.

11.4 Soft-Core Microcontroller Applications

Different from previous chapters, we will refer the reader to successful soft-core micro-controller applications by other researchers in this section. For this purpose, the first repository to look for is by Tracton as mentioned in Sec. 11.2.4. Tracton offers valuable applications on the PicoBlaze microcontroller. As for MicroBlaze-based applications, we can direct the reader to the Digilent website. One good application offered by Digilent is in [43]. There may also be other websites offering good applications of soft-core micro-controllers. It is worth looking at them.

11.5 FPGA Building Blocks Used in Soft-Core Microcontrollers

Soft-core microcontrollers are implemented using standard FPGA building blocks. Therefore, analyzing their implementation details give insight to the reader. Let's start with the PicoBlaze implementation in Sec. 11.2.4. This application requires 233 LUTs and 227 flip-flops. On the other hand, the MicroBlaze MCS implementation in Sec. 11.3.2 requires 654 LUTs, 290 flip-flops, and four block RAMs. As can be seen here, both implementations fit to the FPGA on the Basys3 board without any difficulty. Hence, the user can add extra peripherals to the Artix-7 FPGA besides fundamental implemented blocks. Available resources on the Artix-7 FPGA also allow embedding multiple microcontrollers for parallel operation. If we compare PicoBlaze and MicroBlaze MCS, we can see that PicoBlaze requires almost one-third of the LUT used by MicroBlaze. Both microcontrollers require similar number of flip-flops. Note that the block RAM is inherently added by the MicroBlaze MCS. On the other hand, PicoBlaze has its own RAM block.

11.6   Summary

FPGAs can be used to implement soft-core microcontrollers available either as an IP block or a HDL description. This opens up a way to design custom-made microcontroller systems. Hence, desired or unconventional peripherals can be added to the microcontroller easily via the FPGA design. In this chapter, we explored methods for such implementations. Note that the coverage of the topic in this chapter can be considered introductory. The reader can check available references to master this topic. Xilinx also offers a different platform called system on chip (SoC) which incorporates the hardcore processors and FPGA chips. One recent SoC is Zynq family which has ARM Cortex A9 processors with the Artix-7 FPGAs. This SoC family deserves special consideration. Therefore, we suggest the reader to explore it for more advanced applications.

11.7   Exercises

11.1 Modify the application in Sec. 11.2.4 such that the LED turns on when the corresponding switch goes to logic level 0. The LED turns off when the switch goes to logic level 1.

11.2 Create an IP block for the basic microcontroller in Sec. 11.2.4 such that the same application can be implemented in VHDL.

11.3 Modify the application in Sec. 11.3.2 such that only four LEDs and switches are used in implementation.

11.4 Redo Exercise 11.3 in VHDL.

11.5 The FPGArduino project is a good example of implementing Arduino as a soft-core microcontroller [44]. Follow the steps there to implement Arduino on the Arty board.

11.6 Imagination Technologies offers a soft-core processor on its website [45]. Follow the steps there to implement this microcontroller on the Basys3 board.

11.7 ARM offers the core of its Cortex M0 processor in [46]. Follow the steps there to implement this microcontroller on the Basys3 board.

CHAPTER 12

Digital Interfacing

A digital system communicates with outside world through its analog and digital interface. This chapter focuses on digital interfacing for the field-programmable gate array (FPGA) design. Hence, we will first cover serial communication protocols as universal asynchronous receiver/transmitter (UART), serial peripheral interface (SPI), and inter-integrated circuit I^2C. Then, we will explore video graphics array (VGA) interfacing to connect a display to the FPGA. Afterward, we will cover universal serial bus (USB) and ethernet connections. We will provide Verilog and VHDL descriptions to digital interfacing concepts except ethernet. For it, we will benefit from an available IP block in Vivado. To explain digital interfacing concepts clearly, we provide related applications in this chapter.

12.1 Universal Asynchronous Receiver/Transmitter

A universal asynchronous receiver/transmitter (UART) is a digital communication protocol for two or more devices. We will focus only on UART communication between two devices in this book. Hence, one device will be the transmitter; the other will be the receiver. Communication is done by sending and receiving data asynchronously between the transmitter and receiver. Being asynchronous, the UART does not need a common clock between the transmitter and receiver. Thus, connected devices can work independently. The serial pin of the transmitter is generally called transmit (TX). The corresponding receiver pin is generally called receive (RX). The connection between the transmitter and receiver is established by physically wiring these two pins.

The UART communication can be established between different devices. We will especially focus on the one between the FPGA board of interest (Basys3 or Arty) and PC. To do so, we will develop hardware description language (HDL) description of transmitter and receiver modules. The transmitter module will be basically a shift register that loads parallel data and shifts it in a specific rate through TX pin of the device. The receiver module will convert the received serial data through RX pin into parallel form to be processed by the receiver. Before dealing with HDL descriptions, let's first focus on the working principles of UART.

12.1.1 Working Principles of UART

To use a UART, we should understand how it works. Therefore, we introduce data format, timing, transmission, and reception operations in this section. These will help us forming HDL descriptions in the following section.

12.1.1.1 Data Format

Data is transmitted in terms of packages in the UART. Data framing of a UART package begins with a start bit, followed by seven to eight data bits optionally attached by a parity

259

FIGURE 12.1 Data framing of a UART package with eight-bit data.

bit (explained in Sec. 8.6), and concluded by one or two stop bits. This setup can be seen in Fig. 12.1.

12.1.1.2 Timing

Although the UART works in asynchronous manner, the transmitter and receiver should have same timing values to transmit and receive data. In other words, the data can be transmitted in asynchronous manner. However, as the transmission starts, the receiver should know the duration of each pulse in the UART package. This is set by the baud rate which determines the timing. The baud rate is denoted by bits per second (bps). For example, a 2400-bps indicates a 416-μs bit width (or period) in the UART transmission.

12.1.1.3 Transmission Operation

We can explain the transmission operation in the UART as a state machine. We will explain this state machine in detail in Sec. 12.1.2. Here, let's briefly summarize it. The TX pin should be at logic level 1 when the transmitter is in idle mode. Once transmission starts, a falling edge is created on the data transmit line which wakes up the receiver. Afterward, the clock is set according to the baud rate and all bits are sent one by one in every clock cycle in the transmitter side. The receiver should have the same baud rate for receiving transmitted bits sequentially. As the transmit operation finalizes, the TX pin should be set to logic level 1 for one or two bit widths to inform the receiver that the transmission is done. These are also called stop bit(s). The number of stop bits and usage of parity bit should also be predetermined so that the transmitter and receiver have same settings.

12.1.1.4 Reception Operation

We can explain the reception operation in the UART as another state machine. Although we will explain this state machine in Sec. 12.1.2, let's briefly summarize it here. The receiver will be in ready state initially. When a falling edge signal (start bit) comes to RX pin, it starts receiving data bits sequentially. To do so, the receiver should have an internal timer with the predetermined baud rate as in the transmitter. After receiving start bit, the timer waits for a certain time to sample the first data bit. This offset allows starting the sampling process in the middle of the first data pulse. Note that although data is sent as logic levels 1 and 0 by the transmitter, these are converted to analog pulse signals. Hence, the sampling operation converts the received analog signal to logic level 0 or 1 again. Afterward, we perform the sampling operation at each successive time period to recover data bits. As all bits are received this way, the receiver checks the parity bit within the received data (if the protocol consists one). When stop bit(s) is received, the receiver turns back to ready state waiting to receive the next data packet.

12.1.2 UART in Verilog

We can describe the transmit and receive operations as two separate modules in Verilog. Let's start with the transmitter module.

12.1.2.1 The Transmitter Module

The Verilog description of the transmitter module is presented in Listing 12.1. This module has three inputs as send, data, and clk. send is used to trigger starting a transmit operation. data carries data to be transmitted. clk is used to enter the 100-MHz clock of the FPGA board (Basys3 or Arty) to the module. The transmitter module has two outputs as ready and tx. When ready is at logic level 1, this indicates that the module is ready to transmit data. Output tx should be directly connected to TX pin of the device.

The working principles of the transmitter module (as a state machine) are as follows. Within the module, the baud rate is defined as a parameter and set to 9600 bps by default. Here, baud_timer calculates the number of clock cycles needed for a particular baud rate by dividing the main clock frequency to the baud rate. The transmitter module has three states as RDY, LOAD_BIT, and SEND_BIT. RDY state indicates that the module is ready to send next data package. When in the LOAD_BIT state, the data is loaded to tx output. Finally, SEND_BIT state indicates that the data is being transmitted. Initial state of the module is set as RDY. Hence, it waits for the send trigger. When send is set to logic level 1, the module loads data with a leading zero and a trailing one to txData. Afterward, the module switches to LOAD_BIT state. Here, the first bit to be transmitted (LSB in our configuration) is loaded to txBit. Then, the module waits for bit_index_max clock cycles in SEND_BIT state. Then, it switches back to LOAD_BIT state to load the next bit to be transmitted. This operation is repeated until the last stop bit is transmitted. At the end of the transmission operation, the state is set as RDY. Hence, the transmitter module starts waiting for the next send trigger. In this module, txBit is wired to tx and ready is set as a conditional assignment such that when state equals to RDY, it is at logic level 1, otherwise 0.

12.1.2.2 The Receiver Module

The Verilog description of the receiver module is presented in Listing 12.2. The module has two inputs as clk and rx. clk is used to enter the 100-MHz clock of the FPGA board (Basys3 or Arty) to the module as in the transmitter module. rx should be directly connected to RX pin of the device. Through it, the receiver module listens for a possible incoming package. The receiver module has four outputs as data, parity, ready, and error. data represents the received data. parity shows the received parity bit. ready indicates that the receive operation is complete. Finally, error shows if the data package is received with or without error.

The working principles of the receiver module (as a state machine) are as follows. Within the module, the baud rate is defined as a parameter and set to 9600 bps similar to the transmitter module. As in the transmitter module, baud_timer calculates the number of clock cycles needed for a particular baud rate by dividing the main clock frequency to baud rate. The receiver module has five states as RDY, START, RECEIVE, WAIT, and CHECK. The state machine starts initially at RDY state, which indicates that the module is ready to receive the next data package. Hence, it listens to the RX pin through rx at every rising edge of the clock. When rx goes to logic level 0, the state machine goes to START. There, it waits for half of the baud_timer period where it ends up in the middle of the start signal. First data bit will be ready to be read after waiting for baud_timer period. WAIT state acts as a delay station in which the receiver waits for baud_timer period. Then, it returns to RECEIVE state unless ready is at logic level 1. In RECEIVE state, the incoming data is sampled. Then, bitIndex is incremented by one and checked whether it has reached the maximum value (eight for our case).

Listing 12.1 Verilog Description of the UART Transmitter Module

```verilog
module UART_tx_ctrl(ready,uart_tx,send,data,clk);

input send, clk;
input [7:0] data;
output ready, uart_tx;

parameter baud = 9600;
parameter bit_index_max = 10;

localparam [31:0] baud_timer = 100000000/baud;

localparam RDY = 2'b00, LOAD_BIT = 2'b01, SEND_BIT = 2'b10;

reg [1:0] state = RDY;
reg [31:0] timer = 0;
reg [9:0] txData;
reg [3:0] bitIndex;
reg txBit=1'b1;

always @ (posedge clk)

case (state)
RDY:
        begin
        if (send)
        begin
        txData <= {1'b1,data,1'b0};
        state <= LOAD_BIT;
        end
        timer <= 14'b0;
        bitIndex <= 0;
        txBit <= 1'b1;
        end
LOAD_BIT:
        begin
        state <= SEND_BIT;
        bitIndex <= bitIndex + 1'b1;
        txBit <= txData[bitIndex];
        end
SEND_BIT:
        if (timer == baud_timer)
        begin
        timer <= 14'b0;
        if (bitIndex == bit_index_max)
        state <= RDY;
        else state <= LOAD_BIT;
        end
        else timer <= timer + 1'b1;
default:
        state <= RDY;
endcase

assign uart_tx = txBit;
assign ready = (state == RDY);
endmodule
```

Listing 12.2 Verilog Description of the UART Receiver Module

```verilog
module UART_rx_ctrl(clk,rx,data,parity,ready,error);

input clk, rx;
output reg [7:0] data;
output reg parity;
output reg ready=0;
output reg error=0;

parameter baud=9600;

localparam RDY=3'b000, START=3'b001, RECEIVE=3'b010, WAIT=3'b011,
    CHECK=3'b100;

reg [2:0] state = RDY;

localparam [31:0] baud_timer = 100000000/baud;

reg [31:0] timer = 32'b0;
reg [3:0] bitIndex = 3'b0;
reg [8:0] rxdata;

always @ (posedge clk)

case (state)
RDY:
        if (rx == 1'b0)
        begin
        state <= START;
        bitIndex <= 3'b0;
        end
START:
        if (timer == baud_timer/2)
        begin
        state <= WAIT;
        timer <= 14'b0;
        error <= 1'b0;
        ready <= 1'b0;
        end
        else timer <= timer + 1'b1;
WAIT:
        if (timer == baud_timer)
        begin
        timer <= 14'b0;
        if (ready) state <= RDY;
        else state <= RECEIVE;
        end
        else timer <= timer + 1'b1;
RECEIVE:
        begin
        rxdata[bitIndex] <= rx;
        bitIndex <= bitIndex + 1'b1;
        if (bitIndex == 4'd8) state <= CHECK;
        else state <= WAIT;
        end
```

```
CHECK:
        if (^rxdata[7:0] == rxdata[8])
        begin
        ready <= 1'b1;
        state <= WAIT;
        data <= rxdata[7:0];
        parity <= rxdata[8];
        end
        else
        begin
        ready <= 1'b1;
        data[7:0] <= 8'bx;
        error <= 1'b1;
        state <= RDY;
        end
endcase
endmodule
```

Since we have eight data bits and a parity bit, the state machine has to switch to CHECK state after all bits are received. Even parity check is performed in CHECK state. If the received data package is consistent with parity bit, then ready is set to logic level 1 and the next state is set to WAIT. Received data and parity values in rxdata are written to data and parity outputs. If the parity check fails, then error and ready go to logic level 1. data is filled with logic level 1. Then, the reception operation ends. The receiver turns back to RDY state waiting to receive the next data package.

12.1.3 UART in VHDL

As in Verilog, we can describe the transmit and receive operations in two separate modules in VHDL. Let's start with the transmitter module.

12.1.3.1 The Transmitter Module

The VHDL description of the transmitter module is presented in Listing 12.3. In this description, we tried to keep the input, output definitions, and state names the same as in Listing 12.1. Hence, the reader can associate the working principles of the corresponding Verilog description with the VHDL description here.

12.1.3.2 The Receiver Module

We provide the VHDL description of the receiver module in Listing 12.4. Again, this module has the same working principles as its Verilog version in Listing 12.1.

12.1.4 UART Applications

The UART needs an RS-232 port for communication. Unfortunately, Basys3 and Arty boards do not have such a port. However, they share the micro USB port for the UART communication as mentioned in Chap. 3. To run UART applications in this section, the reader should have a terminal software such as RealTerm on the host PC. Connect your Basys3 board to the PC through its USB cable and turn the board on. Find the assigned COM port number by looking at the device manager. Here, we assume that the application is run on a PC with Microsoft Windows operating system. Please consult related resources for other operating systems. On the terminal, set the baud rate to 9600 bps; the COM port to the one Basys3 board connected; parity bit to "None"; data bits to eight; and stop bits to one. The default demo implemented on the Basys3 board includes a

Listing 12.3 VHDL Description of the UART Transmitter Module

```vhdl
library ieee;
use ieee.std_logic_1164.all;

entity UART_tx_ctrl is
generic(baud : integer := 9600);
port(send : in std_logic;
      clk : in std_logic;
     data : in std_logic_vector (7 downto 0);
    ready : out std_logic;
  uart_tx : out std_logic);
end UART_tx_ctrl;

architecture behavioral of UART_tx_ctrl is

constant baud_timer : integer := 100000000/baud;
constant bit_index_max: integer := 10;

type state_type is (RDY, LOAD_BIT, SEND_BIT);

signal state   : state_type := RDY;
signal timer   : integer := 0;
signal txData  : std_logic_vector (9 downto 0);
signal bitIndex: integer range 0 TO bit_index_max := 0;
signal txBit   : std_logic := '1';

begin
process(clk)
begin
if rising_edge(clk) then

case state is
when RDY =>
      if send then
      txData <= '1' & data & '0';
      state <= LOAD_BIT;
      end if;
      timer <= 0;
      bitIndex <= 0;
      txBit <= '1';
when LOAD_BIT =>
      state <= SEND_BIT;
      bitIndex <= bitIndex + 1;
      txBit <= txData(bitIndex);
when SEND_BIT =>
      if (timer = baud_timer) then
      timer <= 0;
      if (bitIndex = bit_index_max) then
      state <= RDY;
      else state <= LOAD_BIT;
      end if;
      else timer <= timer + 1;
      end if;
end case;
end if;
end process;

uart_tx <= txBit;
ready <= '1' when (state = RDY) else '0';

end behavioral;
```

Listing 12.4 VHDL Description of the UART Receiver Module

```vhdl
library ieee;
use ieee.std_logic_1164.all;

entity UART_rx_ctrl is
generic(baud : integer := 9600);
port(clk : in std_logic;
      rx : in std_logic;
    data : out std_logic_vector (7 downto 0);
  parity : out std_logic;
   ready : out std_logic := '0';
   error : out std_logic := '0');
end UART_rx_ctrl;

architecture behavioral of UART_rx_ctrl is

constant baud_timer  : integer := 100000000/baud;
type state_type is (RDY, START, RECEIVE, WAITING, CHECK);

signal state   : state_type := RDY;
signal timer   : integer := 0;
signal bitIndex: integer range 0 TO 8 := 0;
signal rxdata  : std_logic_vector (8 downto 0);

begin
process(clk)
begin
if rising_edge(clk) then

case state is
when RDY =>
        if rx = '0' then
        state <= START;
        bitIndex <= 0;
        end if;
when START =>
        if timer = baud_timer/2 then
        state <= WAITING;
        timer <= 0;
        error <= '0';
        ready <= '0';
        else timer <= timer + 1;
        end if;
when WAITING =>
        if(timer = baud_timer) then
        timer <= 0;
        if(ready) then state <= RDY;
        else state <= RECEIVE;
        end if;
        else timer <= timer + 1;
        end if;
when RECEIVE =>
        rxdata(bitIndex) <= rx;
        bitIndex <= bitIndex + 1;
        if (bitIndex = 8) then state <= CHECK;
```

```
          else state <= WAITING;
          end if;
  when CHECK =>
          if (xor rxdata(7 downto 0) = rxdata(8)) then
          ready <= '1';
          state <= WAITING;
          data <= rxdata(7 downto 0);
          parity <= rxdata(8);
          else
          ready <= '1';
          data(7 downto 0) <= (others => '-');
          error <= '1';
          state <= RDY;
          end if;
  end case;
  end if;
  end process;
  end behavioral;
```

UART module. Through it, the user can check whether the connection has been established between the PC and Basys3 board. To check it, press the center button (btnC) on the Basys3 board. You should see the sentence "BASYS3 GPIO/UART DEMO!" on the terminal window. If you press any of the remaining four buttons on the board, you should see "Button press detected!" on the terminal window. We will use the same setup for our UART applications next.

12.1.4.1 *Transmitting Data from the Basys3 Board to Host PC*

The first UART application will be on transmitting data from the Basys3 board to host PC. Therefore, connect your board to the host PC and check the status of connection as explained in the previous section. Assuming that everything is set correctly and working properly, we will implement our application. Therefore, we will build a top module which employs the transmitter module and transmits incremental ASCII codes starting from hexadecimal number 41 (corresponding to character A) when center button (btnC) is pressed. The top module in Verilog is presented in Listing 12.5. The VHDL version of the top module is also given in Listing 12.6.

The transmitter top module has two inputs as clock (clk) and center button (btnC). These values will be obtained from the Basys3 board through its XDC file. The top module has one output for the transmitter port of the board as RsTx. The top module uses UART_tx_ctrl (given in Listing 12.1) and debounce (given in Listing 9.20) as submodules. Hence, they should be included to the project. Note that the baud rate for the UART_tx_ctrl module is set to 19200 bps here. The top module has three states as TX_WAIT_BTN, TX_SEND_CHAR, and TX_SEND_WAIT. Therefore, we can explain its working principles as a state machine. The top module is initially at TX_WAIT_BTN state. Here, it waits for a button press. When the button is pressed, the state machine enters TX_SEND_CHAR state. Here, initStr is loaded to uartData to be sent and initStr is incremented by one. Therefore, the next character in ASCII table (Table 6.6) is reached. In TX_SEND_CHAR state, uartSend (connected to send in the transmit module) goes to logic level 1. When the state machine is in state TX_SEND_WAIT, it waits for the transmitter to send loaded data. When uartRdy is at logic level 1, the state machine turns back to state TX_WAIT_BTN.

Listing 12.5 Verilog Description of the Transmitter Top Module

```verilog
module UART_tx_top(btnC,clk,RsTx);

input btnC;
input clk;
output RsTx;

localparam TX_WAIT_BTN=2'b00, TX_SEND_CHAR=2'b01, TX_SEND_WAIT=2'b10;

reg [1:0] state = TX_WAIT_BTN;

wire uartRdy;
wire uartSend;
reg [7:0] uartData = 0;
reg [7:0] initStr = 8'h41; // ASCII 'A' character

wire btnCclr;
reg btnC_prev;

UART_tx_ctrl #(19200) TX(.send(uartSend),.data(uartData),.clk(clk),
    .ready(uartRdy),.uart_tx(RsTx));

debounce dbc(clk,btnC,btnCclr);

always @ (posedge clk)
begin
btnC_prev <= btnCclr;
case (state)
TX_WAIT_BTN:
        if (btnC_prev == 0 && btnCclr == 1'b1) state <= TX_SEND_CHAR;
TX_SEND_CHAR:
        begin
        uartData <= initStr;
        initStr <= initStr + 1'b1;
        state <= TX_SEND_WAIT;
        end
TX_SEND_WAIT:
        if (uartRdy) state <= TX_WAIT_BTN;
default:
        state <= TX_WAIT_BTN;
endcase
end

assign uartSend = (state == TX_SEND_CHAR);

endmodule
```

Listing 12.6   VHDL Description of the Transmitter Top Module

```vhdl
library ieee;
use ieee.std_logic_1164.all;
use ieee.numeric_std.all;

entity UART_tx_top is
port(btnC : in std_logic;
      clk : in std_logic;
     RsTx : out std_logic);
end UART_tx_top;

architecture behavioral of UART_tx_top is

type state_type is (TX_WAIT_BTN, TX_SEND_CHAR, TX_SEND_WAIT);

signal state   : state_type := TX_WAIT_BTN;
signal uartRdy : std_logic;
signal uartSend: std_logic;
signal uartData: std_logic_vector (7 downto 0);
signal initStr : std_logic_vector (7 downto 0) := x"41";
signal btnCclr : std_logic;
signal btnC_prev : std_logic;

component UART_tx_ctrl
generic(baud : integer);
port(send : in std_logic;
      clk : in std_logic;
     data : in std_logic_vector (7 downto 0);
    ready : out std_logic;
  uart_tx : out std_logic);
end component;

component debounce is
port(clk : in std_logic;
     btn : in std_logic;
 btn_clr : out std_logic);
end component;

begin
process(clk)
begin
if rising_edge(clk) then
btnC_prev <= btnCclr;
case state is
when TX_WAIT_BTN =>
        if btnC_prev = '0' and btnCclr = '1' then
        state <= TX_SEND_CHAR;
        end if;
when TX_SEND_CHAR =>
        uartData <= initStr;
        initStr <= std_logic_vector(unsigned(initStr) + 1);
        state <= TX_SEND_WAIT;
when TX_SEND_WAIT =>
        if uartRdy = '1' then state <= TX_WAIT_BTN;
        end if;
```

```
end case;
end if;
end process;

uartSend <= '1' when (state = TX_SEND_CHAR) else '0';

TX : UART_tx_ctrl generic map(baud => 19200)
port map(send => uartSend, data => uartData, clk => clk, ready =>
    uartRdy, uart_tx => RsTx);

dbc : debounce port map(clk => clk, btn => btnC, btn_clr => btnCclr);

end behavioral;
```

To run this application, we should form a project in Vivado containing HDL description files for the top module and submodules used within it. Besides, we should also add the modified XDC file to this project. For more information on this issue, please see Chap. 4. As we generate the bitstream and embed it on the Basys3 board, it is ready to be tested. Now, open the terminal program on the host PC. Set your baud rate to 19200 bps, select no parity bit option and one stop bit. As btnC is pressed on the Basys3 board, you should see character A on the terminal window. As we press btnC again, the next character in the ASCII table should be seen on the terminal window.

12.1.4.2 Receiving Data to the Basys3 Board from Host PC

The next UART application will be on receiving data to the Basys3 board from host PC. Within this application, we will turn on LEDs on the Basys3 board by received data. Top module for this application is presented in Listing 12.7 in Verilog. The corresponding VHDL description is given in Listing 12.8.

The receiver top module has two inputs as clock (clk) and receive (RsRx). It has one output to adjust LEDs on the Basys3 board as the vector led. The input clk will be obtained and the output led will be fed to LEDs on the Basys3 board through its XDC file. The receiver module (UART_rx_ctrl) in Listing 12.2 is used as a submodule here. The baud rate for this submodule is set to 19200 bps. We can represent operations in the top module as a state machine with three states as RX_RDY, RX_WAIT, and RX_DATARDY. When in the RX_RDY state, the state machine waits for data_ready to switch from logic level 1 to 0. As this transition occurs, the state machine goes to RX_WAIT until the module receives the data. When the data is received, the state machine goes to RX_DATARDY. Meanwhile, data_ready goes to logic level 1 again. At the same time, the received data is written to the first eight bits of led. The parity bit is represented as the ninth led entry. If a transmission error occurs, it is indicated in the tenth bit of led. Last six bits of led are used to show how many packages have been received from the host PC. To run this application, please follow the steps explained in the previous application. Since the receiver module is capable of receiving the parity bit, do not forget to select even parity on your terminal settings.

12.2 Serial Peripheral Interface

The serial peripheral interface (SPI) is a digital communication protocol for two or more devices as the UART. In this book, we will focus only on the SPI communication between two devices. Hence, one device will be the transmitter and the other receiver. Different from the UART, the SPI is a synchronous communication protocol. Besides, communication between the transmitter and receiver is duplex. In other words, data is

---

Listing 12.7 Verilog Description of the Receiver Top Module

```verilog
module UART_rx_top(clk,RsRx,led);

input clk, RsRx;
output reg [15:0] led=0;

localparam RX_RDY=2'b00, RX_WAIT=2'b01, RX_DATARDY=2'b10;

reg [1:0] state=RX_RDY;

wire data_ready;
wire [7:0] out;
wire parity;
wire error;

UART_rx_ctrl #(19200) RX(.clk(clk),.rx(RsRx),.data(out),
    .parity(parity),.ready(data_ready),.error(error));

always @ (posedge clk)

case (state)
RX_RDY:
        if (!data_ready) state <= RX_WAIT;
RX_WAIT:
        if (data_ready) state <= RX_DATARDY;
RX_DATARDY:
        begin
        led[9:0] <= {error,parity,out};
        led[15:10] <= led[15:10] + 1'b1;
        state <= RX_RDY;
        end
endcase
endmodule
```

---

transmitted and received at the same time in the SPI. Therefore, the SPI communication uses four wires. Two of these wires are for data transfer. One wire is used for the common clock signal (for synchronization). The fourth wire is used to enable (select) signal to be explained in Sec. 12.2.1.

Being synchronous, the SPI needs a common clock signal generated by either the transmitter or receiver. Clock generating side is called leader. The other side is called follower. These roles are generally called master and slave in literature. However, we prefer leader and follower naming in this book. Therefore, we will use the terms leader-follower instead of master-slave from this point on. As a result we can have leader-transmitter, leader-receiver, follower-transmitter, and follower-receiver options. We will cover all these next.

12.2.1 Working Principles of SPI

The working principles of the SPI are simpler than the UART. To understand them, we introduce the data format, connection diagram, transmission and reception operations, and timing in this section. These will help us forming HDL descriptions for transmission and reception next.

Listing 12.8 VHDL Description of the Receiver Top Module

```vhdl
library ieee;
use ieee.std_logic_1164.all;
use ieee.numeric_std.all;

entity UART_rx_top is
port(clk : in std_logic;
    RsRx : in std_logic;
     led : out std_logic_vector (15 downto 0));
end UART_rx_top;

architecture behavioral of UART_rx_top is

type state_type is (RX_RDY, RX_WAIT, RX_DATARDY);

signal state     : state_type := RX_RDY;
signal data_ready: std_logic;
signal rxout     : std_logic_vector (7 downto 0);
signal parity    : std_logic;
signal error     : std_logic;

component UART_rx_ctrl
generic(baud : integer);
port(clk : in std_logic;
      rx : in std_logic;
    data : out std_logic_vector (7 downto 0);
  parity : out std_logic;
   ready : out std_logic;
   error : out std_logic);
end component;

begin
process(clk)
begin
if rising_edge(clk) then

case state is
when RX_RDY =>
        if (not data_ready) then state <= RX_WAIT; end if;
when RX_WAIT =>
        if (data_ready) then state <= RX_DATARDY; end if;
when RX_DATARDY =>
        led(9 downto 0) <= error & parity & rxout;
        led(15 downto 10) <= std_logic_vector(unsigned(led(15 downto
            10)) + 1);
        state <= RX_RDY;
end case;
end if;
end process;

RX : UART_rx_ctrl generic map (baud => 19200)
port map(clk => clk, rx => RsRx, data => rxout, parity => parity, ready
    => data_ready, error => error);

end behavioral;
```

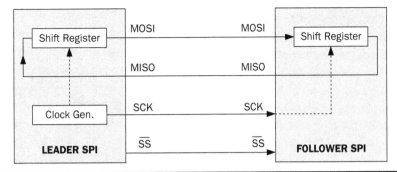

Figure 12.2 SPI connection diagram between the leader and follower.

12.2.1.1 Data Format

Different from the UART, data packet size is not constant in the SPI. This is an advantage since the user can select the packet size as he or she desires. Moreover, the dedicated common clock and enable signals avoid using start and stop bits in the UART. The only requirement here is the need for determining the data packet size. Hence, the transmitter and receiver can understand each other.

12.2.1.2 Connection Diagram

The SPI uses a dedicated clock line, two data lines (one for transmitter, one for receiver), and a select (enable) line as mentioned in the previous section. We provide the connection diagram between two devices using these lines in Fig. 12.2. Here, the clock signal is denoted by SCK. The leader output, follower input is denoted by MOSI. The leader input, follower output is denoted by MISO. Select is denoted by $\overline{SS}$ which is used by the leader to wake up the follower. The select line is also used when more than one follower is connected to a single leader.

12.2.1.3 Transmission and Reception Operations

In the SPI, the data transmission and reception is controlled by the leader through SCK and $\overline{SS}$ signals. When there is no transmission, $\overline{SS}$ stays at logic level 1 and SCK stays either at logic level 0 or 1 depending on the SPI mode. The modes of the SPI and their timing diagrams will be discussed later in Sec. 12.2.1.4. The SPI communication starts when the leader wakes the follower by setting $\overline{SS}$ to logic level 0. Next, the leader and follower start interchanging data in every clock cycle set by SCK. Here, either the leader sends a bit through MOSI line or the follower sends a bit through MISO line. The SPI mode also determines if data will be sent on the rising or falling edge of SCK. After all bits are transferred, the common clock stops and leader deselects the follower by changing $\overline{SS}$ to logic level 1.

12.2.1.4 Timing

As mentioned previously, SCK is generated by the leader and fed to the follower. Here, SCK depends on the maximum data rate of the transmitter and receiver. Hence, the device with the lowest rate defines its limit. Besides frequency, the leader also adjusts the polarity and phase of clock denoted by CPOL and CPHA, respectively. Four possible combinations of CPOL and CPHA are presented in Fig. 12.3. These combinations are called modes of the SPI.

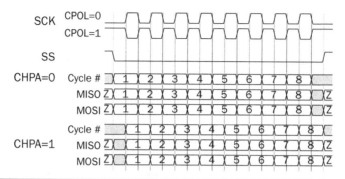

FIGURE 12.3 SPI communication timing diagram.

Modes		Clock	Data	Bits on MISO and MOSI lines		
CPOL	CPHA	base	availability	Placed	Stay	Captured
0	0	0	0 to 1 transition	Clock goes to 0	Clock at 1	0 to 1 transition
0	1	1	0 to 1 transition	Clock goes to 1	Clock at 0	1 to 0 transition
1	0	0	1 to 0 transition	Clock goes to 1	Clock at 0	1 to 0 transition
1	1	1	1 to 0 transition	Clock goes to 0	Clock at 1	0 to 1 transition

TABLE 12.1 SPI Modes in Tabular Form Based on Clock Operation

We can briefly summarize the SPI modes presented in Fig. 12.3 and in Table 12.1. Here, each operation in the corresponding mode is summarized based on the clock. For more information on the SPI modes, please see [47].

12.2.2 SPI in Verilog

We can describe the transmit and receive operations (for the leader and follower) as separate modules in Verilog. Let's start with transmitter modules.

12.2.2.1 Transmitter Modules

As explained previously, either the leader or follower can be a transmitter. Therefore, we should form a different description for each option. The Verilog description of the leader-transmitter setup is presented in Listing 12.9. Inputs to this module are clock (clk), data to be sent (data), and send event trigger (send). The outputs of this module are sck, mosi, ss, and busy. The first three of these are directly associated with the SPI lines. The fourth output shows if the module is busy while transmitting data. Within the module, the data length to be sent is set as a parameter. The frequency division is applied to the input clock so that sck is set at 2 MHz. CPOL and CPHA are set to zero. Thus, all changes are performed on the falling edge of sck.

The working principle of the leader-transmitter module can be explained as follows. The module is a state machine triggered in every falling edge of sck. In other words, the data on the mosi line does not change while sck is at logic level 1. The state machine is initially at RDY state. The transmission of data starts when send goes to logic level 1. The state changes to START while index is set to the first bit of data and busy goes to logic level 1. In START state, the leader module sets ss to logic level 0 to wake up the follower. Then, the first bit of the data is loaded to mosi line and index is decreased by one.

Listing 12.9 Verilog Description of the SPI Leader-Transmitter Module

```verilog
module SPI_leader_transmitter(clk,data,send,sck,ss,mosi,busy);

parameter data_length=8;

input clk;
input [data_length-1:0] data;
input send;
output reg sck=0;
output reg ss=1;
output reg mosi;
output reg busy=0;

localparam RDY=2'b00, START=2'b01, TRANSMIT=2'b10, STOP=2'b11;

reg [1:0] state=RDY;
reg [7:0] clkdiv=0;
reg [7:0] index=0;

always @ (posedge clk)
//sck is set to 2 MHz
if (clkdiv == 8'd24)
        begin
        clkdiv <= 0;
        sck <= ~sck;
        end
else clkdiv <= clkdiv + 1;

always @ (negedge sck)

case(state)
RDY:
        if (send)
        begin
        state <= START;
        busy <= 1;
        index <= data_length-1;
        end
START:
        begin
        ss <= 0;
        mosi <= data[index];
        index <= index - 1;
        state <= TRANSMIT;
        end
TRANSMIT:
        begin
        if (index == 0)
        state <= STOP;
        mosi <= data[index];
        index <= index - 1;
        end
STOP:
        begin
        busy <= 0;
        ss <= 1;
        state <= RDY;
        end
endcase
endmodule
```

Listing 12.10 Verilog Description of the SPI Follower-Transmitter Module

```verilog
module SPI_follower_transmitter(sck,ss,data,miso,busy);

parameter data_length=8;

input sck;
input ss;
input [data_length-1:0] data;
output reg miso;
output reg busy;

localparam RDY=2'b00, TRANSMIT=2'b01, STOP=2'b10;

reg [1:0] state=RDY;
reg [7:0] index=data_length-1;

always @ (negedge sck)

case(state)
RDY:
        if (!ss)
        begin
        miso <= data[index];
        state <= TRANSMIT;
        busy <= 1;
        index <= index - 1;
        end
TRANSMIT:
        begin
        if (index == 0)
        state <= STOP;
        miso <= data[index];
        index <= index - 1;
        end
STOP:
        begin
        index <= data_length-1;
        busy <= 0;
        state <= RDY;
        end
endcase
endmodule
```

The next state is TRANSMIT in which index is decreased by one and the corresponding bit of data vector is sent to the output via mosi step by step. When index equals to zero, iteration ends. The state machine switches to STOP state. Then, busy goes to logic level 0 and ss goes to logic level 1. Next, the state machine turns back to RDY state and waits for another send trigger.

The Verilog description of the follower-transmitter module is presented in Listing 12.10. Since this is the follower module, it has inputs sck, ss, and data. The outputs of the module are miso and busy. Timing modes CPOL and CPHA are selected as zero. Thus, all changes are performed on the falling edge of sck.

The working principle of the follower-transmitter module (as a state machine) will be similar to the leader-transmitter module. However, there are major differences as follows. The state machine is initially at RDY state. It is triggered by the falling edge of sck which is generated by the leader-receiver. If ss is at logic level 0, then busy goes to logic level 1. The first bit of the data vector is loaded to mosi. Then, the state machine goes to the state TRANSMIT. Afterward, the module starts sending data. When index equals to 0, it indicates that all data bits have been transmitted. Next, the state machine goes to STOP state. It resets index and sets busy to logic level 0. Afterward, the state machine turns back to RDY state and waits for ss to go logic level 0 for the next transmission cycle.

12.2.2.2   Receiver Modules

As in the transmitter, either the leader or follower can act as a receiver. Therefore, we will analyze both scenarios in this section. Let's start with the follower-receiver module in Listing 12.11. Inputs to this module are sck, ss, and mosi which are directly related to the corresponding SPI signals. The outputs of the module are data, busy, and ready. Here, busy is at logic level 1 when data is being received. When ready goes to logic level 1, the received data will be available in data vector. Here, the data length is defined as a parameter so that it can be changed depending on the application.

The working principles of the follower-receiver module (as a state machine) are as follows. Initially, data_temp is set to the logic level 0 and index is set to the address of the first bit in data vector (data_length-1). The state machine has three states: RDY, RECEIVE, and STOP. RDY is the initial state in which module checks for ss to become logic level 0 at every rising edge of sck. Once ss goes to logic level 0, data_temp is set to receive the first data bit from mosi. Afterward, index is decreased by one; busy goes to logic level 1; ready goes to logic level 0; and the state machine goes to RECEIVE state. Then, the module receives data bits from mosi in every clock cycle like a shift register. When index reaches zero, the state machine goes to STOP state. In this state, busy goes to logic level 0; the received data is written to the data vector; data_temp is set to logic level 0; index is set to the address of the first bit of data. In the next cycle, the state machine turns back to RDY state and waits for another falling edge on ss to receive the next data package. As a reminder, the follower-receiver module should work together with the leader-transmitter module.

The leader-receiver module is presented in Listing 12.12. Here, the clock (clk), MISO line (miso), and receive trigger (get) are inputs of the module which are directly related to the corresponding SPI signals. The outputs of the module are data, sck, ss, busy, and ready. The data length is defined as a parameter of flexibility. The leader-receiver module starts listening the follower-transmitter when get goes to logic level 1. While receiving data, busy stays at logic level 1. Once all bits are received, ready goes to logic level 1. Then, the received data can be obtained from the data vector.

The working principles of the leader-receiver module (as a state machine) are as follows. Within the module, the frequency division is applied to the main clock to have 2-MHz sck. The state machine has three states: RDY, RECEIVE, and STOP. In RDY state, the module checks for a receive trigger in every rising edge of sck. Once the module is triggered, the state machine goes to RECEIVE state. Then, ss goes to logic level 0; busy goes to logic level 1; ready goes to logic level 0; all bits of data_temp are set to logic level 0; and (data_length-1) is loaded to index. In RECEIVE state, data bits are received from the follower in every rising edge of sck. Meanwhile, index is decreased until it reaches zero. Afterward, the state machine goes to STOP state. Here, busy goes to

---

Listing 12.11 Verilog Description of the SPI Follower-Receiver Module

```verilog
module SPI_follower_receiver(sck,ss,mosi,data,busy,ready);

parameter data_length=8;

input sck;
input ss;
input mosi;
output reg [data_length-1:0] data;
output reg busy=0;
output reg ready=0;

localparam RDY=2'b00, RECEIVE=2'b01, STOP=2'b10;
reg [1:0] state=RDY;

reg [data_length-1:0] data_temp=0;
reg [7:0] index=data_length-1;

always @ (posedge sck)

case(state)
RDY:
        if (!ss)
        begin
        data_temp[index] <= mosi;
        index <= index - 1;
        busy <= 1;
        ready <= 0;
        state <= RECEIVE;
        end
RECEIVE:
        begin
        if (index == 0)
        state <= STOP;
        else index <= index - 1;
        data_temp[index] <= mosi;
        end
STOP:
        begin
        busy <= 0;
        ready <= 1;
        data_temp <= 0;
        data <= data_temp;
        index <= data_length-1;
        state <= RDY;
        end
endcase
endmodule
```

Listing 12.12 Verilog Description of the SPI Leader-Receiver Module

```verilog
module SPI_leader_receiver(clk,miso,get,data,sck,ss,busy,ready);

parameter data_length=8;

input clk;
input miso;
input get;
output reg [data_length-1:0] data;
output reg sck=0;
output reg ss=1;
output reg busy=0;
output reg ready=0;

localparam RDY=2'b00, RECEIVE=2'b01, STOP=2'b10;

reg [1:0] state=RDY;
reg [data_length-1:0] data_temp=0;
reg [7:0] clkdiv=0;
reg [7:0] index=0;

always @ (posedge clk)
//sck is set to 2 MHz
        if (clkdiv == 8'd24)
        begin
        clkdiv <= 0;
        sck <= ~sck;
        end
        else
        clkdiv <= clkdiv + 1;

always @ (posedge sck)

case(state)
RDY:
        if (get)
        begin
        ss <= 0;
        state <= RECEIVE;
        busy <= 1;
        ready <= 0;
        data_temp <= 0;
        index <= data_length-1;
        end
RECEIVE:
        begin
        if (index == 0) state <= STOP;
        data_temp[index] <= miso;
        index <= index - 1;
        end
STOP:
        begin
        busy <= 0;
        ready <= 1;
        ss <= 1;
        data <= data_temp;
        state <= RDY;
        end
endcase
endmodule
```

logic level 0; `ready` goes to logic level 1; `ss` goes to logic level 1; and the received data is written to the `data` vector. In the next clock cycle, state machine turns back to RDY state and waits for another `get` trigger to receive the next incoming data. As a reminder, the leader-receiver module should work together with the follower-transmitter module.

12.2.3   SPI in VHDL

As in Verilog, we can describe the transmit and receive operations (for the leader and follower) as separate modules in VHDL. Let's start with the transmitter modules.

12.2.3.1   Transmitter Modules

The VHDL description of the leader-transmitter and follower-transmitter modules is presented in Listings 12.13 and 12.14. In both the descriptions, we tried to keep the input-output definitions and state names the same as in the corresponding Verilog descriptions presented in Listings 12.9 and 12.10. Hence, the reader can associate the working principles of the VHDL and Verilog descriptions.

12.2.3.2   Receiver Modules

The VHDL description of the leader-receiver and follower-receiver modules is presented in Listings 12.15 and 12.16. As in the transmitter modules, we tried to keep the input-output definitions and state names the same as in the corresponding Verilog descriptions in Listings 12.12 and 12.11. Hence, the reader can associate the working principles of the VHDL and Verilog descriptions.

12.2.4   SPI Application

To provide an actual application, we connect Digilent's ambient light sensor (PmodALS) module to the Basys3 board. The sensor module returns eight-bit data based on the illumination level through the SPI. In our application, we will receive this data and convert it to BCD form to show it on the seven-segment display of the Basys3 board.

The sensor module is connected to the top six pins (1 to 6) of the JB PMOD port on the Basys3 board. Necessary adjustments are done on the XDC file to use this pin. Since the sensor module is set to work in the follower-transmitter mode, we set Basys3 to work in the leader-receiver mode. We provide the Verilog description of the top module of the SPI application in Listing 12.17. Unfortunately, the naming on the sensor module is different from the standard SPI applications. Please see the Digilent website for more information on this issue [48]. Therefore, $\overline{SS}$, MOSI, and SCK lines are denoted by `cs`, `sdo`, and `sck` in the top module. The sensor module works with a clock frequency between 1 and 4 MHz. This is in line with our `spi_leader_transmitter` module since the master clock on the Basys3 board (with 100 MHz rate) is divided to obtain a 2-MHz clock. Therefore, we can directly integrate it.

The top module in Listing 12.17 uses submodules SPI_leader_ receiver, binaryto2BCD, and `sevenseg_driver`. The binary to BCD converter and seven-segment display driver have been mentioned in Sec. 10.6. Every unit of the BCD converter is connected directly to the seven-segment display driver. The top module can be represented as a state machine with two states. These are TRACK and GETIN. In TRACK state, `delay` is decreased by one in every master clock cycle until it reaches zero. This time slot is referred to as the delay time. It is specifically introduced to keep the coherency of digits in the seven-segment display. When `get` goes to logic level 1, the state changes to GETIN and the SPI module starts working. The top module waits until the SPI module does its job and takes `ready` to logic level 1. Afterward, `get` goes to logic level 0;

Listing 12.13 VHDL Description of the SPI Leader-Transmitter Module

```vhdl
library ieee;
use ieee.std_logic_1164.all;

entity SPI_leader_transmitter is
generic (data_length : integer := 8);
port(clk : in std_logic;
    data : in std_logic_vector (data_length-1 downto 0);
    send : in std_logic;
     sck : out std_logic := '0';
      ss : out std_logic := '1';
    mosi : out std_logic;
    busy : out std_logic := '0');
end SPI_leader_transmitter;

architecture behavioral of SPI_leader_transmitter is

type state_type is (RDY, START, TRANSMIT, STOP);

signal state  : state_type := RDY;
signal clkdiv : integer := 0;
signal index  : integer := 0;

begin
process(clk)
begin
if rising_edge(clk) then
--sck is set to 2 MHz
        if (clkdiv = 24) then
        clkdiv <= 0;
        sck <= not sck;
        else
        clkdiv <= clkdiv + 1;
        end if;
end if;
end process;

process(clk)
begin
if rising_edge(clk) then

case state is
when RDY =>
        if send = '1' then
        state <= START;
        busy <= '1';
        index <= data_length - 1;
        end if;
when START =>
        ss <= '0';
        mosi <= data(index);
        index <= index - 1;
        state <= TRANSMIT;
when TRANSMIT =>
        if (index = 0) then
        state <= STOP;
        mosi <= data(index);
        index <= index - 1;
        end if;
```

```
when STOP =>
        busy <= '0';
        ss <= '1';
        state <= RDY;
end case;
end if;
end process;
end behavioral;
```

Listing 12.14 VHDL Description of the SPI Follower-Transmitter Module

```
library ieee;
use ieee.std_logic_1164.all;

entity SPI_follower_transmitter is
generic (data_length : integer := 8);
port(sck : in std_logic;
     ss : in std_logic;
   data : in std_logic_vector (data_length-1 downto 0);
   miso : out std_logic;
   busy : out std_logic);
end SPI_follower_transmitter;

architecture behavioral of SPI_follower_transmitter is

type state_type is (RDY, TRANSMIT, STOP);

signal state  : state_type := RDY;
signal index  : integer := data_length-1;

begin
process(sck)
begin
if falling_edge(sck) then

case state is
when RDY =>
        if (ss = '0') then
        miso <= data(index);
        state <= TRANSMIT;
        busy <= '1';
        index <= index - 1; end if;
when TRANSMIT =>
        if (index = 0) then
        state <= STOP; end if;
        miso <= data(index);
        index <= index - 1;
when STOP =>
        index <= data_length-1;
        busy <= '0';
        state <= RDY;
end case;
end if;
end process;
end behavioral;
```

Listing 12.15   VHDL Description of the SPI Leader-Receiver Module

```vhdl
library ieee;
use ieee.std_logic_1164.all;

entity SPI_leader_receiver is
generic(data_length : integer := 8);
port(clk : in std_logic;
    miso : in std_logic;
     get : in std_logic;
    data : out std_logic_vector (data_length-1 downto 0);
     sck : out std_logic := '0';
      ss : out std_logic := '1';
    busy : out std_logic := '0';
   ready : out std_logic := '0');
end SPI_leader_receiver;

architecture behavioral of SPI_leader_receiver is

type state_type is (RDY, RECEIVE, STOP);

signal state     : state_type := RDY;
signal data_temp : std_logic_vector (data_length-1 downto 0);
signal clkdiv    : integer := 0;
signal index     : integer := 0;

begin
process(clk)
begin
if rising_edge(clk) then
--sck is set to 2 MHz
        if (clkdiv = 24) then
        clkdiv <= 0;
        sck <= not sck;
        else clkdiv <= clkdiv + 1;
        end if;
end if;
end process;

process(sck)
begin
if rising_edge(sck) then

case state is
when RDY =>
        if get then
        ss <= '0';
        state <= RECEIVE;
        busy <= '1';
        ready <= '0';
        data_temp <= (others => '0');
        index <= data_length-1;
        end if;
when RECEIVE =>
        if (index = 0) then
        state <= STOP;
```

```
                    end if;
                    data_temp(index) <= miso;
                    index <= index - 1;
       when STOP =>
                    busy <= '0';
                    ready <= '1';
                    ss <= '1';
                    data <= data_temp;
                    state <= RDY;
       end case;
       end if;
       end process;
       end behavioral;
```

Listing 12.16 VHDL Description of the SPI Follower-Receiver Module

```
library ieee;
use ieee.std_logic_1164.all;

entity SPI_follower_receiver is
generic(data_length : integer := 8);
port(sck : in std_logic;
      ss : in std_logic;
    mosi : in std_logic;
    data : out std_logic_vector (7 downto 0);
    busy : out std_logic;
   ready : out std_logic);
end SPI_follower_receiver;

architecture behavioral of SPI_follower_receiver is

type state_type is (RDY, RECEIVE, STOP);

signal state     : state_type := RDY;
signal data_temp : std_logic_vector (data_length-1 downto 0);
signal index     : integer := data_length-1;

begin
process (sck)
begin
if rising_edge(sck) then

case state is
when RDY =>
        if (ss = '0') then
        data_temp(index) <= mosi;
        index <= index - 1;
        end if;
when RECEIVE =>
        if (index = 0) then
        state <= STOP;
        else index <= index - 1;
        end if;
        data_temp(index) <= mosi;
when STOP =>
```

```
            busy <= '0';
            ready <= '1';
            data_temp <= (others => '0');
            data <= data_temp;
            index <= data_length-1;
            state <= RDY;
    end case;
    end if;
    end process;

    end behavioral;
```

Listing 12.17 Verilog Description of the Top Module in SPI Application on the Basys3 Board

```verilog
module SPI_lightsensor(clk,sdo,ss,sck,led,seg,an);

input clk;
input sdo;
output ss;
output sck;
output [15:0] led;
output [6:0] seg;
output [3:0] an;

parameter delaytime=500; //in msec

localparam TRACK=1'b0, GETIN=1'b1;

reg state=TRACK;
reg [31:0] delay=2000*delaytime;
reg [15:0] lightdata;

wire [3:0] thos, huns, tens, ones;

wire [15:0] data;
wire busy;
wire ready;
reg get=0;

always @ (posedge sck)
case(state)
TRACK:
    if (delay == 0)
    begin
    get <= 1;
    state <= GETIN;
    end
    else
    delay <= delay - 1'b1;
GETIN:
    if (ready == 1)
    begin
    get <= 0;
    lightdata <= data;
    delay <= 2000*delaytime;
```

```
        state <= TRACK;
    end
endcase

assign led = {thos,huns,tens,ones};

SPI_leader_receiver #(16) spi(clk,sdo,get,data,sck,ss,busy,ready);
binarytoBCD_0 bcd({4'b0000,lightdata[12:5]},thos,huns,tens,ones);
sevenseg_driver_0 segdriver(clk,0,thos,huns,tens,ones,seg,an);

endmodule
```

delay is set to its initial value again; data from the SPI module is written to the light-data vector; and the state turns back to TRACK.

We provide the VHDL version of the top module in Listing 12.18. As in previous applications, we kept the naming convention the same between Verilog and VHDL descriptions. Hence, the reader can easily follow the VHDL description.

12.3   Inter-Integrated Circuit

Inter-integrated circuit (I^2C) is the third and final digital communication protocol we will cover in this chapter. As in the UART and SPI, we will first cover the working principles of the I^2C. Then, we will implement its Verilog and VHDL descriptions. Finally, we will provide a sample application using the I^2C.

12.3.1   Working Principles of I²C

The I^2C is a multi-leader, multi-follower, serial communication protocol between digital devices. We will focus only on the I^2C communication between two devices in this book. In this section, we will cover the working principles of the I^2C in terms of the data format, connection diagram, and transmission and reception operations. More information on these topics can be found in [49].

12.3.1.1   Data Format

In the I^2C, every follower has a unique address. Data transfer starts with this address. When the follower wakes up and acknowledges back the leader, transfer continues with the pointer/address and data or directly data is transferred depending on the protocol. The address of a follower is usually composed of seven bits. However, in some cases the address can be either eight or ten bits. Independent of the address, pointer, and data size, the transfer is performed in terms of eight-bit packages. Each package has seven-bit address, pointer, and data and one-bit acknowledge value. The receiver merges packets to extract data.

12.3.1.2   Connection Diagram

The I^2C data bus has two wires called serial data line (SDA) and serial clock line (SCL). Besides, all connected devices need a common ground and power line. As a result, the I^2C will need four wires for communication. The connection diagram of a generic I^2C is presented in Fig. 12.4. The SDA and SCL are bidirectional lines. Both the lines are connected to V_{DD} by a pull-up resistor. This means they are at logic level 1 when idle.

Listing 12.18 VHDL Description of the Top Module in SPI Application on the Basys3 Board

```vhdl
library ieee;
use ieee.std_logic_1164.all;
use ieee.std_logic_unsigned.all;

entity SPI_lightsensor is
generic(delaytime : integer := 500);   --msec
port(clk : in std_logic;
     sdo : in std_logic;
      ss : out std_logic;
     sck : out std_logic;
     led : out std_logic_vector (15 downto 0);
     seg : out std_logic_vector (6 downto 0);
      an : out std_logic_vector (3 downto 0));
end SPI_lightsensor;

architecture behavioral of SPI_lightsensor is

type state_type is (TRACK, GETIN);
signal state   : state_type := TRACK;
signal delay   : integer := 2000*delaytime;
signal lightdata: std_logic_vector (15 downto 0);
signal thos    : std_logic_vector (3 downto 0);
signal huns    : std_logic_vector (3 downto 0);
signal tens    : std_logic_vector (3 downto 0);
signal ones    : std_logic_vector (3 downto 0);
signal data    : std_logic_vector (15 downto 0);
signal busy    : std_logic;
signal ready   : std_logic;
signal get     : std_logic := '0';

component SPI_leader_receiver is
generic(data_length : integer);
port(clk : in std_logic;
    miso : in std_logic;
     get : in std_logic;
    data : out std_logic_vector (data_length-1 downto 0);
     sck : out std_logic;
      ss : out std_logic;
    busy : out std_logic;
   ready : out std_logic);
end component;

component binarytoBCD is
port(binary : in std_logic_vector (11 downto 0);
       thos : out std_logic_vector (3 downto 0);
       huns : out std_logic_vector (3 downto 0);
       tens : out std_logic_vector (3 downto 0);
       ones : out std_logic_vector (3 downto 0));
end component;

component sevenseg_driver is
port(clk : in std_logic;
     clr : in std_logic;
     in1 : in std_logic_vector (3 downto 0);
```

```vhdl
        in2 : in std_logic_vector (3 downto 0);
        in3 : in std_logic_vector (3 downto 0);
        in4 : in std_logic_vector (3 downto 0);
        seg : out std_logic_vector (6 downto 0);
         an : out std_logic_vector (3 downto 0));
end component;

begin
process(sck)
begin
if rising_edge(sck) then
case state is
        when TRACK =>
        if delay = 0 then
        get <= '1';
        state <= GETIN;
        else delay <= delay - 1;
        end if;
        when GETIN =>
        if ready = '1' then
    get <= '0';
    lightdata <= data;
    delay <= 2000*delaytime;
    state <= TRACK;
        end if;
end case;
end if;
end process;

led <= thos & huns & tens & ones;

spi : spi_leader_receiver generic map (data_length => 16) port map (clk
    => clk,miso => sdo,get => get,data => data,sck => sck,ss => ss,
    busy => busy,ready => ready);

bcd : binarytoBCD port map (binary => "0000"&lightdata (12 downto 5),
    thos => thos,huns => huns,tens => tens,ones => ones);

segdriver : sevenseg_driver port map (clk => clk,clr => '0',in1 => thos
    ,in2 => huns,in3 => tens,in4 => ones,seg => seg,an => an);

end behavioral;
```

Different from the SPI, every follower has a unique address in the I^{2}C. Therefore, the follower and leader can be chosen over the serial data line without the need of a select signal. Thus, other than power and ground signals, the I^{2}C bus has only two wires connected to all the devices. This advantage saves the pin usage compared to the SPI.

12.3.1.3 Transmission and Reception Operations

As mentioned in the previous section, data on the I^{2}C communication is carried by eight-bit packages. The leader starts the transmission by sending the follower address and read/write decision bit. The follower with this address on the network wakes up and acknowledges the leader that it is alive and ready to talk. Then, depending on the

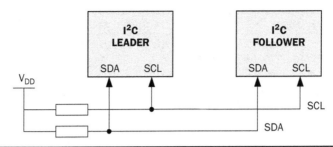

Figure 12.4 Connection diagram of a generic I²C.

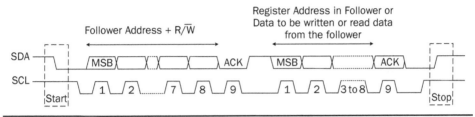

Figure 12.5 I²C timing diagram.

decision bit, the leader writes or reads data from the follower. The leader ends the talk by sending a stop signal. Figure 12.5 shows the complete timing diagram of the I²C communication.

As can be seen in Fig. 12.5, the leader starts transmission by a logic level 1 to 0 transition on SDA while SCL stays at logic level 1. We can call this as the start signal. The transmission ends by a logic level 0 to 1 transition on the SDA while SCL is at logic level 1. We can call this as the stop signal. The address of the device and data is transmitted between start and stop signals. After the start signal, the leader sends the seven-bit address of the follower. Then $R/\overline{W}$ signal is sent, which tells the follower if the leader is going to read or write the data to/from the follower. This is concluded by an acknowledge signal from the follower. Next, the leader starts sending or receiving data (with the MSB first) followed by an acknowledge signal. There are no restrictions on the number of successively transmitted data bits. The communication continues until the leader sends the stop signal. Note that during the acknowledge signal the transmitter releases the SDA line and the receiver pulls the line to logic level 0 while SCL is at logic level 1. We will use these descriptions while implementing the I²C module next.

12.3.2   I²C in Verilog

We provide the Verilog description of the I²C leader module in Listing 12.19. This module has six inputs as `clk`, `reset_n`, `ena`, `addr`, `rw`, and `data_wr`. `clk` corresponds to the clock signal to be used in the module. `reset_n` indicates the active low reset. `ena` is for the active-high enable signal. `addr` represents the address of the follower to be connected to. `rw` stands for the read/write input. When it is at logic level 1, the module reads data. Otherwise, the module writes data. The module has four outputs as `busy`, `data_rd`, `ack_error`, and `eop`. If the module is transmitting or receiving data, `busy` will be at logic level 1. `data_rd` indicates the read data from the follower.

Listing 12.19   Verilog Description of the I²C Leader Module

```verilog
module I2C_leader(clk,reset_n,ena,addr,rw,data_wr,busy,data_rd,
    ack_error,eop,sda,scl);

input clk;
input reset_n;
input ena;
input [6:0] addr;
input rw;
input [7:0] data_wr;

output reg busy;
output reg [7:0] data_rd;
output reg ack_error;
output reg eop=0;

inout sda;
inout scl;

parameter input_clk=100000000;
parameter bus_clk=400000;

localparam READY=4'b0000, START=4'b0001, COMMAND=4'b0010, SLV_ACK1=4'
    b0011, WR=4'b0100, RD=4'b0101, SLV_ACK2=4'b0110, MSTR_ACK=4'b0111,
    STOP=4'b1000;

reg [3:0] state;
integer divider=(input_clk/bus_clk)/4;
reg data_clk;
reg data_clk_prev;
reg scl_clk;
reg scl_ena=0;
reg sda_int=1;
reg sda_ena_n;
reg [7:0] addr_rw;
reg [7:0] data_tx;
reg [7:0] data_rx;
reg [2:0] bit_cnt=3'd7;
reg [31:0] count;

always @ (posedge clk, negedge reset_n)
begin
if (reset_n == 0)
count <= 0;
else
begin
data_clk_prev <= data_clk;
if(count == divider*4-1) count <= 0;
else
begin
count <= count + 1;
if (count >= 0 && count < divider-1)
begin
scl_clk <= 0;
data_clk <= 0; end
```

```verilog
    else if (count >= divider && count < divider*2-1)
    begin
    scl_clk <= 0;
    data_clk <= 1; end
    else if (count >= divider*2 && count < divider*3-1)
    begin
    scl_clk <= 1;
    data_clk <= 1;
    end
    else if (count >= divider*3 && count < divider*4-1) begin
    scl_clk <= 1;
    data_clk <= 0;
    end
    else begin
    scl_clk <= scl_clk;
    data_clk <= data_clk;
    end
    end
    end
    end

    always @ (posedge clk, negedge reset_n)
    begin
    if (reset_n == 0)
    begin
    state <= READY;
    busy <= 1;
    scl_ena <= 0;
    sda_int <= 1;
    ack_error <= 0;
    bit_cnt <= 3'd7;
    data_rd <= 8'b0;
    end
    else
    begin
    if (data_clk == 1 && data_clk_prev == 0)

    case (state)
    READY:
            if (ena == 1)
            begin
            busy <= 1;
            addr_rw <= {addr,rw};
            data_tx <= data_wr;
            state <= START;
            end
            else
            begin
            busy <= 0;
            state <= READY;
            end
    START:
            begin
            busy <= 1;
            sda_int <= addr_rw[bit_cnt];
```

```
                              state <= COMMAND;
                              end
            COMMAND:
                              if (bit_cnt == 0)
                              begin
                              sda_int <= 1;
                              bit_cnt <= 3'd7;
                              state <= SLV_ACK1;
                              end
                              else begin
                              bit_cnt <= bit_cnt - 1;
                              sda_int <= addr_rw[bit_cnt-1];
                              state <= COMMAND;
                              end
            SLV_ACK1:
                              if (addr_rw[0] == 0)
                              begin
                              sda_int <= data_tx[bit_cnt];
                              state <= WR;
                              end
                              else
                              begin
                              sda_int <= 1;
                              state <= RD;
                              end
            WR:
                              begin
                              busy <= 1;
                              if (bit_cnt == 0)
                              begin
                              sda_int <= 1;
                              bit_cnt <= 3'd7;
                              state <= SLV_ACK2;
                              end
                              else if (bit_cnt == 1)
                              begin
                              bit_cnt <= bit_cnt - 1;
                              sda_int <= data_tx[bit_cnt-1];
                              eop <= 1;
                              end
                              else begin
                              bit_cnt <= bit_cnt - 1;
                              sda_int <= data_tx[bit_cnt-1];
                              state <= WR;
                              end
                              end
            RD:
                              begin
                              busy <= 1;
                              if (bit_cnt == 0) begin
                              if (ena == 1 && addr_rw == {addr,rw}) sda_int <= 0;
                              else sda_int <= 1;
                              bit_cnt <= 3'd7;
                              data_rd <= data_rx;
                              state <= MSTR_ACK;
```

```
            end
            else if (bit_cnt == 1)
            begin
            bit_cnt <= bit_cnt - 1;
            eop <= 1;
            end
            else
            begin
            bit_cnt <= bit_cnt - 1;
            state <= RD;
            end
            end
SLV_ACK2:
            if (ena == 1)
            begin
            eop <= 0;
            busy <= 0;
            addr_rw <= {addr,rw};
            data_tx <= data_wr;
            if (addr_rw == {addr,rw})
            begin
            sda_int <= data_wr[bit_cnt];
            state <= WR;
            end
            else state <= START;
            end
            else
            begin
            eop <= 0;
            state <= STOP;
            end
MSTR_ACK:
            if (ena == 1)
            begin
            eop <= 0;
            busy = 0;
            addr_rw <= {addr,rw};
            data_tx <= data_wr;
            if (addr_rw == {addr,rw})
            begin
            sda_int <= 1;
            state <= RD;
            end
            else state <= START;
            end
            else
            begin
            eop <= 0;
            state <= STOP;
            end
STOP:
            begin
            busy <= 0;
            state <= READY;
            end
```

```
        endcase
        else if (data_clk == 0 && data_clk_prev == 1)

        case (state)
        START:
                if (scl_ena == 0)
                begin
                scl_ena <= 1;
                ack_error <= 0;
                end
        SLV_ACK1:
                if (sda != 0 || ack_error == 1) ack_error <= 1;
        RD:
                data_rx[bit_cnt] <= sda;
        SLV_ACK2:
                if (sda != 0 || ack_error == 1) ack_error <= 1;
        STOP:
                scl_ena <= 0;
        endcase
        end
        end

        always @ (clk)
        case (state)
        START   : sda_ena_n <= data_clk_prev;
        STOP    : sda_ena_n <= ~data_clk_prev;
        default : sda_ena_n <= sda_int;
        endcase

        assign scl = (scl_ena == 1 && scl_clk == 0) ? 0 : 1'bz;
        assign sda = (sda_ena_n == 0) ? 0 : 1'bz;

        endmodule
```

ack_error stands for the active-high acknowledge error. eop indicates the end of package. The module also has inout signals sda and scl. These are the serial data and serial clock I^{2}C signals, respectively.

We can explain the working principles of the I^{2}C leader module as a state machine. There are nine states in the machine: READY, START, COMMAND, SLV_ACK1, WR, RD, SLV_ACK2, MSTR_ACK, and STOP. The machine starts in READY state and waits for ena to go to logic level 1. When this happens, busy goes to logic level 1; the follower address with the rw bit is written to addr_rw vector; data_wr is registered to the data_tx vector; and state changes to START. At this state, the MSB of the addr_rw vector is loaded to sda_int. This directly controls the sda port. Then, the state changes to COMMAND. In this state, addr_rw is loaded to sda line bit by bit. At the end of this operation, the leader module waits for an acknowledgement from the follower by going to SLV_ACK1 state. This is when the follower wakes up and says hello to the leader by taking sda line to logic level 0. If the follower does not send an acknowledgement, then ack_error goes to logic level 1. The next state is determined by rw input. If rw is at logic level 1, the state machine switches to RD state. Otherwise, it switches to WR state. In RD state, bits are received one by one. They are stored in the `data_rd` vector. If the bit counter equals to 1, `eop` goes to logic level 1 to indicate that the package is received. After all bits are received (bit counter goes to zero), the state changes to `MSTR_ACK` where the leader acknowledges that the package is received. Similarly in `WR` state, bits on `data_tx` vector are loaded to `sda_int` one by one. When bit counter goes to 1, `eop` goes to logic level 1. When all bits are transferred, the state changes to `SLV_ACK2` where the state machine waits for an acknowledgement from the follower. In both `SLV_ACK2` and `MSTR_ACK` states, if `ena` is still at logic level 1, the new follower address and `rw` bits are written to `addr_rw` vector. If these are same as in previous values, then the state machine goes to the state `START` for a repeated start. Otherwise, the state changes to `STOP` where a stop signal is applied to the I²C line. Then, the state machine goes to `READY` state waiting for the next transmission to start.

Within the I²C leader module in Listing 12.19, we had to use conditional statements in the dataflow modeling as the last two lines of the description. The structure of such a statement is as `assign variable = condition ? 0/1 : value_to_be_assigned`. This representation can be used in other Verilog descriptions as well.

12.3.3   I²C in VHDL

We provide the VHDL version of the I²C leader module in Listing 12.20. This description is the modified version of the one available in [50]. It has been included with their permission. This module shares the same input, output, inout, and state descriptions as its Verilog version in Listing 12.19. To avoid repetition, we direct the reader to the previous section for the working principles of the VHDL description.

12.3.4   I²C Application

In this application, we will use the PMOD three-axis digital compass module which communicates by the I²C. The website of the compass module is provided in [51]. The module uses the Honeywell HMC5883L 3-axis digital compass which measures magnetic field and gives data in three dimensions with 16 bits two's complement form. The range of data varies from the hexadecimal number F800 to 07FF. There are 13 registers inside the chip each having eight-bit data length. The address of the chip is the hexadecimal number 1E in seven bits. We have to configure the chip for continuous measurement mode which is not set by default. This configuration register is located at the address 02 in hexadecimal form. The value for the the continuous mode is 00 in hexadecimal form. The measurements in x, y, and z axes are stored in the register addresses between 03 and 08 in the hexadecimal form. Within the application, we will read compass data and send them to PC terminal via the UART interface.

To run the compass module, we will start the I²C transmission in write mode and set the chip to continuous mode. Then, we will continue with the read mode to read stored measurements in register addresses. The compass chip is designed with an internal address counter which increments the register address after every operation. When the counter reaches the address 08, it turns back to 03. So, once we start reading from 03, we do not have to set the internal register again. The chip needs some time to make measurements. We can observe this time by watching its `drdy` port which goes to logic level 0 when data gets ready in registers. You can find detailed information about the Honeywell HMC5883L 3-axis digital compass chip in its data sheet.

Listing 12.20 VHDL Description of the I²C Leader Module

```vhdl
library ieee;
use ieee.std_logic_1164.all;
use ieee.std_logic_unsigned.all;

entity I2C_leader is
generic(input_clk : integer := 100000000;
           bus_clk : integer := 400000);
port(clk : in std_logic;
  reset_n : in std_logic;
      ena : in std_logic;
     addr : in std_logic_vector (6 downto 0);
       rw : in std_logic;
  data_wr : in std_logic_vector (7 downto 0);
     busy : out std_logic;
  data_rd : out std_logic_vector (7 downto 0);
ack_error: buffer std_logic;
      eop : out std_logic := '0';
      sda : inout std_logic;
      scl : inout std_logic);
end I2C_leader;

architecture behavioral of I2C_leader is
constant divider : integer := (input_clk/bus_clk)/4;

type state_type is (READY, START, COMMAND, SLV_ACK1, WR, RD, SLV_ACK2,
    MSTR_ACK, STOP);

signal state     : state_type;
signal data_clk : std_logic;
signal data_clk_prev: std_logic;
signal scl_clk   : std_logic;
signal scl_ena   : std_logic := '0';
signal sda_int   : std_logic := '1';
signal sda_ena_n: std_logic;
signal addr_rw   : std_logic_vector (7 downto 0);
signal data_tx   : std_logic_vector (7 downto 0);
signal data_rx   : std_logic_vector (7 downto 0);
signal bit_cnt   : integer range 0 TO 7 := 7;

begin
process(clk, reset_n)
variable count : integer range 0 to divider*4;
begin
if (reset_n = '0') then
count := 0;
elsif(clk'event and clk = '1') then
data_clk_prev <= data_clk;
if (count = divider*4-1) then
count := 0;
else
count := count + 1;
case count is
when 0 to divider-1 =>
        scl_clk <= '0';
        data_clk <= '0';
when divider to divider*2-1 =>
```

```
                    scl_clk <= '0';
                    data_clk <= '1';
        when divider*2 to divider*3-1 =>
                    scl_clk <= '1';
                    data_clk <= '1';
        when others =>
                    scl_clk <= '1';
                    data_clk <= '0';
        end case;
        end if;
        end if;
        end process;

        process(clk, reset_n)
        begin
        if (reset_n = '0') then
        state <= READY;
        busy <= '1';
        scl_ena <= '0';
        sda_int <= '1';
        ack_error <= '0';
        bit_cnt <= 7;
        data_rd <= "00000000";
        elsif (clk'event and clk = '1') then
        if (data_clk = '1' and data_clk_prev = '0') then

        case state is
        when READY =>
                    if (ena = '1') then
                    busy <= '1';
                    addr_rw <= addr & rw;
                    data_tx <= data_wr;
                    state <= START;
                    else
                    busy <= '0';
                    state <= READY;
                    end if;
        when START =>
                    busy <= '1';
                    sda_int <= addr_rw(bit_cnt);
                    state <= COMMAND;
        when COMMAND =>
                    if (bit_cnt = 0) then
                    sda_int <= '1';
                    bit_cnt <= 7;
                    state <= SLV_ACK1;
                    else
                    bit_cnt <= bit_cnt - 1;
                    sda_int <= addr_rw(bit_cnt-1);
                    state <= COMMAND;
                    end if;
        when SLV_ACK1 =>
                    if (addr_rw(0) = '0') then
                    sda_int <= data_tx(bit_cnt);
                    state <= WR;
                    else
                    sda_int <= '1';
```

```
                      state <= RD;
                   end if;
         when WR =>
                   busy <= '1';
                   if (bit_cnt = 0) then
                   sda_int <= '1';
                   bit_cnt <= 7;
                   state <= SLV_ACK2;
                   elsif (bit_cnt = 1) then
                   bit_cnt <= bit_cnt - 1;
                   sda_int <= data_tx(bit_cnt-1);
                   eop <= '1';
                   else
                   bit_cnt <= bit_cnt - 1;
                   sda_int <= data_tx(bit_cnt-1);
                   state <= WR;
                   end if;
         when RD =>
                   busy <= '1';
                   if (bit_cnt = 0) then
                   if (ena = '1' and addr_rw = addr & rw) then
                   sda_int <= '0';
                   else
                   sda_int <= '1';
                   end if;
                   bit_cnt <= 7;
                   data_rd <= data_rx;
                   state <= MSTR_ACK;
                   elsif (bit_cnt = 1) then
                   bit_cnt <= bit_cnt - 1;
                   eop <= '1';
                   else
                   bit_cnt <= bit_cnt - 1;
                   state <= RD;
                   end if;
         when SLV_ACK2 =>
                   if (ena = '1') then
                   eop <= '0';
                   busy <= '0';
                   addr_rw <= addr & rw;
                   data_tx <= data_wr;
                   if (addr_rw = addr & rw) then
                   sda_int <= data_wr(bit_cnt);
                   state <= WR;
                   else
                   state <= START;
                   end if;
                   else
                   eop <= '0';
                   state <= STOP;
                   end if;
         when MSTR_ACK =>
                   if (ena = '1') then
                   eop <= '0';
                   busy <= '0';
                   addr_rw <= addr & rw;
                   data_tx <= data_wr;
```

```vhdl
                  if (addr_rw = addr & rw) then
                        sda_int <= '1';
                        state <= RD;
                  else
                        state <= START;
                  end if;
                  else
                        eop <= '0';
                        state <= STOP;
                  end if;
      when STOP =>
                  busy <= '0';
                  state <= READY;
      end case;

      elsif (data_clk = '0' and data_clk_prev = '1') then

      case state is
      when START =>
                  if (scl_ena = '0') then
                        scl_ena <= '1';
                        ack_error <= '0';
                  end if;
      when SLV_ACK1 =>
                  if (sda /= '0' or ack_error = '1') then
                        ack_error <= '1';
                  end if;
      when RD =>
                  data_rx(bit_cnt) <= sda;
      when SLV_ACK2 =>
                  if (sda /= '0' or ack_error = '1') then
                        ack_error <= '1';
                  end if;
      when STOP =>
                  scl_ena <= '0';
      when others =>
                  null;
      end case;
      end if;
      end if;
      end process;

with state select
sda_ena_n <= data_clk_prev when START,
not data_clk_prev when STOP,
sda_int when others;

scl <= '0' when (scl_ena = '1' and scl_clk = '0') else 'Z';
sda <= '0' when sda_ena_n = '0' else 'Z';

end behavioral;
```

The Verilog description of the top module for our application is presented in Listing 12.21. This module has three inputs as follows. clk is the master clock on the Basys3 board. btnC is the center push-button on the Basys3 board which is used as reset button. drdy is the data-ready signal coming from the compass module. The outputs of the module are led corresponding to 16 LEDS on the Basys3 board and RsTx which is the UART transmitter port of the Basys3 board. Since we will have the I^{2}C communication, there will be inout signals scl and sda.

The top module in Listing 12.21 can be described as a state machine which basically works with eop signals introduced in Sec. 12.3.2. If btnC is pressed on the Basys3 board, reset_n, package counter (pck_cnt) and enable ena go to logic level 0. reset_n is connected to the reset input in the I2C_leader module. When pck_cnt equals to zero, the state machine goes to START state and pulls ena to logic level 1; writes the follower address to addr vector; sets rw to write mode; and loads data to be written in data_wr vector. In this case, the first data to be written is the address of the configuration register in the compass chip which is 02 in hexadecimal form. Then, the state machine waits for eop signal coming from the I2C_leader module. A rising edge on eop indicates that communication is established with the follower and the first eight-bit data is almost sent. Then, pck_cnt is incremented by one and state changes to WRITE1. Here, data for the configuration register is loaded to data_wr vector. After the second eop signal, we can understand that the compass chip is configured. The state goes to WAITDATA where it waits for drdy to go to logic level 1 to start reading data from the compass registers. During this time, ena goes to logic level 0. Once we are informed that data is ready (drdy is at logic level 0), ena goes to logic level 1 again and the state changes to READDATA by incrementing pck_cnt by one. In this state, rw goes to logic level 1, which corresponds to the read mode. With the next eop signal, the state machine will be ready to read data from the data_rd vector. Before the last bit of the first register is read, eop goes to logic level 1 and the state changes to READXH. When eop goes back to logic level 0, the eight-bit data on the data_rd vector corresponds to the first eight bits of the x-axis on the compass. Therefore, we write it to the MSB eight bit of the x_axis vector. Then, all data bits are received similarly. At READZL state, ena goes to logic level 0. The state machine waits drdy to go to logic level 0 to avoid repeated reads. When drdy goes to logic level 0, pck_cnt is reset to START state and the operation starts again.

Once we have all the data, the UART transmitter module is triggered with the rising edge of drdy signal. The module responsible for this operation is UART_word_tx. We provide the Verilog description of this module in Listing 12.22. This module sends data stored at x_axis, y_axis, and z_axis to host PC. The reader can observe these values through a terminal program as explained in Sec. 12.1.

We provide the VHDL description of the top module of our application in Listing 12.23. This description can be understood by the corresponding Verilog description since the descriptions share the same naming conventions as in previous sections. We also provide the VHDL version of the UART_word_tx module in Listing 12.24.

Listing 12.21 Verilog Description of the Top Module for the I²C Application

```verilog
module I2C_compass(clk,btnC,drdy,scl,sda,RsTx);

input clk;
input btnC;
input drdy;
inout scl;
inout sda;
output RsTx;

parameter [6:0] slv_addr = 7'b0011110;

localparam START=4'b0000, WRITE1=4'b0001, WAITDATA=4'b0010, READDATA=4'
    b0011, READXH=4'b0100, READXL=4'b0101, READYH=4'b0110, READYL=4'
    b0111, READZH=4'b1000, READZL=4'b1001;

reg reset_n=0;
reg ena=0;
reg [6:0] addr;
reg rw;
reg [7:0] data_wr;
wire busy;
reg [3:0] pck_cnt=0;
wire [7:0] data_rd;
wire ack_error;
wire eop;
reg eop_prev;
wire btnCclr;

reg [15:0] x_axis=0;
reg [15:0] y_axis=0;
reg [15:0] z_axis=0;

I2C_leader I2C_comp(clk,reset_n,ena,addr,rw,data_wr,busy,data_rd,
    ack_error,eop,sda,scl);

debounce_0 db1(clk,btnC,btnCclr);

always @ (posedge clk)
begin
  if (btnCclr == 1'b1)
  begin
  reset_n <= 0;
  ena <= 0;
  pck_cnt <= 0;
  end
  else reset_n <= 1'b1;
eop_prev <= eop;
  if (eop_prev == 0 && eop == 1) pck_cnt <= pck_cnt + 1'b1;

case (pck_cnt)
START:
      begin
      ena <= 1;
      addr <= slv_addr;
```

```
                rw <= 0;
                data_wr <= 8'h02;
                end
        WRITE1:
                data_wr <= 8'h00;
        WAITDATA:
                if (drdy == 0)
                begin
                ena <= 1;
                pck_cnt <= pck_cnt + 4'b0001;
                end
                else ena <= 0;
        READDATA:
                rw <= 1;
        READXH:
                if (eop == 0) x_axis[15:8] <= data_rd;
        READXL:
                if (eop == 0) x_axis[7:0] <= data_rd;
        READYH:
                if (eop == 0) y_axis[15:8] <= data_rd;
        READYL:
                if (eop == 0) y_axis[7:0] <= data_rd;
        READZH:
                if (eop == 0) z_axis[15:8] <= data_rd;
        READZL:
                begin
                ena <= 0;
                if (eop == 0) z_axis[7:0] <= data_rd;
                if (drdy == 1'b1) pck_cnt <= START;
                end
endcase
end

reg [255:0] word;
reg wordsend=0;
reg [5:0] wordlength;
reg drdy_prev;

UART_word_tx_0 TX(clk,wordsend,word,wordlength,RsTx);

always @ (posedge clk)
begin
drdy_prev <= drdy;
if (drdy_prev == 1'b0 && drdy == 1'b1)
  begin
  word <= {x_axis[15:8],x_axis[7:0],y_axis[15:8],y_axis[7:0],z_axis
      [15:8],z_axis[7:0]};
  wordlength <= 6'b000110;
  wordsend <= 1'b1;
  end
else wordsend <= 1'b0;
end
endmodule
```

Listing 12.22 Verilog Description of the UART_word_tx Module

```verilog
module UART_word_tx(clk,send,word,count,tx);

input clk;
input send;
input [255:0] word; //max 32 characters
input [5:0] count; //number of characters (max 32)
output tx;

localparam RDY=2'b00, SEND=2'b01, WAITING=2'b10;

reg [1:0] state=RDY;

integer index;
reg send_prev=0;
reg txsend=0;
reg [7:0]txdata;
wire txready;
reg txready_prev;

UART_tx_ctrl_0 TX(txready,tx,txsend,txdata,clk);

always @ (posedge clk)
begin
txready_prev <= txready;
send_prev <= send;

case (state)
RDY:
        if (send_prev == 0 && send == 1)
        begin
        index <= count;
        state <= SEND;
        end
SEND:
        begin
        txsend <= 1'b1;
        txdata <= word[index*8-1 -: 8];
        state <= WAITING;
        end
WAITING:
        if (txready_prev == 1 && txready == 0)
        begin
        txsend <= 1'b0;
        index <= index - 1'b1;
        end
        else if (txready_prev == 0 && txready == 1)
        begin
        if (index == 0)
        state <= RDY;
        else state <= SEND;
        end
endcase
end

endmodule
```

Listing 12.23 VHDL Description of the Top Module for the I²C Application

```vhdl
library ieee;
use ieee.std_logic_1164.all;
use ieee.numeric_std.all;

entity I2C_compass is
generic(slv_addr : std_logic_vector (6 downto 0) := "0011110");
port(clk : in std_logic;
    btnC : in std_logic;
    drdy : in std_logic;
     scl : inout std_logic;
     sda : inout std_logic;
    RsTx : out std_logic);
end I2C_compass;

architecture behavioral of I2C_compass is

signal reset_n  : std_logic := '0';
signal ena      : std_logic := '0';
signal addr     : std_logic_vector (6 downto 0);
signal rw       : std_logic;
signal data_wr  : std_logic_vector (7 downto 0);
signal busy     : std_logic;
signal pck_cnt  : integer range 0 to 15 := 0;
signal data_rd  : std_logic_vector (7 downto 0);
signal ack_error: std_logic;
signal eop      : std_logic;
signal eop_prev : std_logic;
signal btnCclr  : std_logic;
signal x_axis   : std_logic_vector (15 downto 0) := (others => '0');
signal y_axis   : std_logic_vector (15 downto 0) := (others => '0');
signal z_axis   : std_logic_vector (15 downto 0) := (others => '0');
signal fillzeros: std_logic_vector (207 downto 0) := (others => '0');

--UART signals
signal wordsend  : std_logic := '0';
signal word      : std_logic_vector (255 downto 0);
signal wordlength: std_logic_vector (5 downto 0);
signal drdy_prev : std_logic;

component I2C_leader
generic(input_clk : integer;
         bus_clk : integer);
port(clk : in std_logic;
 reset_n : in std_logic;
     ena : in std_logic;
    addr : in std_logic_vector (6 downto 0);
      rw : in std_logic;
 data_wr : in std_logic_vector (7 downto 0);
    busy : out std_logic;
 data_rd : out std_logic_vector (7 downto 0);
ack_error: buffer std_logic;
     eop : out std_logic;
     sda : inout std_logic;
     scl : inout std_logic);
```

```vhdl
end component;

component debounce_0
port(clk : in std_logic;
     btn : in std_logic;
 btn_clr : out std_logic);
end component;

component UART_word_tx_0
port(clk : in std_logic;
    send : in std_logic;
    word : in std_logic_vector (255 downto 0);
   count : in std_logic_vector (5 downto 0);
      tx : out std_logic);
end component;

begin
process(clk)
begin
if rising_edge(clk) then
  if btnCclr='1' then
  reset_n <= '0';
  ena <= '0';
  pck_cnt <= 0;
  else
  reset_n <= '1';
  end if;
eop_prev <= eop;
if eop_prev = '0' and eop = '1' then
  pck_cnt <= pck_cnt + 1;
  end if;

case pck_cnt is
when 0 => --START
        ena <= '1';
        addr <= slv_addr;
        rw <= '0';
        data_wr <= x"02";
when 1 => --WRITE1
        data_wr <= x"00";
when 2 => --WAITDATA
        if drdy = '0' then
        ena <= '1';
        pck_cnt <= pck_cnt + 1;
        else
        ena <= '0';
        end if;
when 3 => --READDATA
        rw <= '1';
when 4 => --READXH
        if eop = '0' then x_axis(15 downto 8) <= data_rd; end if;
when 5 => --READXL
        if eop = '0' then x_axis(7 downto 0) <= data_rd; end if;
when 6 => --READYH
        if eop = '0' then y_axis(15 downto 8) <= data_rd; end if;
```

```vhdl
    when 7 => --READYL
          if eop = '0' then y_axis(7 downto 0) <= data_rd; end if;
    when 8 => --READZH
          if eop = '0' then z_axis(15 downto 8) <= data_rd; end if;
    when 9 => --READZL
          ena <= '0';
          if eop = '0' then z_axis(7 downto 0) <= data_rd; end if;
          if drdy = '1' then pck_cnt <= 0; end if; --RETURNS TO START
    when others =>
          pck_cnt <= 0;
    end case;
    end if;
    end process;

    process (clk)
    begin
    if rising_edge(clk) then
      drdy_prev <= drdy;
      if drdy_prev = '0' and drdy = '1' then
      word <= fillzeros & x_axis & y_axis & z_axis;
      wordlength <= "000110";
      wordsend <= '1';
    else
      wordsend <= '0';
    end if;
    end if;
    end process;

    I2C_comp : I2C_leader
    generic map(input_clk => 100000000, bus_clk => 400000)
    port map(clk => clk, reset_n => reset_n, ena => ena, addr => addr, rw
        => rw, data_wr => data_wr, busy => busy, data_rd => data_rd,
        ack_error => ack_error, eop => eop, sda => sda, scl => scl);

    db1 : debounce_0 port map(clk => clk, btn => btnC, btn_clr => btnCclr);

    TX : UART_word_tx_0 port map(clk => clk, send => wordsend, word =>
        word, count => wordlength, tx => RsTx);

    end behavioral;
```

Listing 12.24 VHDL Description of the UART_word_tx Module

```vhdl
library ieee;
use ieee.std_logic_1164.all;
use ieee.std_logic_unsigned.all;

entity UART_word_tx is
port(clk : in std_logic;
    send : in std_logic;
     word : in std_logic_vector (255 downto 0);
   count : in integer;
      tx : out std_logic);
end UART_word_tx;
```

```vhdl
architecture behavioral of UART_word_tx is

type state_type is (RDY, SENDX, WAITING);
signal state     : state_type := RDY;
signal index     : integer;
signal send_prev: std_logic;
signal txsend    : std_logic;
signal txdata    : std_logic_vector (7 downto 0);
signal txready   : std_logic;
signal txready_prev: std_logic;

component UART_tx_ctrl_0 is
port(ready : out std_logic;
    uart_tx: out std_logic;
      send : in std_logic;
      data : in std_logic_vector (7 downto 0);
       clk : in std_logic);
end component;

begin
process(clk)
begin
if rising_edge(clk) then
txready_prev <= txready;
send_prev <= send;
case state is
when RDY =>
    if send_prev = '0' and send = '1' then
    index <= count;
    state <= SENDX;
    end if;
when SENDX =>
    txsend <= '1';
    txdata <= word(index*8-1 downto (index-1)*8);
    state <= WAITING;
when WAITING =>
    if txready_prev = '1' and txready = '0' then
    txsend <= '0';
    index <= index - 1; end if;
    if txready_prev = '0' and txready = '1' then
        if index = 0 then
        state <= RDY;
        else state <= SENDX;
        end if;
    end if;
end case;
end if;
end process;

TXW : UART_tx_ctrl_0 port map (ready => txready,uart_tx => tx,send =>
    txsend,data => txdata,clk => clk);

end behavioral;
```

12.4 Video Graphics Array

The video graphics array (VGA) is a display standard used in the CRT and LCD monitors. The Basys3 board has a VGA port as mentioned in Chap. 3. We will use it to develop projects in Verilog and VHDL in this section. Let's start with the working principles of the VGA.

12.4.1 Working Principles of VGA

In the VGA, the display is formed of pixels (picture elements). These are grouped into horizontal lines. Horizontal lines placed on the screen form a frame. Therefore, a pixel location has both horizontal and vertical coordinates. One standard VGA display size is 640×480 pixels. This should be read as follows. The display is formed of 480 horizontal lines each holding 640 pixels.

The time needed to display a single pixel is determined by a pixel clock. Hence, pixels in a horizontal line are displayed by the successive clock signals. When end of the line is reached, the display should continue with a new line. This is set by the horizontal synchronization signal. When all lines in a frame are displayed, a new frame should be formed. This is set by the vertical synchronization signal which also defines the refresh rate of display. The horizontal and vertical synchronization signals depend on pixel clock by definition. Moreover, the monitor needs some time before applying horizontal and vertical synchronization signals. This is called front porch. Similarly, we should wait for a certain amount of time after displaying pixels in a horizontal line and frame. This is called back porch. More information on the VGA timing can be found in [52].

Every pixel has red, green, and blue (RGB) values in the VGA. As mentioned in Chap. 3, the VGA connector on the Basys3 board allows these RGB values to be represented by at most 12 bits. In this scenario, the RGB values get four bits each. Hence, a pixel can have one of $2^4 \times 2^4 \times 2^4 = 4096$ different colors. One can also use eight bits to represent the RGB values. Then, the RGB values get three, three, and two bits, respectively. Hence, a pixel can have one of $2^3 \times 2^3 \times 2^2 = 256$ different colors.

12.4.2 VGA in Verilog

We can display an image using the VGA connection of the Basys3 board. To do so, we first provide the Verilog description of the VGA module in Listing 12.25. This module works in connection with the distributed ROM and clock divider modules. We will introduce such a complete application in Sec. 12.4.4.

The inputs to the VGA module are `clk25`, `pixel_data`, `sx`, and `sy`. `clk25` represents the clock signal fed to the VGA module. For our case, it will be 25 MHz. `pixel_data` represents the vector holding RGB pixel values to be displayed. The VGA module is set to work with eight-bit data. Hence, the RGB values get three, three, and two bits, respectively, as mentioned before. `sx` and `sy` represent the image size to be displayed. Although the default display size in VGA module is 640×480 pixels, it is not possible to keep such an image in the Artix-7 FPGA block or distributed ROM. Therefore, we can set `sx` and `sy` to 80 and 87 pixels at most. The outputs of the VGA module are `red`, `green`, `blue`, `Hsync`, `Vsync`, and `pixel_address`. The outputs `red`, `green`, and `blue` represent the pixel color values each being three, three, and two bits, respectively. `Hsync` and `Vsync` represent horizontal and vertical synchronization signals. Finally, `pixel_address` represents the address of the pixel to be fed to the accompanying ROM module.

Listing 12.25 Verilog Description of the VGA Module

```verilog
module VGA_module (clk25,pixel_data,sx,sy,red,green,blue,Hsync,Vsync,
    pixel_addr);

input clk25;
input [7:0] pixel_data;
input [7:0] sx, sy;
output reg [2:0] red, green, blue;
output reg Hsync, Vsync;
output reg [12:0] pixel_addr;

localparam HDISP=640;
localparam HFP=16;
localparam HPW=96;
localparam HLIM=800;

localparam VDISP=480;
localparam VFP=10;
localparam VPW=2;
localparam VLIM=525;

reg [10:0] hcount=0;
reg [10:0] vcount=0;
reg enable=0;

always@(posedge clk25)
begin

if (hcount < HLIM-1)
    hcount <= hcount+1;
    else
    begin
    hcount<=0;
    if (vcount < VLIM-1)
    vcount <= vcount+1;
    else
    vcount <= 0;
    end

if (vcount > sy)
    begin
    pixel_addr<=-1;
    enable <= 0;
    end
    else
    begin
    if (hcount < sx)
    begin
    enable <= 1;
    pixel_addr<=pixel_addr+1;
    end
    else
    enable <= 0;
    end
```

```
if (enable==1)
   begin
   red <= pixel_data[2:0];
   green <= pixel_data[5:3];
   blue  <= pixel_data[7:6];
   end
   else
   begin
   red   <= 3'b000;
   green <= 2'b00;
   blue  <= 2'b00;
   end

if (hcount > (HDISP+HFP) && hcount <=(HDISP+HFP+HPW))
   Hsync <= 0;
   else
   Hsync <= 1;

if (vcount >= (VDISP+VFP) && vcount < (VDISP+VFP+VPW))
   Vsync <= 0;
   else
   Vsync <= 1;

end
endmodule
```

The working principles of the VGA module in Listing 12.25 are as follows. The module is set to work with 640×480 pixels. These values are represented as local parameters in the module. Similarly, front and back porch values for horizontal and vertical synchronization signals are set as local parameters. Based on these values, the maximum horizontal and vertical display limits are set as 800 and 525 pixels as local parameters within the module. We assume that the input clock to the module (clk25) has 25-MHz frequency. Based on these values, the refresh time of a frame is $800 \times 600/25 \times 10^{-6}$ s. Hence, the refresh rate of the display becomes 52 Hz, which is a suitable value. The VGA module calculates the pixel address to be displayed, whether to generate horizontal and vertical synchronization signals, and RGB values to be used in display. All these operations depend on accurate and synchronous timing calculations. Besides, no detailed operation is performed.

12.4.3 VGA in VHDL

The VHDL version of the VGA module introduced in the previous section is presented in Listing 12.26. Within this module, we set all input, output, and parameter names the same as presented in Listing 12.25. Besides, the working principles of both modules are also the same. Hence, the reader can follow the explanation in the previous section for the VHDL version of the VGA module as well.

12.4.4 VGA Application

We can use the VGA module (in Verilog or VHDL) in a simple application to show its working principles. This application displays an RGB image on the display connected to VGA port of the Basys3 board. We first provide the Verilog version of the top module for the application in Listing 12.27.

Listing 12.26 VHDL Description of the VGA Module

```vhdl
library ieee;
use ieee.std_logic_1164.all;
use ieee.numeric_std.all;

entity VGA_module is
port(clk25 : in std_logic;
pixel_data : in std_logic_vector (7 downto 0);
        sx : in std_logic_vector (9 downto 0);
        sy : in std_logic_vector (9 downto 0);
       red : out std_logic_vector (2 downto 0);
     green : out std_logic_vector (2 downto 0);
      blue : out std_logic_vector (1 downto 0);
     Hsync : out std_logic;
     Vsync : out std_logic;
pixel_addr : out std_logic_vector (12 downto 0));
end VGA_module;

architecture behavioral of VGA_module is
constant HDISP : integer := 640;
constant HFP   : integer := 16;
constant HPW   : integer := 96;
constant HLIM  : integer := 800;

constant VDISP : integer := 480;
constant VFP   : integer := 10;
constant VPW   : integer := 2;
constant VLIM  : integer := 525;

signal hcount : unsigned (9 downto 0) := (others => '0');
signal vcount : unsigned (9 downto 0) := (others => '0');

signal enable : std_logic := '0';
signal pixcount : unsigned (12 downto 0);

begin
pixel_addr <= std_logic_vector (pixcount);

process(clk25)
begin
if rising_edge(clk25) then

if hcount < HLIM-1 then
    hcount <= hcount+1;
    else
    hcount<=(others => '0');
    if vcount < VLIM-1 then
    vcount <= vcount+1;
    else
    vcount <= (others => '0');
    end if;
end if;

if vcount > unsigned(sy) then
    pixcount<="1111111111111";
```

```
        enable <= '0';
        else

if hcount < unsigned(sx) then
        enable <= '1';
        pixcount<=pixcount+1;
        else enable <= '0';
        end if;
end if;

if enable = '1' then
        red <= pixel_data(2 downto 0);
        green <= pixel_data(5 downto 3);
        blue <= pixel_data(7 downto 6);
        else
        red <= (others => '0');
        green <= (others => '0');
        blue  <= (others => '0');
end if;

if hcount> (HDISP+HFP) and  hcount <=(HDISP+HFP+HPW) then
        Hsync <= '0';
        else
        Hsync <= '1';
end if;

if vcount >= (VDISP+VFP) and vcount < (VDISP+VFP+VPW) then
        Vsync <= '0';
        else
        Vsync <= '1';
end if;

end if;
end process;
end behavioral;
```

The top module in Listing 12.27 has input and output values directly set for the Basys3 board. Hence, XDC file namings are used there. Besides, the top module has three submodules: clock, memory, and VGA. The VGA submodule is the one in Listing 12.25. The clock submodule is for dividing 100-MHz clock of the Basys3 board to 25 MHz to be used in the VGA submodule. To do so, we used "Clocking Wizard IP" which can be found in IP Catalog → FPGA Features and Design → Clocking. With its simple interface, this IP block allows generating a frequency divider. The image to be displayed is kept in the distributed ROM as explained in Sec. 9.5. Here, the image is saved in ROM as an initialization file in coe format. Here, we suggest a two-step operation. We provide the MATLAB file to generate a text file from a given TIFF image on this book's companion website, www.mhprofessional.com/1259837904. First, the reader can convert the image file to suitable text format via this file. Second, this file should be converted to coe format as explained in Sec. 9.5. This way, the image of interest can be included to the ROM module. As the project is built and the VGA port is connected to the display, the image should be seen on it.

We also provide the VHDL version of the top module for the VGA application in Listing 12.28. This module has the same working principles as presented in Listing 12.27. Therefore, the explanations there directly apply to it as well.

Listing 12.27 Top Module of the VGA Application in Verilog

```verilog
module VGA_top_module(clk,vgaRed,vgaGreen,vgaBlue,Hsync,Vsync);

input clk;
output [2:0] vgaRed;
output [2:0] vgaGreen;
output [1:0] vgaBlue;
output Hsync;
output Vsync;

//image size
parameter sx=80;
parameter sy=80;

wire clk_25;
wire [12:0] pixel_addr;
wire [7:0] pixel_data;

clk_wiz_0 clock (clk25, clk);

dist_mem_gen_0 memory(.a(pixel_addr),.spo(pixel_data));

VGA_module VGA(
.clk25(clk25),
.pixel_data(pixel_data),
.sx(sx),
.sy(sy),
.red(vgaRed),
.green(vgaGreen),
.blue(vgaBlue),
.Hsync(Hsync),
.Vsync(Vsync),
.pixel_addr(pixel_addr));

endmodule
```

Listing 12.28 Top Module of the VGA Application in VHDL

```vhdl
library ieee;
use ieee.std_logic_1164.all;

entity VGA_top_module is
port(clk : in std_logic;
  vgaRed : out std_logic_vector (2 downto 0);
vgaGreen : out std_logic_vector (2 downto 0);
 vgaBlue : out std_logic_vector (1 downto 0);
   Hsync : out std_logic;
   Vsync : out std_logic);
end VGA_top_module;

architecture behavioral of VGA_top_module is
--image size sx = 80, sy = 80;
constant sx : std_logic_vector (9 downto 0) :="0001010000";
constant sy : std_logic_vector (9 downto 0) :="0001010000";
```

```vhdl
signal clk_25 : std_logic;
signal pixel_data : std_logic_vector (7 downto 0) ;
signal pixel_addr : std_logic_vector (12 downto 0);

component clk_wiz_0
port(clk_in1 : in std_logic;
    clk_out1 : out std_logic);
end component;

component dist_mem_gen_0
port(a : in std_logic_vector (12 downto 0);
   spo : out std_logic_vector (7 downto 0));
end component;

component VGA_module
port(clk25 : in std_logic;
pixel_data : in std_logic_vector (7 downto 0);
       sx : in std_logic_vector (9 downto 0);
       sy : in std_logic_vector (9 downto 0);
      red : out std_logic_vector (2 downto 0);
    green : out std_logic_vector (2 downto 0);
     blue : out std_logic_vector (1 downto 0);
    Hsync : out std_logic;
    Vsync : out std_logic;
pixel_addr : out std_logic_vector (12 downto 0));
end component;

begin

clock: clk_wiz_0 PORT MAP (
clk_in1 => clk, clk_out1 => clk_25);

memory: dist_mem_gen_0 PORT MAP(
a =>pixel_addr,spo =>pixel_data);

VGA: VGA_module port map(
clk25 => clk_25,
pixel_data => pixel_data,
sx => sx, sy => sy,
red => vgaRed,
green => vgaGreen,
blue => vgaBlue,
Hsync => Hsync,
Vsync => Vsync,
pixel_addr => pixel_addr);

end behavioral;
```

12.5 Universal Serial Bus

The Universal Serial Bus (USB) is an industry standard developed to unify connection, communication, and power supply between digital devices. It can be used between a PC and keyboard, mouse, external hard drive, printer, and digital camera. The Basys3 and Arty boards have USB ports used for powering and programming purposes as explained in Chap. 3. The Basys3 board also has an extra USB port which can be used to connect peripherals such as keyboard. In low level, the USB operations are not easy to manage. Fortunately, the Basys3 board has a PIC24FJ128 chip which provides USB HID host capability as mentioned in Chap. 3. We will use this option to develop the HDL projects here. Specifically, we will focus on interfacing a keyboard to Basys3 board since the PIC24FJ128 chip available on the board converts the USB input to standard PS/2 signals to communicate with a mouse or keyboard. Here, the Basys3 board will be the receiver. The keyboard will be the transmitter. Therefore, we will focus only on the USB-receiving module next.

12.5.1 USB-Receiving Module in Verilog

The Verilog description of the USB-receiving module is given in Listing 12.29. The inputs of this module are `ps2data` and `ps2clk`. The outputs of the module are `data` and `ready`. The serial data on `ps2data` is received in every falling edge of `ps2clk`. Data is stored in `data` output and `ready` goes to logic level 1 once all bits are received and verified by parity check.

The USB-receiving module (represented as a state machine) has four states: RDY, RECEIVE, PARITY, and STOP. The state machine starts at RDY state initially. In every falling edge of `ps2clk`, it checks whether `ps2data` is at logic level 0, which indicates the start of a new transmission. If `ps2data` is at logic level 0, `ready` goes to logic level 0; index is reset; and state changes to RECEIVE. After the start bit, in every falling edge of `ps2clk`, data at `ps2data` is received serially by incrementing `index`. Data is stored temporarily in vector `received`. When `index` reaches seven, the last bit is received and state changes to PARITY where the parity bit is stored in `prty`. Then, in STOP state if the `ps2data` is at logic level 1, `ready` goes to logic level 1. Here, parity check is performed, such as if the parity of received data matches received parity bit, data is written to the output vector. Otherwise the hexadecimal number EE is written to the output vector to indicate a parity error.

12.5.2 USB-Receiving Module in VHDL

The VHDL version of the USB-receiving module is given in Listing 12.30. Here, the working principles of both Verilog and VHDL descriptions are the same. Therefore, the reader can check the Verilog description to understand the VHDL version.

12.5.3 USB Keyboard Application

As a USB application, we will read numeric data (1 to 8) coming from a keyboard connected to the USB port of the Basys3 board. The received data (as a keyscan code) will be displayed on eight LEDs (from 15 to 8) of the board. If any number from 1 to 8 is pressed on the keyboard, the corresponding LED (7 to 0) will toggle. Hence, this application aims to show how a keyscan code can be read from a USB keyboard.

The keyscan code table of the PS/2 keyboard can be found in [53]. It is straightforward to get data from the keyboard and check if it matches a number between one to

Listing 12.29 Verilog Description of the USB Receiver Module

```verilog
module USB_keyboard(ps2data,ps2clk,data,ready);

input ps2data;
input ps2clk;
output reg [7:0] data;
output reg ready=0;

parameter    RDY=2'b00, RECEIVE=2'b01, PARITY=2'b10, STOP=2'b11;

reg [2:0] state = RDY;
reg [7:0] received;
reg prty;
integer index=0;

always @ (negedge ps2clk)

case (state)
RDY:
        if (ps2data == 1'b0)
        begin
        state <= RECEIVE;
        ready <= 1'b0;
        index <= 0; end
RECEIVE:
        begin
        if (index == 7) state <= PARITY;
        received[index] <= ps2data;
        index <= index + 1;
        end
PARITY:
        begin
        prty <= ps2data;
        state <= STOP;
        end
STOP:
        if (ps2data == 1'b1) begin
        state <= RDY;
        ready <= 1'b1;
        if (prty != ~(^received)) data <= 8'hEE;
        else data <= received; end
endcase
endmodule
```

eight. However, reading a keyboard button data requires some processing. Therefore, let's first focus on this operation. Whenever a button is pressed on the keyboard, its make code should be sent. The make code is the eight-bit code you see on the reference keyscan code table. Once the button is released, a break code is transmitted. The break code has a specific eight-bit code (the hexadecimal number F0 for characters and numbers) followed by the same make code of the button. Let's explain how this is done by an example. The hexadecimal keyscan code of the button for number one is 16. When the button is pressed, the make code of the button is sent immediately. Once the button is released, the keyboard sends the break code as hexadecimal number F0 and 16 again.

Listing 12.30 VHDL Description of the USB Receiver Module

```vhdl
library ieee;
use ieee.std_logic_1164.all;

entity USB_keyboard is
port(ps2data : in std_logic;
     ps2clk : in std_logic;
       data : out std_logic_vector (7 downto 0);
      ready : out std_logic);
end USB_keyboard;

architecture behavioral of USB_keyboard is

type state_type is (RDY, RECEIVE, PARITY, STOP);

signal state   : state_type := RDY;
signal received: std_logic_vector (7 downto 0);
signal prty    : std_logic;
signal index   : integer := 0;

begin
process(ps2clk)
begin
if falling_edge(ps2clk) then

case state is
when RDY =>
        if (ps2data = '0') then
        state <= RECEIVE;
        ready <= '0';
        index <= 0;
        end if;
when RECEIVE =>
        if (index = 7) then state <= PARITY; end if;
        received(index) <= ps2data;
        index <= index + 1;
when PARITY =>
        prty <= ps2data;
        state <= STOP;
when STOP =>
        if (ps2data = '1') then
        state <= RDY;
        ready <= '1';
        if (prty /= xnor received) then data <= x"EE";
        else data <= received; end if;
        end if;
end case;
end if;
end process;

end behavioral;
```

Therefore, for the normal button press and release, we should get the corresponding hexadecimal make code, number F0, and make code again. If the button is pressed and held down, this key is called typematic. If the button press exceeds the typematic delay time of the keyboard, it continues sending the make code repeatedly until the button is released. Once the button is released, the transmission concludes with the break code. This should be taken into account in reading a button press from the keyboard.

We provide the Verilog description of the top module of the application in Listing 12.31. The inputs of the top module are clk, PS2Clk, and PS2Data. PS2Clk and PS2Data are directly connected to the USB port of the Basys3 board via its XDC file. clk is the master clock of the board. The output of the module is 16-element led vector. Again, this output is connected to LEDs on the Basys3 board via its XDC file.

Listing 12.31 Verilog Description of the USB Keyboard Application

```verilog
module USB_keyboard_app(clk,PS2Clk,PS2Data,led);

input clk;
input PS2Clk;
input PS2Data;
output [15:0] led;

localparam PRESS=2'b00, EXTEND=2'b01, RLS=2'b10, CHECK=2'b11;

reg [1:0] state = PRESS;
reg [23:0] received;
wire [7:0] data;
reg [7:0] ledreg=0;
wire ready;
reg ready_prev;

USB_keyboard kb1(.ps2data(PS2Data),.ps2clk(PS2Clk),.data(data),.ready
    (ready));

always @ (posedge clk)
begin
ready_prev <= ready;

case (state)
PRESS:
        if (ready_prev==0 && ready==1)
        begin
        received[23:16] <= data;
        state <= EXTEND;
        end
EXTEND:
        if (ready_prev==0 && ready==1)
        begin
        received[15:8] <= data;
        state <= RLS;
        end
RLS:
        if (received[15:8] != 8'hF0)
        state <= EXTEND;
        else if (ready_prev==0 && ready==1)
```

```
          begin
          received[7:0] <= data;
          state <= CHECK;
          end
CHECK:
          begin
          case (received[7:0])
          8'h16:   ledreg[0] <= ~ledreg[0];   // toggle if 1 is pressed
          8'h1E:   ledreg[1] <= ~ledreg[1];   // toggle if 2 is pressed
          8'h26:   ledreg[2] <= ~ledreg[2];   // toggle if 3 is pressed
          8'h25:   ledreg[3] <= ~ledreg[3];   // toggle if 4 is pressed
          8'h2E:   ledreg[4] <= ~ledreg[4];   // toggle if 5 is pressed
          8'h36:   ledreg[5] <= ~ledreg[5];   // toggle if 6 is pressed
          8'h3D:   ledreg[6] <= ~ledreg[6];   // toggle if 7 is pressed
          8'h3E:   ledreg[7] <= ~ledreg[7];   // toggle if 8 is pressed
          endcase
          state <= PRESS;
      end
endcase
end

assign led = {received[7:0],ledreg};

endmodule
```

Top module for the USB application can be taken as a state machine with four states: PRESS, EXTEND, RLS, and CHECK. Based on these, we can explain the working principles of the top module as follows. In every rising edge of clk, ready signal from the USB_keyboard module is checked to detect its rising edge. In PRESS state, if ready goes to logic level 1 (when a button is pressed) the eight-bit receiving data (make code) is stored in the most significant eight bits of 24-bit received vector and the state changes to EXTEND. The top module waits for another rising edge of ready which can be a repeated make code (when the button is still pressed) or the first byte of the break code which is the hexadecimal number F0 if the button is released. Here, the received data is written to the middle eight bits of received vector and the state is changed to RLS. In this state, the previously received data is checked for whether the button is released or not before waiting for the next ready signal. If the last received data is the hexadecimal number F0, this means that the button is released and we are ready to check if the pressed button was one of the numbers from one to eight. Otherwise, we should understand that the button is still pressed and we should go back to EXTEND state to wait for the button release. This depends on the designer. For a continuous press, you can send multiple characters or wait for the button release and send the character only once. We selected the latter option in our module. Hence, if the button is released, the last eight-bit data will be received and written to the least significant eight bits of the received vector. In CHECK state, the corresponding LED on the Basys3 board is toggled if the pressed button was one of the numbers in the keyboard from 1 to 8. The received data is also displayed on LEDs 15 to 8. The state of the machine turns back to PRESS unconditionally and waits for another ready signal.

The VHDL description of the USB keyboard application is given in Listing 12.32. This description is in line with its Verilog counterpart. Therefore, the explanation there applies to this description as well.

Listing 12.32 VHDL Description of the USB Keyboard Application

```vhdl
library ieee;
use ieee.std_logic_1164.all;

entity USB_keyboard_app is
port(clk : in std_logic;
  PS2Clk : in std_logic;
 PS2Data : in std_logic;
     led : out std_logic_vector (15 downto 0));
end USB_keyboard_app;

architecture behavioral of USB_keyboard_app is

type state_type is (PRESS, EXTEND, RLS, CHECK);

signal state     : state_type := PRESS;
signal received  : std_logic_vector (23 downto 0);
signal data      : std_logic_vector (7 downto 0);
signal ledreg    : std_logic_vector (7 downto 0);
signal ready     : std_logic;
signal ready_prev: std_logic;

component USB_keyboard is
port(ps2data : in std_logic;
       ps2clk : in std_logic;
         data : out std_logic_vector (7 downto 0);
        ready : out std_logic);
end component;

begin
process (clk)
begin
if rising_edge(clk) then
ready_prev <= ready;

case state is
when PRESS =>
  if (ready_prev='0' and ready='1') then
  received(23 downto 16) <= data;
  state <= EXTEND;
  end if;
when EXTEND =>
  if (ready_prev='0' and ready='1') then
  received(15 downto 8) <= data;
  state <= RLS;
  end if;
when RLS =>
  if (received(15 downto 8) /= x"F0") then
  state <= EXTEND;
  elsif (ready_prev='0' and ready='1') then
  received(7 downto 0) <= data;
  state <= CHECK;
  end if;
when CHECK =>
case received(7 downto 0) is
```

```
when x"16" => ledreg(0) <= not ledreg(0); --toggle if 1 is pressed
when x"1E" => ledreg(1) <= not ledreg(1); --toggle if 2 is pressed
when x"26" => ledreg(2) <= not ledreg(2); --toggle if 3 is pressed
when x"25" => ledreg(3) <= not ledreg(3); --toggle if 4 is pressed
when x"2E" => ledreg(4) <= not ledreg(4); --toggle if 5 is pressed
when x"36" => ledreg(5) <= not ledreg(5); --toggle if 6 is pressed
when x"3D" => ledreg(6) <= not ledreg(6); --toggle if 7 is pressed
when x"3E" => ledreg(7) <= not ledreg(7); --toggle if 8 is pressed
when others => ledreg <= ledreg;
end case;
    state <= PRESS;
end case;
end if;
end process;

led <= received(7 downto 0) & ledreg;

kb1 : USB_keyboard port map(ps2data => PS2Data, ps2clk => PS2Clk, data
    => data, ready => ready);

end behavioral;
```

12.6   Ethernet

Ethernet is an industry standard for local area network (LAN). Each device in the network has a unique address called internet protocol (IP) address. IP is the the main communication protocol in networking and it essentially forms the Internet. The data communication is established on IP address headers followed by data signals. Unfortunately, IP has a complicated structure which makes it hard to implement it at the lowest level.

To explain the working principles of ethernet and how it can be implemented on an FPGA, we benefit from the application in [54]. Here, an echo server is developed on the Arty board since it has an integrated ethernet connector as explained in Chap. 3. As all the steps explained in the mentioned website are followed, and the project is implemented on the board, every character you write to the terminal echoes back to your IP terminal. This application can be expanded further for more advanced applications.

12.7   FPGA Building Blocks Used in Digital Interfacing

The digital interfacing concepts introduced in this section are both complex and diverse. Therefore, we can assume that almost all the FPGA blocks introduced in previous chapters are used in either one or two applications. Note that the digital interfacing applications introduced in this chapter are not unique. In other words, HDL modules given in this chapter are not unique for a given interfacing option. Hence, comparing different digital interfacing options based on their FPGA resource usage will not be fair. As a result, we ask the reader to check his or her digital interfacing application's FPGA usage.

12.8 Summary

The digital interfacing is becoming more and more important with the introduction of the Internet of things in which communication of two or more embedded devices is a fundamental necessity. Therefore, we started with serial communication methods as UART, SPI, and I^2C as well as advanced ethernet protocol in this chapter. These communication methods can also be used in interfacing the FPGA with sensor chips using them. Then, we introduced other important interfacing methods, the VGA and USB. The VGA allows the user to display images on a monitor. Hence, it opens up a way for more advanced applications. In the same way, using a USB connection allows the reader to benefit from more advanced peripheral devices such as a keyboard or mouse. As a result, methods introduced in this chapter can improve the quality of the application to be developed on the FPGA. We will provide such examples in the following chapters.

12.9 Exercises

12.1 Implement the provided UART applications on the Basys3 board in Sec. 12.1 to check how they work.

12.2 Repeat Exercise 12.1 using the Arty board.

12.3 One method of testing the UART transmitter and receiver modules is implementing both on the same FPGA. If these modules are also appropriately connected via their TX and RX pins in a top module, then the UART communication can be simulated on one FPGA chip. Implement such an application on the Basys3 board such that when btnC is pressed, the transmitter module sends a character to the receiver module. As this character is received, led[0] on the Basys3 board toggles.

12.4 Modify the application in Exercise 12.3 to work on the Arty board.

12.5 Implement the provided SPI application on the Basys3 board in Sec. 12.2 to check how it works.

12.6 Repeat Exercise 12.5 using the Arty board.

12.7 Repeat Exercise 12.3 using the SPI communication on the Basys3 or Arty board.

12.8 Repeat Exercise 12.3 using the I^2C communication on the Basys3 or Arty board.

12.9 Implement the provided VGA application on the Basys3 board in Sec. 12.4 to check how it works.

12.10 Modify the VGA application in Sec. 12.4 such that when a color image is given, its
  a. red band is displayed only.
  b. green band is displayed only.
  c. blue band is displayed only.

12.11 Implement the provided USB application on the Basys3 board in Sec. 12.5 to check how it works.

12.12 Modify the application in Exercise 12.11 such that when a button is pressed on the keyboard, it is displayed on the rightmost seven-segment display of the Basys3 board. Note that numbers can be displayed easily on the seven-segment display. However, characters on the keyboard should be limited such that only a subset of them can be displayed such as "E", "R", "U", and so on.

12.13 Modify the application in Exercise 12.11 such that when a button is pressed on the keyboard, the corresponding character is displayed on the terminal window of the host PC via the UART communication.

12.14 Implement the provided ethernet application on the Arty board in Sec. 12.6 to check how it works.

CHAPTER 13

Advanced Applications

We have used the field-programmable gate array (FPGA) to implement both basic digital systems and sample applications up to this chapter. The FPGA can also be used to develop more functional and advanced digital systems using tools introduced in previous chapters. This chapter will be on such advanced applications. Therefore, we will start with integrated logic analyzer IP core analyzer first. This IP core will allow us to analyze a working digital system on the FPGA chip. Afterward, we will focus on the XADC block usage to process analog signals on the FPGA. Then, we will provide 22 applications of which nine have implementation details here. Remaining applications will only have their description such that they can be implemented on the FPGA.

13.1 Integrated Logic Analyzer IP Core Usage

When designing a digital system on an FPGA, we may need to observe internal signals of the design. Vivado offers a way to achieve this by integrated logic analyzer (ILA) IP core usage. This core acts as an actual logic analyzer for monitoring signals in the FPGA. Since this is a very important topic for actual digital system realization, we focus on it in this section. We pick a simple Verilog project on the Basys3 board to show how ILA can be used. For more information on this issue, please see [55, 56].

Let's take an example project to blink the rightmost LED on the Basys3 board every second. We provide the Verilog description for this project in Listing 13.1. In this module, there is a clock divider producing output clk1 from the main clock of the Basys3 board. Assume that we would like to observe this signal using ILA.

We can observe the signal clk1 using ILA applying the following steps before synthesizing the project. In Flow Navigator, select Synthesis → Set up Debug. In the opening Set Up Debug window, add the signal clk1 to the list. This can be done in several ways. The easiest way is dragging and dropping it from the Netlist window in Project Explorer. Make sure that the probe type is selected as "Data and Trigger". As we press next, a new window titled ILA Core Options appears asking for the ILA features. The user can select the sample data size in this tab. We should also tick Capture control box in this tab. As we press Next, the final window appears titled as Set up Debug Summary. Pressing Finish button at this window finalizes the debug setup. Afterward, we should generate bitstream and embed it on the FPGA of the Basys3 board. Different from steps explained in Sec. 4.6.2, there will be two files to be embedded now, one for the actual implementation (Bitstream file), the other for debugging (Debug probes file). Afterward, the ILA window appears in the project explorer window. As we press "Run trigger for this ILA core" button, ILA starts working and the result is displayed on the screen. The user can also export this result by right clicking (and selecting Export ILA data) on the signal of

---

Listing 13.1 Verilog Description of the Example Project for ILA Usage

```verilog
module ILA_usage(led,clk);

input clk;
output reg [0:0] led =0;

reg [2:0] Fout;
reg clk1;

reg [23:0] count;

initial
begin
Fout=3'b000;
count=24'h000000;
end;

//clock divider
always @ (negedge clk)
begin
Fout <= Fout + 1'b1;
clk1 <=Fout[2];
end

//blink the LED
always @ (posedge clk1)
begin
count <= count + 1'b1;
if (count ==24'd12500000)
begin
led[0]<=~led[0];
count <=0;
end;

end
endmodule
```

---

interest in the ILA window. In the opening window, the user should enter the target location and file format type for the export operation. We suggest using VCD format for exporting data. Hence, it can be opened as a text file.

13.2   The XADC Block Usage

Processing an analog signal in digital system requires analog-to-digital conversion as the first step. The Artix-7 FPGA has a specific XADC block for this purpose as mentioned in Sec. 2.2.8. This block is connected to the JXADC Pmod port of the Basys3 board. It is capable of converting four external differential signals to digital form since the port has four differential pins. Also, the XADC block has an internal temperature sensor which can be selected to read its output. In this section, we will focus on the usage of the XADC block with two applications. The first application will be on reading temperature value from the internal sensor of the FPGA chip. The second application will be on measuring voltage level of a battery connected to ports of the Basys3 board.

We will benefit from an IP block (XADC Wizard) available in Vivado to use the XADC block. Therefore, we should first create a project and add XADC Wizard to it. To do so, we should select XADC Wizard from the IP Catalog → FPGA Features and Design → XADC. By double-clicking on the IP block, we can open its configuration window. For our two applications, we will remove reset_in and change Startup Channel Selection mode to Channel Sequencer in the Basic tab. We can select which channels to add to the XADC block from the Channel Sequencer tab. For our first application (temperature sensing), we should tick the TEMPERATURE box. For our second application (measuring voltage), we should tick the vauxp6/vauxn6 box. Now, we are ready to integrate the IP block to our project.

After integrating the IP block to the project, you can go ahead and check its instantiation file. Although there are many inputs and outputs, we do not have to use all of them for a basic ADC operation. To be more specific, we will only need the address register (daddr_in), clock in (dclk_in), enable (den_in), data out (do_out), end of conversation signal (eoc_out), and data ready signal (drdy_out) for our two applications. The address register will have value 00 and 16 in hexadecimal form for TEMPERATURE and vauxp6/vauxn6 inputs, respectively. For more information on the usage of the XADC Wizard, please see [5, 57].

As the first application, we will read analog values from the internal temperature sensor of the FPGA. We will show the result on seven-segment display and LEDs of the Basys3 board. Top module for the first application is given in Listing 13.2. This module

Listing 13.2 Verilog Description of the Top Module to Convert Analog Temperature Value to Digital Form

```verilog
module XADC_temperature(clk,led,an,seg);

input clk;
output reg [15:0] led;
output [3:0] an;
output [6:0] seg;

wire enable;
wire ready;
wire [15:0] data;
reg [11:0] temper;

xadc_wiz_0  XLXI_7 (.daddr_in(7'h00),.dclk_in(clk),.den_in(enable),
    .do_out(data),.eoc_out(enable),.drdy_out(ready));

wire [3:0] thos,huns,tens,ones;
binarytoBCD_0 bcd(.binary(temper),.thos(thos),.huns(huns),.tens(tens),
    .ones(ones));
sevenseg_driver_0 seg1(.clk(clk),.clr(1'b0),.in1(thos),.in2(huns),
    .in3(tens),.in4(ones),.seg(seg),.an(an));

always @( posedge(clk))
if(ready == 1'b1)
    begin
    led <= data;
    temper <= ((data[15:4]*504)/4096)-273;
    end
endmodule
```

has one input as clk (the main clock signal on the Basys3 board). The outputs of the module are 16-bit led, seven-segment display outputs seg and an associated with XDC file of the Basys3 board.

In Listing 13.2, we integrate the xadc_wiz_0 module with den_in connected to eoc_out which ensures that the device operates in continuous mode. We provide the temperature value in raw digital form on 16 LEDs of the Basys3 board. We also show the temperature value in Celsius on seven-segment display. Raw digital data can be converted into Celsius form with the help of the conversion formula in [5]. Note that although the output of the XADC module is 16 bits, only most significant 12 bits are valid since the XADC block has the 12-bit resolution.

As the second application, we will read voltage level on an alkaline battery connected to the Basys3 board. Alkaline batteries have 1.5 V when they are fully charged. This value goes down to 1 V when the battery is dead. To measure voltage level on the battery, we should have a voltage divider circuit since analog input of the XADC block accepts at most 1 V. To do so, connect two high-valued resistors in series with one end at the positive terminal of the battery and the other end at the negative terminal. Then, connect the common node of resistors to XA1_P port while connecting negative terminal of the battery to XA1_A of the Basys3 board.

We will show the result on the seven-segment display and LEDs of the Basys3 board. Top module for the second application is given in Listing 13.3. This module has three inputs: clk (the main clock signal on the Basys3 board) and auxiliary inputs vauxp6, vauxn6 which are connected to XA1_P and XA1_A via the XDC file of the Basys3 board. The outputs of the module are 16-bit led, seven-segment display values seg and an associated with the XDC file of the Basys3 board.

In Listing 13.3, we integrate the xadc_wiz_0 module with den_in connected to eoc_out, which ensures that the device operates in continuous mode. We provide the voltage value in raw digital form on 16 LEDs of the Basys3 board. We also show the voltage value (in millivolts) on the seven-segment display. Note that since we applied voltage division, we should have seen only half of the voltage value. However, since we modified the read value in the description, we will see actual voltage levels. Therefore, if the voltage value is around 1500 (mV), this means the battery is full. If the reading is around 1000 (mV), this means the battery is about to die.

13.3 Adding Two Floating-Point Numbers

We have introduced arithmetic operations on floating-point numbers in Chap. 6. We have mentioned there that operations on the floating-point numbers are complex. In this section, we will handle addition operation on two floating-point numbers in half form. We provide the Verilog description of the corresponding module in Listing 13.4. This module has four inputs as follows. clk represents the main clock to be fed to the module. add is for starting the addition operation. number1 and number2 stand for floating-point numbers to be added. The module has two outputs as result and ready. The first one holds the result of the operation. The second one indicates that the operation has ended.

The floating point adder module in Listing 13.4 uses two 42-bit vectors to perform shifting operations on number1 and number2. The adder module is a state machine. When add is at logic level 1, the state machine starts working. Hence, ready is set to logic level 0, and state goes to START. Here, first sign bit of numbers are compared.

Listing 13.3 Verilog Description of the Top Module to Convert External Voltage Value to Digital Form

```verilog
module XADC_voltage(clk,vauxp6,vauxn6,led,an,seg);

input clk;
input vauxp6;
input vauxn6;
output reg [15:0] led;
output [3:0] an;
output [6:0] seg;

wire enable;
wire ready;
wire [15:0] data;
reg [11:0] volt;

xadc_wiz_0  XLXI_7 (.daddr_in(7'h16),.dclk_in(clk),.den_in(enable),
    .vauxp6(vauxp6),.vauxn6(vauxn6),.do_out(data),
    .eoc_out(enable),.drdy_out(ready));

wire [3:0] thos,huns,tens,ones;
binarytoBCD_0 bcd(.binary(volt),.thos(thos),.huns(huns),.tens(tens),
    .ones(ones));
sevenseg_driver_0 seg1(.clk(clk),.clr(1'b0),.in1(thos),.in2(huns),
    .in3(tens),.in4(ones),.seg(seg),.an(an));

always @ (posedge(clk))
if(ready == 1'b1)
  begin
  led <= data;
  volt <= (data[15:4]*1500)/4095;
  end
endmodule
```

Listing 13.4 Adding Two Floating-Point Numbers in Verilog

```verilog
module floating_point_adder(clk,add,number1,number2,result,ready);

input clk;
input add;
input [15:0] number1;
input [15:0] number2;
output reg [15:0] result;
output reg ready;

localparam RDY=3'b000,START=3'b001,NEGPOS=3'b010,OP=3'b011,
    SHIFT=3'b100,WRITE=3'b101,RSLT=3'b110;

reg [3:0] state = RDY;
reg [41:0] bigreg;
reg [41:0] smallreg;
reg [41:0] resultreg;
reg resultsign;
```

```verilog
reg [4:0] resultshift;
reg [9:0] resultfrac;
reg [5:0] pos=0;

integer bigshift;
integer smallshift;
reg [5:0] i;

always @ (posedge clk)

case (state)
RDY:
        if (add)
        begin
        bigreg <= {16'b0,1'b1,25'b0};
        smallreg <= {16'b0,1'b1,25'b0};
        pos <= 6'b0;
        i <= 6'd41;
        ready <= 0;
        state <= START;
        end
START:
        begin
        if (number1[15] != number2[15])
        if (number1[14:10] > number2[14:10])
        begin
        bigreg[24:15] <= number1[9:0];
        bigshift <= number1[14:10] - 4'b1111;
        resultsign <= number1[15];
        smallreg[24:15] <= number2[9:0];
        smallshift <= number2[14:10] - 4'b1111;
        end
        else if (number2[14:10] > number1[14:10])
        begin
        bigreg[24:15] <= number2[9:0];
        bigshift <= number2[14:10] - 4'b1111;
        resultsign <= number2[15];
        smallreg[24:15] <= number1[9:0];
        smallshift <= number1[14:10] - 4'b1111;
        end
        else
        begin
        if (number1[9:0] > number2[9:0])
        begin
        bigreg[24:15] <= number1[9:0];
        bigshift <= number1[14:10] - 4'b1111;
        resultsign <= number1[15];
        smallreg[24:15] <= number2[9:0];
        smallshift <= number2[14:10] - 4'b1111;
        end
        else if (number2[9:0] > number1[9:0])
        begin
        bigreg[24:15] <= number2[9:0];
        bigshift <= number2[14:10] - 4'b1111;
        resultsign <= number2[15];
```

```
            smallreg[24:15] <= number1[9:0];
            smallshift <= number1[14:10] - 4'b1111;
            end
            else result <= {2'b00,4'b1111,10'b0};
            end
            else
            begin
            bigreg[24:15] <= number1[9:0];
            bigshift <= number1[14:10] - 4'b1111;
            resultsign <= number1[15];
            smallreg[24:15] <= number2[9:0];
            smallshift <= number2[14:10] - 4'b1111;
            end
            state <= NEGPOS;
            end
NEGPOS:
            begin
            if (bigshift > 0)
            bigreg <= bigreg << bigshift;
            else bigreg <= bigreg >> ((~bigshift)+1);
            if (smallshift > 0)
            smallreg <= smallreg << smallshift;
            else smallreg <= smallreg >> ((~smallshift)+1);
            state <= OP;
            end
OP:
            begin
            if (number1[15] != number2[15])
            resultreg = bigreg - smallreg;
            else
            resultreg = bigreg + smallreg;
            state <= SHIFT;
            end
SHIFT:
            if (resultreg[i] == 1'b1 || i == 0) begin
            pos = i;
            state <= WRITE;
            end
            else i <= i - 1'b1;
WRITE:
            begin
            resultshift <= pos - 10;
            if (pos >= 11)
            resultfrac <= resultreg[pos-1 -: 10];
            state <= RSLT;
            end
RSLT:
            begin
            result <= {resultsign,resultshift,resultfrac};
            ready <= 1'b1;
            state <= RDY;
            end
        endcase

    endmodule
```

If these are different, the one with the bigger absolute value is determined. Therefore, exponential and fractional parts of numbers are compared successively. The aim here is subtracting the smaller number from the bigger one and keeping sign of the bigger number. Afterward, the state machine goes to NEGPOS state. Here, shifting operations are done and state of the machine goes to OP. Here, sign bits are considered again to decide whether to perform addition or subtraction operation. If sign of both numbers are the same, we add them and go to the SHIFT state to perform another shifting operation. Then, the state goes to WRITE in which the machine forms result arrays. The final state is RST, in which the final result is prepared and ready is set to logic level 1. Hence, the state machine goes to RDY state for a new addition operation.

The user can check how the floating-point adder module works by using its testbench file given at this book's companion website, www.mhprofessional.com/1259837904. We strongly suggest that the reader check floating-point addition operations considered in Chap. 6 to cross-check the results there. Moreover, the module introduced in this section can be expanded further to handle subtraction, division, and multiplication operations on floating-point numbers as well. Xilinx also offers an IP block for floating-point operations under IP catalog → Math Functions → Floating Point. The reader can check it for efficient floating-point calculations.

13.4 Calculator

The calculator application has been improved up to this chapter. Now, it is time to finalize it. Therefore, we modify it such that two-digit decimal numbers can be taken as input. A USB keyboard can be used for this purpose. The result (which can go up to 4096) will be seen on seven-segment display of the Basys3 board. We provide the top module for the calculator application in Listing 13.5. As can be seen here, the top module uses several IP modules developed in previous chapters. These are the seven-segment display driver, calculator, debounce, and binary to BCD modules. When the calculator IP module is used, the number length is set as seven.

Listing 13.5 Calculator in Final Form Implemented on the Basys3 Board in Verilog

```verilog
module calculator_topmodule(clk,btnC,btnU,btnD,btnR,btnL,sw,PS2Clk,
    PS2Data,seg,an);

input clk;
input btnC,btnU,btnD,btnR,btnL;
input [0:0] sw;
input PS2Clk,PS2Data;
output [6:0] seg;
output [3:0] an;

reg [3:0] state = 4'b0;

reg [6:0] number1;
reg [6:0] number2;
reg [3:0] num1ones, num1tens, num2ones, num2tens;

reg [3:0] in1,in2,in3,in4;
```

```
sevenseg_driver_0 seg1(clk,sw[0],in1,in2,in3,in4,seg,an);

wire [3:0] keyout;
wire numready;
reg numready_prev;

keypad_app_0 key1(clk,PS2Clk,PS2Data,keyout,numready);

reg [1:0] op;
wire [15:0] result;

calculator_0 cal1(number1,number2,op,result);

wire btnCclk, btnUclr, btnDclr, btnLclr, btnRclr;
reg btnC_prev, btnU_prev, btnD_prev, btnL_prev, btnR_prev;

debounce_0 dbnC(clk,btnC,btnCclr);
debounce_0 dbnU(clk,btnU,btnUclr);
debounce_0 dbnD(clk,btnD,btnDclr);
debounce_0 dbnR(clk,btnR,btnRclr);
debounce_0 dbnL(clk,btnL,btnLclr);

reg [11:0] binary;
wire [3:0] thos,huns,tens,ones;

binarytoBCD_0 bcd1(binary,thos,huns,tens,ones);

always @ (posedge clk)
begin
if (sw[0])
begin
number1 <= 7'b0;
number2 <= 7'b0;
{num1ones,num1tens,num2ones,num2tens} <= 16'b0;
{in1,in2,in3,in4} <= 16'b0;
state <= 4'b0;
end
else
begin
numready_prev <= numready;

case (state)
4'b0000:
        if (numready_prev == 1'b0 && numready == 1'b1)
        begin
        num1tens <= keyout;
        in4 <= keyout;
        state <= state +1'b1;
        end
4'b0001:
        if (numready_prev == 1'b0 && numready == 1'b1)
        begin
        num1ones <= keyout;
        in4 <= keyout;
        in3 <= num1tens;
```

```verilog
                        number1 <= num1tens*10+keyout;
                        state <= state +1'b1;
                        end
            4'b0010:
                        begin
                        if (btnUclr) begin op <= 2'b00; state <= state +1'b1; end
                        if (btnDclr) begin op <= 2'b01; state <= state +1'b1; end
                        if (btnRclr) begin op <= 2'b10; state <= state +1'b1; end
                        if (btnLclr) begin op <= 2'b11; state <= state +1'b1; end
                        end
            4'b0011:
                        begin
                        in3 <= 4'b0;
                        in4 <= 4'b0;
                        state <= state +1'b1;
                        end
            4'b0100:
                        if (numready_prev == 1'b0 && numready == 1'b1)
                        begin
                        num2tens <= keyout;
                        in4 <= keyout;
                        state <= state +1'b1;
                        end
            4'b0101:
                        if (numready_prev == 1'b0 && numready == 1'b1)
                        begin
                        num2ones <= keyout;
                        in4 <= keyout;
                        in3 <= num2tens;
                        number2 <= num2tens*10+keyout;
                        state <= state +1'b1;
                        end
            4'b0110:
                        if (btnCclr)
                        begin
                        binary <= result[11:0];
                        state <= state +1'b1;
                        end
            4'b0111:
                        begin
                        in4 <= ones;
                        in3 <= tens;
                        in2 <= huns;
                        in1 <= thos;
                        end
            endcase

        end
end
endmodule
```

The top module also uses the keyboard keypad controller module. Before explaining the top module, let's focus on this module first. We provide the Verilog description of the keyboard controller module in Listing 13.6. The working principles of this module are very similar to the USB keyboard application in Chap. 12. There, we processed the scancode to toggle LEDs on the Basys3 board. Here, we convert the scancode of numbers 0 to 9 to the corresponding binary code. Hence, we can easily use a USB keyboard as keypad.

In Listing 13.5, the direction buttons on the Basys3 board are used as operation entries. Hence, btnU is used for addition; btnD is used for subtraction; btnR is used for multiplication; btnL is used for division; and btnC is used as the equal sign. The top module uses the main clock of Basys3 (clk) and communicates with the USB keyboard by PS2Clk and PS2Data ports. The first switch of the Basys3 board (sw[0]) acts as reset input (which will be used after an operation). The seven-segment display ports, seg and an, are also integrated in the top module.

The working principles of the calculator top module (as a state machine) are as follows. The reset input sw[0] is checked at every positive edge of the main clock. If it is at logic level 1, then all numbers, seven-segment display digits, and state of the machine go to zero. When sw[0] is at logic level 0, the state machine checks for a ready signal from the USB keyboard. When the first number is entered via keyboard, it is written to the rightmost digit on the seven-segment display. The user should enter the number as he or she is using an actual calculator. Hence, the first entry will be the tens digit of the first number. Then, ones digit of the first number should be entered by keyboard. After the first number is entered, the state machine waits for operator selection. This can be done by pressing one of the direction buttons as mentioned before. Afterward, the second number should be entered similar to the first one. Pressing the center button (designated as the equal sign) will generate the result of the operation and show it on the seven-segment display. The user should reset the calculator (by using sw[0]) for a new operation. Note that reset can be applied in any phase of the calculation.

13.5 Home Alarm System

We can use sensors instead of switches to realize an actual home alarm system. Therefore, we replace switches representing windows and door by proximity sensors. The proximity sensor we picked works as follows. If someone goes in front of the sensor, it provides the output of logic level 0. Otherwise, the output of the sensor is at logic level 1.

Listing 13.6 Verilog Description of the Keyboard Keypad Controller Module

```
module keypad_app(clk,PS2Clk,PS2Data,keyout,ready);

input clk;
input PS2Clk;
input PS2Data;
output reg [3:0] keyout;
output reg ready=0;

localparam PRESS=2'b00, EXTEND=2'b01, RLS=2'b10, CHECK=2'b11;

reg [1:0] state = PRESS;
reg [23:0] received;
```

```verilog
wire [7:0] keydata;
wire keyready;
reg keyready_prev;

USB_keyboard kb1(.ps2data(PS2Data),.ps2clk(PS2Clk),.data(keydata),
    .ready(keyready));

always @ (posedge clk)
begin
keyready_prev <= keyready;
case (state)
PRESS:
        if (keyready_prev==0 && keyready==1)
        begin
        ready <= 0;
        received[23:16] <= keydata;
        state <= EXTEND;
        end
EXTEND:
        if (keyready_prev==0 && keyready==1)
        begin
        received[15:8] <= keydata;
        state <= RLS;
        end
RLS:
        if (received[15:8] != 8'hF0)
        state <= EXTEND;
        else if (keyready_prev==0 && keyready==1)
        begin
        received[7:0] <= keydata;
        state <= CHECK;
        end
CHECK:
        begin
        case (received[7:0])
        8'h16 : keyout <= 4'b0001;
        8'h1E : keyout <= 4'b0010;
        8'h26 : keyout <= 4'b0011;
        8'h25 : keyout <= 4'b0100;
        8'h2E : keyout <= 4'b0101;
        8'h36 : keyout <= 4'b0110;
        8'h3D : keyout <= 4'b0111;
        8'h3E : keyout <= 4'b1000;
        8'h46 : keyout <= 4'b1001;
        8'h45 : keyout <= 4'b0000;
        endcase
        state <= PRESS;
        ready <= 1;
        end
endcase
end
endmodule
```

Listing 13.7 Home Alarm System in Final Form Implemented on the Basys3 Board in Verilog

```verilog
module home_alarm_topmodule(clk,sw,btnC,btnU,led,seg,an,JA,JB,JC);

input clk;
input [7:0] sw;
input btnC, btnU;
output [4:0] led;
output [6:0] seg;
output [3:0] an;
input [3:0] JA;
input [3:7] JB;
output [3:0] JC; //bluetooth JC[2]:rx, JC[3]:tx

wire movement, proximity, sound;

assign movement = JB[3];
assign proximity = ~JB[7];
assign sound = ~JA[3];
assign led[4:2] = {JB[3],JB[7],JA[3]};

wire btnCclr, btnUclr;

debounce_0 dbC(clk,btnC,btnCclr);
debounce_0 dbU(clk,btnU,btnUclr);

home_alarm2_0 alarm1(.clk(clk),.pass(sw),.act(btnCclr),.door(btnUclr),
    .win1(movement),.win2(proximity),.win3(sound),.blinkled(led[1]),
    .alarmled(led[0]),.seg(seg),.an(an),.buzzer(JC[0]));

reg [255:0] word= {"A","L","A","R","M",8'h0A};

UART_word_tx_0 TX(clk,led[0],word,6'b000110,JC[2]);

endmodule
```

Besides these sensors, we also added a movement (PIR) sensor and sound detector to the home alarm system. The output of the movement sensor is at logic level 0. If it senses a movement, this output goes to logic level 1. If the sound detector detects a sound higher than its sensitivity value (threshold), then its output goes to logic level 0. Otherwise, its output stays at logic level 1. Based on these improvements, we provide the final form of the home alarm system in Listing 13.7.

Movement, proximity, and sound sensors all have three pins: VCC, GND, and OUT. They are all supplied by 3.3 V from the Basys3 board. The output of the movement sensor is connected to JB[3]. The proximity sensor output is connected to JB[7]. The sound sensor output is connected to JA[3].

13.6 Digital Safe System

We can finalize the digital safe system by adding a USB keyboard to it. Besides, the digital safe will work as explained in Chap. 10. We provide the modified and final form of the digital safe in Listing 13.8.

Listing 13.8 Digital Safe System in Final Form Implemented on the Basys3 Board in Verilog

---

```verilog
module digital_safe_topmodule(clk,btnU,btnC,btnD,btnR,PS2Data,PS2Clk,
    seg,an,led);

input clk;
input btnC,btnU,btnD,btnR;
input PS2Data;
input PS2Clk;
output [3:0] an;
output [6:0] seg;
output [15:0] led;

wire btnCclr,btnDclr,btnUclr,btnRclr;

debounce_0 dbc(clk,btnC,btnCclr);
debounce_0 dbu(clk,btnU,btnUclr);
debounce_0 dbd(clk,btnD,btnDclr);
debounce_0 dbr(clk,btnR,btnRclr);

reg [3:0] disp1;
reg [3:0] disp2;
reg [3:0] disp3;
reg [3:0] disp4;

sevenseg_driver_0 seg7(clk,1'b0,disp1,disp2,disp3,disp4,seg,an);

wire [1:0] safestate;
reg [15:0] password;

digital_safe2_0 safe1(.clk(clk),.passinput(password),.pass_set
    (btnUclr),.pass_reg(btnDclr),.pass_lock(btnCclr),.safestate(
        safestate));

wire keyready;
reg keyready_prev;
reg [1:0] keystate=0;
wire [7:0] keyout;

keypad_app_0 key1(.clk(clk),.PS2Clk(PS2Clk),.PS2Data(PS2Data),.keyout
    (keyout),.ready(keyready));

always @ (posedge clk)
begin
keyready_prev <= keyready;
if (btnRclr) keystate <= 2'b0;
else if (btnCclr == 1'b1) password <= 16'b0;
else

case (keystate)
2'b00:
        if (keyready_prev == 0 && keyready == 1) begin
        password[15:12] <= keyout;
        keystate <= keystate + 1'b1;
        end
```

```
    2'b01:
            if (keyready_prev == 0 && keyready == 1) begin
            password[11:8] <= keyout;
            keystate <= keystate + 1'b1;
            end
    2'b10:
            if (keyready_prev == 0 && keyready == 1) begin
            password[7:4] <= keyout;
            keystate <= keystate + 1'b1;
            end
    2'b11:
            if (keyready_prev == 0 && keyready == 1) begin
            password[3:0] <= keyout;
            keystate <= 2'b00;
            end
endcase
end

always @ (posedge clk)
case(safestate)
        2'b00: {disp1,disp2,disp3,disp4} <= {4{4'b1100}}; //C
        2'b01: {disp1,disp2,disp3,disp4} <= {4{4'b0000}}; //0
        2'b10: {disp1,disp2,disp3,disp4} <= password;
        2'b11: {disp1,disp2,disp3,disp4} <= {4{4'b0101}}; //S
endcase

assign led = password;

endmodule
```

Let's explain the working principles of the digital safe system (as a state machine) step by step. The system starts with a default password 1234. When the user enters it, the safe opens. Here, user has two options. The first one is changing the password. The second one is locking the safe again. When btnC on the Basys3 board is pressed, the safe locks again. If the user presses btnU, digital safe goes to the password changing state. Here, it expects the user to enter a new password. This can be done by using numbers on the keyboard. Since this is a prototype system, the entered password is also shown on the seven-segment display (and LEDs) of Basys3. When a new password is entered, the user should press btnD to save it. Afterward, btnC should be pressed to lock the safe again. While entering the password digits, the user may press btnR anytime to restart again.

13.7 Car Park Occupied Slot Counting System

We can finalize the car park occupied slot counting system in several ways. First, we can add a bluetooth module such that the user can open garage gate by using his or her cell phone. Here, a simple Android application developed under MIT App Inventor may be sufficient [58]. Also, we can add a proximity sensor to the garage gate. Hence, we can detect whether the car is passing through the gate. We can also add a stepper motor to open and close the garage gate.

Let's start with the stepper motor. The stepper motor used in this application is 24BYJ48. This stepper motor has five terminals. The four of them drive coils and the last one is ground as can be seen in Fig. 13.1. To run this motor, we will use the Digilent

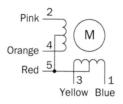

FIGURE 13.1 Stepper motor terminals.

Pmod STEP: stepper motor driver [59]. We will connect terminals of the stepper motor from left to right like pink, yellow, orange, and blue to the driver. The red wire is the ground. For connection properties, please see the website appearing in [59].

We provide the Verilog description of the stepper motor driver in Listing 13.9. There are five inputs in the module. These are clk (clock signal), rst (active-high reset), en (active-high enable), trig (active-high trigger), and dir (direction; logic level 1: clockwise, logic level 0: counter clockwise). The output of the module is a four-bit vector driver.

We can explain the working principles of the stepper motor driver (as a state machine) presented in Listing 13.9 as follows. The module has two parameters. motorfreq determines the frequency in which the motor will be driven. degree is used to tell the module how much it will turn in degrees (360 for a full spin). The state machine has three states as follows. In RDY state, the driver waits for a trigger. In TURN state, the motor turns depending on the degree parameter. The last state, MOTORSTATE is used to drive the motor. Within the stepper motor driver module, the input clock signal, clk, is divided to generate motorclk in frequency set by the parameter motorfreq. This frequency will be used to run the motor. Since motorfreq is around 100 Hz at most, it is hard to catch the trigger signal. To avoid this issue, trig signal is sampled in every clk signal. If its rising edge is catched, trig_int goes to logic level 1, and the state machine waits until the state changes to TURN. Hence, we make sure that turning of the motor has begun. In every rising edge of motorclk, if the machine is in RDY state, the module checks if en and trig_int are both at logic level 1. If this is the case, the state changes to TURN and the motor is driven by four-bit outputs depending on the dir bit until the predetermined degree is satisfied.

We provide the top module for the final car park occupied slot counting system in Listing 13.10. Different from previous versions of the application, the proximity sensor is located in the garage gate which controls if the car is still passing through. Switches on the Basys3 board imitate the output of the proximity sensors located in each parking slot. Once the user approaches the gate, he or she sends the character O via cell phone to the bluetooth adapter connected to the Basys3 board via UART communication. Then, the gate opens and waits for 12 seconds to close unless the car is still passing through. Since the steppermotor module is integrated as an IP block to the project, it asks for the rotation degree in initialization window. If you do not integrate your module in your IP library, then you have to add 90 degree as a parameter to module initialization.

Let's briefly explain the hardware used in this application. The output of the proximity sensor is connected to JC[3] port of the Basys3 board, the one we have used before. The bluetooth adapter HC-06 is connected to JB port, such that the receiving port of the module is connected to JB[3]. The clock used by the module is the master clock on the Basys3 board. The stepper motor introduced in this section is used at the garage gate.

Listing 13.9 Verilog Description of the Stepper Motor Driver Module

```verilog
module stepper_motor(clk,rst,en,trig,dir,driver);

input clk;
input rst;
input en;
input trig;
input dir; //1:clockwise, 0:counterclockwise
output reg [3:0] driver;

parameter motorfreq=100; //in Hz
parameter degree=180;

localparam motordegree = (820*degree)/360;
integer degreecount;

localparam RDY=1'b0, TURN=1'b1;

reg state=RDY;

localparam A=2'b00, B=2'b01, C=2'b10, D=2'b11;
reg [1:0] motorstate=A;

localparam motorcount = 100000000/(motorfreq*2); //26'd500000;
integer count;
reg motorclk=0;

reg trig_prev,trig_int;

always @ (posedge clk)
begin
if (rst == 1'b1)
begin
count <= 0;
motorclk <= 1'b0;
end
else if (count == motorcount)
begin
count <= 0;
motorclk <= ~motorclk;
end
else
count <= count + 1;
end

always @ (posedge clk)
begin
trig_prev <= trig;
if (trig_prev == 1'b0 && trig == 1'b1)
trig_int <= 1'b1;
else if (state == TURN)
trig_int <= 1'b0;
end
```

```verilog
always @ (posedge motorclk)

case (state)
RDY:
  if (en == 1'b1 && trig_int == 1'b1)
  state <= TURN;
  else
  begin
  driver <= 4'b0000;
  degreecount <= 0;
  end
TURN:
  if (degreecount == motordegree)
  state <= RDY;
  else
  begin
  degreecount <= degreecount + 1;
  case (motorstate)
       A:
       begin
       driver = 4'b1000;
       if (dir) motorstate <= B;
       else motorstate <= D;
       end
       B:
       begin
       driver = 4'b0100;
       if (dir) motorstate <= C;
       else motorstate <= A;
       end
       C:
       begin
       driver = 4'b0010;
       if (dir) motorstate <= D;
       else motorstate <= B;
       end
       D:
       begin
       driver = 4'b0001;
       if (dir) motorstate <= A;
       else motorstate <= C;
       end
       endcase
       end
  endcase
endmodule
```

Listing 13.10 Car Park Occupied Slot Counting System in Final Form Implemented on the Basys3 Board in Verilog

```verilog
module carpark_topmodule(clk,led,sw,JC,seg,an,JB,JA);

input clk;
input [15:0] sw;
input [3:3] JC;
```

```
output [15:0] led;
output [6:0] seg;
output [3:0] an;
input [3:3] JB; //bluetooth JB[2]:tx, JB[3]:rx
output [7:4] JA;

wire [3:0] tens, ones;
wire [4:0] cars;
reg gate=1'b0;//1:gate open, 0:gate close
reg gate_prev;

integer passcounter=0;
localparam secondtime=100000000; //1 second
reg [7:0] seconds=0;

car_park3_0 park1(.c(cars),.s(sw[15:0]));

binarytoBCD_0 bcd(.binary({7'b0,cars}),.thos(),.huns(),.tens(tens),
    .ones(ones));

sevenseg_driver_0 seg1(.clk(clk),.clr(1'b0),.in1(4'b0),.in2(4'b0),
    .in3(tens),.in4(ones),.seg(seg),.an(an));

wire [7:0] rxdata;
wire rxready;
reg rxready_prev;

UART_rx_ctrl_0 bleUART(.clk(clk),.rx(JB[3]),.data(rxdata),.parity(),
    .ready(rxready),.error());

reg dir=0,trig=0;

stepper_motor_0 gatemotor(.clk(clk),.rst(1'b0),.en(1'b1),.trig(trig),
    .dir(dir),.driver(JA[7:4]));

always @ (posedge clk)
begin
rxready_prev <= rxready;
if (rxready_prev == 1'b0 && rxready == 1'b1) begin
    if (cars < 5'b10000)
    if (rxdata == 8'h4F) gate <= 1'b1; //if O is received
    end
else if (gate == 1'b1 && JC[3] == 1'b1)
    if (passcounter == secondtime)
    begin
    seconds <= seconds + 1'b1;
    passcounter <= 0;
    end
    else if (seconds == 8'b0000_1100) begin
    seconds <= 0;
    gate <= 1'b0;
    end
    else passcounter <= passcounter + 1;
end
```

```
always @ (posedge clk)
begin
gate_prev <= gate;
if (gate_prev == 1'b0 && gate == 1'b1) //gate opens
    begin
    dir <= 1'b0;
    trig <= 1'b1;
    end
else if (gate_prev == 1'b1 && gate == 1'b0) //gate closes
    begin
    dir <= 1'b1;
    trig <= 1'b1;
    end
else
    trig <= 1'b0;
end
endmodule
```

13.8 Vending Machine

We can finalize the vending machine by adding a bluetooth module to send a signal if one of the products runs out of stock. We provide the final form of the vending machine on the Basys3 board in Listing 13.11.

In Listing 13.11, we use the bluetooth module via the UART communication. The bluetooth adapter HC-06 is connected to JC port. Hence, the transmitter port of the module is connected to JC[2] port of the Basys3 board. As a reminder, outofstock vector indicates if one of the products has gone out of stock in the vending machine. Therefore, we check a rising edge in this vector. Assume that the second item went out of stock. Then, the phrase "2 OUT OF STOCK" is loaded to the word vector of vending_machine and wordsend is set to logic level 1 for the UART transmission (to the bluetooth module) to start. Within the top module, the money entrance is imitated by btnR for 25 cents and btnL for 1 dollar again. First four switches on the Bassy3 board are used to select the product and btnC stands for the final buy command. A stepper motor (connected to JA port of the Basys3 board) is integrated to the top module to realize the exact vending machine behavior. As in the previous application, the steppermotor module is integrated as an IP block to the project; it asks for the rotation degree in the initialization window. Set it to 360 degrees for this application. For ease of implementation, only the first product is handled this way. Therefore, when the customer buys the first product, the stepper motor does a full turn in clockwise direction.

13.9 Digital Clock

We can finalize the digital clock by adding alarm and chronometer modules to it. We provide the Verilog description of the alarm module in Listing 13.12. This module is the simplified version of the digital clock module such that it only increments hour and minute digits with the button press.

We provide the top module for the final digital clock application in Listing 13.13. Within this module, sw input determines the mode of the system. Hence, if sw is 00, the regular clock operation is done. When sw is set to 01, the system enters the alarm mode and the user can set the alarm with btnU and btnR buttons. When the alarm is

Listing 13.11 Vending Machine in Final Form Implemented on the Basys3 Board in Verilog

```verilog
module vending_machine_topmodule(clk,btnC,btnR,btnL,sw,led,seg,an,JC,
    JA);

input clk;
input btnC,btnR,btnL;
input [7:0] sw;
output [7:0] led;
output [6:0] seg;
output [3:0] an;
output [2:2] JC;
output [7:4] JA;

wire [11:0] money;

wire btnCclk,btnLclr,btnRclr;

debounce_0 dbnC(clk,btnC,btnCclr);
debounce_0 dbnR(clk,btnR,btnRclr);
debounce_0 dbnL(clk,btnL,btnLclr);

wire [3:0] thos,huns,tens,ones;

binarytoBCD_0 bcd1(money,thos,huns,tens,ones);

sevenseg_driver_0 seg1(.clk(clk),.clr(1'b0),.in1(thos),.in2(huns),
    .in3(tens),.in4(ones),.seg(seg),.an(an));

wire [3:0] outofstock;
wire [3:0] products;
reg product1_prev;
reg [3:0] outofstock_prev;

vending_machine_0 vm1(.clk(clk),.coin1(btnRclr),.coin2(btnLclr),
    .select(sw[3:0]),.buy(btnCclr),.load(sw[7:4]),.money(money),
    .products(products),.outofstock(outofstock));

reg [255:0] word;
reg wordsend=0;
reg [5:0] wordlength;
wire transmit;

UART_word_tx_0 TX(clk,wordsend,word,wordlength,transmit);

reg dir=1,trig=0; //always clockwise
stepper_motor_0 sm(.clk(clk),.rst(1'b0),.en(1'b1),.trig(trig),
    .dir(dir),.driver(JA[7:4]));

always @ (posedge clk)
begin
outofstock_prev <= outofstock;
if (outofstock_prev[0] == 1'b0 && outofstock[0] == 1'b1)
    begin
    word <= {"1",8'h20,"O","U","T",8'h20,"O","F",8'h20,"S","T","O","C",
        "K",8'h0A};
```

```
      wordlength <= 6'b01111;
      wordsend <= 1'b1;
      end
  else if (outofstock_prev[1] == 1'b0 && outofstock[1] == 1'b1)
      begin
      word <=   {"2",8'h20,"O","U","T",8'h20,"O","F",8'h20,"S","T","O","C"
         ,"K",8'h0A};
      wordlength <= 6'b01111;
      wordsend <= 1'b1;
      end
  else if (outofstock_prev[2] == 1'b0 && outofstock[2] == 1'b1)
      begin
      word <=   {"3",8'h20,"O","U","T",8'h20,"O","F",8'h20,"S","T","O","C"
         ,"K",8'h0A};
      wordlength <= 6'b01111;
      wordsend <= 1'b1;
      end
  else if (outofstock_prev[3] == 1'b0 && outofstock[3] == 1'b1)
      begin
      word <=   {"4",8'h20,"O","U","T",8'h20,"O","F",8'h20,"S","T","O","C"
         ,"K",8'h0A};
      wordlength <= 6'b01111;
      wordsend <= 1'b1;
      end
  else wordsend <= 1'b0;
  end


always @ (posedge clk)
begin
product1_prev <= products[0];
if (product1_prev == 1'b0 && products[0] == 1'b1) //product1 is bought
    trig <= 1'b1;
else
    trig <= 1'b0;
end

assign led[7:4] = outofstock;
assign led[3:0] = products;
assign JC[2] = transmit;


endmodule
```

Listing 13.12 Verilog Description of the Alarm Module for Digital Clock

```
module alarm(clk,rst,hrup,minup,m1,m2,h1,h2);

input clk;
input rst;
input hrup,minup;
output [3:0]m1,m2,h1,h2;

reg [5:0] hour=0,min=0;
reg hrup_prev,minup_prev;

always @ (posedge clk)
```

```
begin
hrup_prev <= hrup;
minup_prev <= minup;
if (minup_prev == 1'b0 && minup == 1'b1)
  if (min == 6'd59)
  min<=0;
  else
  min <= min+1'd1;
else if (hrup_prev == 1'b0 && hrup == 1'b1)
  if (hour == 6'd23)
  hour<=0;
  else
  hour <= hour+1'd1;
end

binarytoBCD_0 mins(.binary(min),.thos(),.huns(),.tens(m2),.ones(m1));

binarytoBCD_0 hours(.binary(hour),.thos(),.huns(),.tens(h2),.ones(h1));


endmodule
```

Listing 13.13 Digital Clock in Final Form Implemented on the Basys3 Board in Verilog

```
module digital_clock_topmodule(clk,sw,btnC,btnU,btnD,btnR,btnL,led,seg,
    an);

input clk;
input [1:0] sw;
input btnC,btnU,btnD,btnR,btnL;
output reg [15:0] led;
output [6:0] seg;
output [3:0] an;

reg btnC_prev,btnU_prev,btnR_prev;

wire [3:0] s1,s2,m1,m2,h1,h2; //clock variables
wire [3:0] alm1,alm2,alh1,alh2; //alarm variables
wire [3:0] chs1,chs2,chm1,chm2; //chronometer variables

reg [3:0]dig1,dig2,dig3,dig4;

sevenseg_driver_0 seg7(clk,1'b0,dig1,dig2,dig3,dig4,seg,an);

wire btnCclr,btnUclr,btnRclr;

debounce_0 dbC(clk,btnC,btnCclr);
debounce_0 dbU(clk,btnU,btnUclr);
debounce_0 dbR(clk,btnR,btnRclr);

reg clockrst=0,chen=0,chrst=0,alarmrst=0;
reg hrup,minup,alhrup,alminup;

digital_clock_0 tm(clk,1'b1,clockrst,hrup,minup,s1,s2,m1,m2,h1,h2);

digital_clock_0 chr(clk,chen,chrst,1'b0,1'b0,chs1,chs2,chm1,chm2, , );

alarm_0 alm(clk,alarmrst,alhrup,alminup,alm1,alm2,alh1,alh2);
```

```verilog
always @ (posedge clk)
begin
btnC_prev <= btnCclr;
btnU_prev <= btnUclr;
btnR_prev <= btnRclr;

led[7:0] <= {s2,s1};

case (sw)
2'b00:
        begin
        if (btnC_prev == 1'b0 && btnCclr == 1'b1)
        clockrst <= 1'b1;
        else clockrst <= 1'b0;
        if (btnU_prev == 1'b0 && btnUclr == 1'b1)
        hrup <= 1'b1;
        else hrup <= 1'b0;
        if (btnR_prev == 1'b0 && btnRclr == 1'b1)
        minup <= 1'b1;
        else minup <= 1'b0;
        {dig4,dig3,dig2,dig1} <= {m1,m2,h1,h2};
        end
2'b01:
        begin
        alarmrst <= btnCclr;
        alhrup <= btnUclr;
        alminup <= btnRclr;
        {dig4,dig3,dig2,dig1} <= {alm1,alm2,alh1,alh2};
        end
2'b11:
        begin
        if (btnU_prev == 1'b0 && btnUclr == 1'b1)
        chen <= ~chen;
        else if (btnC_prev == 1'b0 && btnCclr == 1'b1)
        begin
        chen <= 1'b1;
        chrst <= 1'b1;
        end
        else if (btnC_prev == 1'b1 && btnCclr == 1'b0)
        begin
        chen <= 1'b0;
        chrst <= 1'b0;
        end
        {dig4,dig3,dig2,dig1} <= {chs1,chs2,chm1,chm2};
        end
endcase

if (alm1==m1 && alm2==m2 && alh1==h1 && alh2==h2)
led [15:8] <= {8{s1[0]}};
else led [15:8] <= 8'b0;

end
endmodule
```

activated, the leftmost eight LEDs on the Basys3 board flash for 60 seconds. When sw is set to 11, the system enters the chronometer mode. Here, the digital clock module is used again without its outputs. With every btnU press, the chronometer counts or pauses. When the user presses btnC, the module clears the output and gets ready for the next count. All three modes can work at the same time. Therefore, if the user wants to use the chronometer he or she can do so without disturbing the digital clock and alarm operations.

13.10 Moving Wave via LEDs

In this application, we will form a moving wave application via LEDs on the Basys3 board. To do so, we will benefit from the pulse width modulation (PWM) which forms digital periodic pulses with varying width (duty cycle). Hence, the PWM is used to obtain analog signals from a digital system most of the times [32]. We provide the Verilog description to generate a PWM signal in Listing 13.14. This module has two inputs. These are clk (main clock signal) and dutyc (duty cycle). The output of the module is pwm_out which is the PWM signal.

The working principles of the PWM generator module in Listing 13.14 are as follows. There is a four-bit pwmc vector in the module. What we do is basically incrementing pwmc in every clock cycle and comparing it with dutyc. If pwmc is smaller than or equal to dutyc, then the output will be at logic level 1, otherwise it will be at logic level 0. We divide the full period of a square wave into 16 parts. The output starts at logic level 1. With dutyc, we decide when it goes to logic level 0. Hence, for a 50% duty cycle, we should set dutyc to seven as half of its maximum value. Here, we use the Clock Wizard IP for frequency division. Hence, clock-based operations within the module are done appropriately.

We provide the top module for the moving wave application in Listing 13.15. This module uses the PWM_generator module to drive all 16 LEDs on the Basys3 board with

Listing 13.14 Verilog Description of PWM Module

```verilog
module PWM_generator(clk,dutyc,pwm_out);

input clk;
input [3:0] dutyc; //from 0 to 14 in decimal form
output reg pwm_out;

reg [3:0] pwmc=0;
wire clk25MHz;

clk_wiz_0 clk_divider(clk,clk25MHz);

always @ (posedge clk25MHz)
begin
pwmc <= pwmc + 1'b1;
if (pwmc <= dutyc)
  pwm_out <= 1'b1;
else
  pwm_out <= 1'b0;
end
endmodule
```

Listing 13.15 Moving Wave Application Implemented on the Basys3 Board in Verilog

```verilog
module moving_wave_topmodule(clk,sw,led);

input clk;
input [2:0] sw;
output reg [15:0] led=0;

reg [3:0] dutyc=0;

wire led_out;

PWM_generator_0 pwm1(.clk(clk),.dutyc(dutyc),.pwm_out(led_out));

reg [27:0] clock_div=0;
reg [3:0] led_index=0;

always @ (posedge clk)
begin
clock_div <= clock_div + 1'b1;
led[led_index-1'b1] <= 0;
led[led_index] <= led_out;

case (led_index)
    4'b0000 : dutyc <= 4'b0000;
    4'b0001 : dutyc <= 4'b0001;
    4'b0010 : dutyc <= 4'b0010;
    4'b0011 : dutyc <= 4'b0011;
    4'b0100 : dutyc <= 4'b1000;
    4'b0101 : dutyc <= 4'b0101;
    4'b0110 : dutyc <= 4'b0110;
    4'b0111 : dutyc <= 4'b0111;
    4'b1000 : dutyc <= 4'b1000;
    4'b1001 : dutyc <= 4'b1001;
    4'b1010 : dutyc <= 4'b1010;
    4'b1011 : dutyc <= 4'b1011;
    4'b1100 : dutyc <= 4'b1100;
    4'b1101 : dutyc <= 4'b1101;
    4'b1110 : dutyc <= 4'b1110;
    4'b1111 : dutyc <= 4'b1110;
endcase
end

always @ (posedge clock_div[20+sw])
led_index <= led_index + 1'b1;

endmodule
```

different duty cycles. To have a moving wave effect visible to our eyes, we apply the frequency division (with a counter having 28 bits) to the main clock of the Basys3 board. First three switches on the Basys3 board can be used to adjust speed of the moving wave.

We can further improve the moving wave application by adding a joystick as input medium. Hence, the user can decide on the wave movement direction using it. To do so, the XADC module should also be used in connection with the joystick.

13.11 Translator

We can design a digital system to translate voice commands from English to Spanish (or another language) and show them on a 16×2 LCD. The system will have two parts. The first part will recognize the spelled out English word. We can use the EasyVR shield for this purpose [60]. This module has predefined speaker-independent word sets. Also, you can create your own speaker-dependent word set. In the second part of the translator system, we will get the recognition result and form a state machine in the FPGA to provide the translated word corresponding to the recognized one. Then, this word is shown on the LCD.

The LCD we have used in our system is WH1602N with a built-in controller ST7066 or equivalent. To use the LCD display, we need a Verilog description. We provide such an LCD driver module in Listing 13.16. This module has four inputs: clk, reset, wr_en, and data_in. The main clock signal, clk, is expected to be 100 MHz. Active-high reset signal resets the operation and the module waits for an active-high write enable

Listing 13.16 Verilog Description of LCD Driver Module

```verilog
module LCD_driver(clk,reset,wr_en,data_in,data_out,en,rs);

input clk;
input reset;
input wr_en;
input [7:0] data_in;
output reg [7:0] data_out;
output reg en;
output reg rs;

parameter clk_param = 100000;

localparam INIT=2'b00,WAIT=2'b01,WRITE=2'b10;
reg [2:0] state = INIT;

localparam init_index=3;
localparam char_index=15;
reg [3:0] init_count=0;
integer limit_count=0;
reg [7:0] init [3:0];
reg [3:0] clear=0;

initial begin
init[0]=8'h30; // 1 Line, 5x8 Dots
init[1]=8'h01; // Clear display
init[2]=8'h06; // Increment cursor (Shift cursor to right)
init[3]=8'h0F; // Display on cursor blinking
rs=1'b1;
end

always@(negedge clk)
begin
rs <= 1'b1;
en <= 1'b1;
if (reset)
```

```verilog
    begin
    state <= INIT;
    init_count <= 0;
    limit_count <= 0;
    clear <= 0;
    end
else
    begin
    case (state)
    INIT:
    begin
    rs <= 0;
    data_out <= init[init_count];
    if (limit_count == clk_param)
        begin
        en <= 0;
        limit_count <= 0;
        init_count <= init_count + 1'b1;
        if (init_count == init_index)
            begin
            init_count <= 0;
            state <= WAIT;
            end
        end
    else
        limit_count <= limit_count + 1;
    end
    WAIT:
    if (wr_en)
        state <= WRITE;
    WRITE :
    begin
    data_out <= data_in;
    if (limit_count == clk_param)
        begin
        en <= 0;
        limit_count <= 0;
        clear <= clear + 1'b1;
        if (clear == char_index)
            begin
            state <= INIT;
            clear <= 0;
            end
        else
            state <= WAIT;
        end
    else
        limit_count <= limit_count + 1;
    end
    endcase
    end
end
endmodule
```

wr_en signal. data_in is eight-bit data that will be transmitted to the display. The module has three outputs: data_out (eight-bit data output), en (enable signal to drive the LCD), and rs (data/instruction selection signal for LCD).

The working principle of the LCD driver module (as a state machine) is as follows. The module starts in INIT state where it initializes the display by the predefined eight-bit commands. Using them, we set the display for one line and 5×8 dots; then cleared the display screen; set the cursor direction; and changed the cursor to blinking mode. After initialization, the machine goes to WAIT state where it waits for wr_en signal to go to logic level 1. Once this happens, the machine goes to WRITE state. Here, it transfers data_in to data_out and waits for 1 microsecond. Afterward, the machine turns back to WAIT state by incrementing clear vector by one and waits for another wr_en signal. Once clear reaches 15 in decimal form, that means it reached the end of the line and it turns back to INIT state and clears the display.

We can connect the LCD to the Basys3 board as follows. The LCD's eight-bit data bus line (DB0 to DB7) should be connected to JA port of the board (starting from JA[0] to JA[7]). Enable signal of the LCD (E) should be connected to JB[0]. Similarly, RS signal of the LCD should be connected to JB[2]. You can connect the R/W port of the LCD to ground since we will always be in write mode. Also, do not forget to supply V_{DD} port of LCD with 5 V and connect V_{SS} to ground. There is a contrast port V_O on the LCD. This port can be connected to ground for maximum contrast. Finally, A and K ports control the back light of the LCD screen. You can set A to 5 V or 3.3 V, and K goes to the ground to lit your LCD screen.

The EasyVR shield communicates through the UART interface. Hence, we can use the UART transmitter and receiver blocks introduced in Chap. 12. After activating the EasyVR shield, it recognizes words in its predefined word set 1 as default. This word set includes English words Action, Move, Turn, Run, Look, Attack, Stop, and Hello. The Spanish translation of these words are Accion, Movimiento, Giro, Correr, Mirar, Ataque, Detener, and Hola, respectively.

Assuming that the reader does not have an EasyVR module, we simulate the translation operation by feeding input signals via the first eight switches of the Basys3 board. We provide the Verilog description of the top module for the translator constructed this way in Listing 13.17. This module has three inputs: clk (main clock of Basys3), reset

---

Listing 13.17 Translator Implemented on the Basys3 Board in Verilog

```verilog
module translator_topmodule(clk,reset,sw,rs,en,data_out);

input clk;
input reset;
input [7:0] sw;
output rs;
output en;
output [7:0] data_out;

parameter clk_param =16000000;

reg [7:0] data [15:0];
reg [7:0] character;
wire [7:0] data_in;
integer counter=0;
reg [3:0] index=0;
```

```verilog
reg wr_en=0;

LCD_driver_0 lcd1(.clk(clk),.reset(reset),.wr_en(wr_en),.data_in
    (data_in),
.data_out(data_out),.en(en),.rs(rs));

assign data_in = character;

always @ (posedge clk)
begin
if (reset)
    counter <= 0;
else
    begin
    if (counter == clk_param)
        counter <= 0;
    else
        counter <= counter + 1;
    end
end

always @ (posedge clk)
begin
if (reset)
    begin
    wr_en <= 0;
    index <= 0;
    end
else
    begin
    if (counter == clk_param)
        begin
        wr_en <= 1'b1;
        character <= data[index];
        index <= index + 1'b1;
        end
    else
        wr_en <= 0;
    end
end

always @ (posedge clk)
case (sw)
8'h01:
begin // ACTION - ACCION
data[0]<="A"; data[1]<="C"; data[2]<="C"; data[3]<="I";
data[4]<="O"; data[5]<="N"; data[6]<=" "; data[7]<=" ";
data[8]<=" "; data[9]<=" "; data[10]<=" "; data[11]<=" ";
data[12]<=" "; data[13]<=" "; data[14]<=" "; data[15]<=" ";
end
8'h02:
begin // MOVE - MOVIMIENTO
data[0]<="M"; data[1]<="O"; data[2]<="V"; data[3]<="I";
data[4]<="M"; data[5]<="I"; data[6]<="E"; data[7]<="N";
data[8]<="T"; data[9]<="O"; data[10]<=" "; data[11]<=" ";
```

```verilog
data[12]<=" "; data[13]<=" "; data[14]<=" "; data[15]<=" ";
end
8'h04:
begin // TURN - GIRO
data[0]<="G"; data[1]<="I"; data[2]<="R"; data[3]<="O";
data[4]<=" "; data[5]<=" "; data[6]<=" "; data[7]<=" ";
data[8]<=" "; data[9]<=" "; data[10]<=" "; data[11]<=" ";
data[12]<=" "; data[13]<=" "; data[14]<=" "; data[15]<=" ";
end
8'h08:
begin // RUN - CORRER
data[0]<="C"; data[1]<="O"; data[2]<="R"; data[3]<="R";
data[4]<="E"; data[5]<="R"; data[6]<=" "; data[7]<=" ";
data[8]<=" "; data[9]<=" "; data[10]<=" "; data[11]<=" ";
data[12]<=" "; data[13]<=" "; data[14]<=" "; data[15]<=" ";
end
8'h10:
begin // LOOK - MIRAR
data[0]<="M"; data[1]<="I"; data[2]<="R"; data[3]<="A";
data[4]<="R"; data[5]<=" "; data[6]<=" "; data[7]<=" ";
data[8]<=" "; data[9]<=" "; data[10]<=" "; data[11]<=" ";
data[12]<=" "; data[13]<=" "; data[14]<=" "; data[15]<=" ";
end
8'h20:
begin // ATTACK - ATAQUE
data[0]<="A"; data[1]<="T"; data[2]<="A"; data[3]<="Q";
data[4]<="U"; data[5]<="E"; data[6]<=" "; data[7]<=" ";
data[8]<=" "; data[9]<=" "; data[10]<=" "; data[11]<=" ";
data[12]<=" "; data[13]<=" "; data[14]<=" "; data[15]<=" ";
end
8'h40:
begin // STOP - DETENER
data[0]<="D"; data[1]<="E"; data[2]<="T"; data[3]<="E";
data[4]<="N"; data[5]<="E"; data[6]<="R"; data[7]<=" ";
data[8]<=" "; data[9]<=" "; data[10]<=" "; data[11]<=" ";
data[12]<=" "; data[13]<=" "; data[14]<=" "; data[15]<=" ";
end
8'h80:
begin // HELLO - HOLA
data[0]<="H"; data[1]<="O"; data[2]<="L"; data[3]<="A";
data[4]<=" "; data[5]<=" "; data[6]<=" "; data[7]<=" ";
data[8]<=" "; data[9]<=" "; data[10]<=" "; data[11]<=" ";
data[12]<=" "; data[13]<=" "; data[14]<=" "; data[15]<=" ";
end
default:
begin // MAKE A SELECTION - HAS UNA ELECCION
data[0]<="H"; data[1]<="A"; data[2]<="S"; data[3]<=" ";
data[4]<="U"; data[5]<="N"; data[6]<="A"; data[7]<=" ";
data[8]<="E"; data[9]<="L"; data[10]<="E"; data[11]<="C";
data[12]<="C"; data[13]<="I"; data[14]<="O"; data[15]<="N";
end
endcase

endmodule
```

(active-high reset signal), and sw vector (first eight switches on the Basys3 board). The outputs of the module are rs, en, and data_out, all of which are LCD driving signals.

The top module in Listing 13.17 uses the LCD_driver module to show the translation results. The top module has an internal counter which counts up to 160 milliseconds. If reset signal goes to logic level 1, then counter, wr_en, and index values will be equal to logic level 0. When counter reaches clk_param (corresponding to 160 milliseconds), index of data memory is loaded into eight-bit character vector. This is directly connected to data_in of the LCD_driver module. There is a case statement at the end of the top module which loads Spanish translation corresponding to the given command (or English word). For our application, depending on which switch is at logic level 1, the corresponding word is loaded to data memory. Hence, each character of this word is displayed with the help of the LCD_driver module. The reader can modify this section if translation to another language is desired.

13.12   Air Freshener Dispenser

We can modify the air freshener dispenser system developed for the MSP430 microcontroller to work on the Basys3 board [32]. The system will have four different programs to spray fresh odor in 5-, 10-, 15-, and 20-second intervals. These values should be in minutes in an actual system. However, we set such values to observe the system output. The system should have a counter for these operations. When counter reaches the designated time value, the kit sprays the fresh odor and restarts counting again. We can use two switches to select among four programs. Besides, there should be an instant spray button. When it is pressed, the fresh odor should be sprayed and the counter should be reset. When the user selects another program, the counter should restart again. There should also be an on/off switch for the system. Spraying fresh odor can be indicated by blinking an LED on the board for three seconds.

13.13   Obstacle-Avoiding Tank

We can modify the obstacle-avoiding tank system developed for the MSP430 microcontroller to work on the Basys3 board [32]. Hence, we will build a tank which is driven by two stepper motors. The proximity sensors on the front edges of the tank will be used to sense obstacles on the way. The tank will change its direction by controlling motor speeds accordingly. The proximity sensor we have used in previous applications can also be employed for this application. By using the tuning screw on the sensor, the designer can adjust the distance the tank will turn when it faces an obstacle.

The sensors can be connected to JB or JC ports of the Basys3 board. The motor driver should also be connected to the JA port of the board. Since this application will be integrated on a tank, the board itself can be powered by a battery to ensure autonomy of the tank. Hence, 5 V has to be applied to external power pins of the board. If the battery's voltage is above 5 V, a regulator should be used.

13.14   Intelligent Washing Machine

We can modify and improve the washing machine system developed for the MSP430 microcontroller to work on the Basys3 board [32]. The washing machine will be simulated by a stepper motor. Hence, the reader should check how it works in the car park occupied slot counting system.

The washing machine is controlled by five buttons. The two of them are for the main on/off and rotation speed. The remaining three buttons are for program selection as follows:

- Prewash: 30 rotations in one direction, then 30 rotations in the other direction
- Normal wash: 100 rotations in one direction, then 100 rotations in the other direction
- Final spin: 50 rotations in one direction, but faster than prewash and normal wash

The normal wash program can be improved by adding intelligence to it. To do so, we can include an IR transmitter and receiver LED pair. The IR transmitter emits IR light when fed with voltage. The IR receiver LED produces voltage when it absorbs IR light. We can form a structure by using the IR transmitter and receiver such that when water passing through them is dirty, no light transmission occurs. Hence, the output of the receiver LED can be taken as logic level 0. When the water passing through these LEDs is clean, the light transmission occurs. Hence, the output of the receiver LED can be taken as logic level 1. Therefore, when the water is dirty, normal wash program is repeated again. This program ends when water becomes clean.

When the main on/off button is pressed, the system is activated. To indicate this, the rightmost LED on the Basys3 board will turn on. In this state, all programs (prewash, normal wash, and final spin) can be performed. Each program can be selected by a specific button. There is an extra button for adjusting the rotation speed as slow and fast. Depending on the selection, the leftmost LED on the board will be either on or off. When the main on/off button is pressed again, the system will be deactivated. To indicate this, the rightmost LED will turn off.

13.15   Non-Touch Paper Towel Dispenser

We can modify the non-touch paper towel dispenser system developed for the MSP430 microcontroller to work on the Basys3 board [32]. The system has a light-dependent resistor (LDR). When the user crosses his or her hand by an LDR, this will indicate that the paper towel is needed. This should start the timing module. The rightmost LED on the Basys3 board will turn on for four seconds to indicate that the paper towel is fed. During this time, no other paper towel request is accepted. When the waiting time is over, the LED will turn off. The system will wait for a new paper towel request.

We can also put a DC motor instead of an LED. To do so, we should set the PWM frequency to 5 kHz. The duty cycle of the PWM signal should be 50%. The DC motor will rotate for four seconds to simulate the feeding of the paper towel. Again, no other paper towel request is accepted during this time. After the waiting time is over, the motor will stop.

13.16   Traffic Lights

We can modify the traffic light system developed for the MSP430 microcontroller to work on the Basys3 board [32]. The traffic light is located on a road which has two-sided car traffic and a crosswalk for pedestrians. There are green and red lights for both cars and pedestrians. Also there are buttons on each side of the crosswalk for pedestrians. The green light duration for cars is 60 seconds. During this time, if any of the buttons are pressed, light turns to red after 60 seconds for cars. Then, the green light turns on after two seconds for pedestrians. The green light duration for pedestrians is 20 seconds. If the button is not pressed, that means there are no pedestrian. Hence, the green light for cars stays on. If a pedestrian presses the button in any time after 60 seconds, it turns to red for cars. Then the system waits for two seconds and the green light turns on for pedestrians to cross.

To implement this system on the Basys3 board, we should use the available LEDs on it to simulate red and green lights. Moreover, we should also form a counter module to indicate one second as we have done in digital clock and car park occupied slot counting systems.

13.17   Car Parking Sensor System

We can modify the car parking sensor system developed for the MSP430 microcontroller to work on the Basys3 board [32]. The system will start working when the proximity sensor reads a value corresponding to one meter. This value will be shown by the seven-segment display and turning on all 16 LEDs on the board. Afterward, as the distance between the car and obstacle decreases, LEDs start to turn off with respect to distance. As the distance falls lower than 50 cm, then the buzzer starts working. The frequency of the sound produced by the buzzer will increase with respect to proximity such that when the car is five centimeters close to the obstacle, the buzzer will have the highest frequency value. Note that the proximity sensor used in this application will be different from the previously used ones. Here, we will need a proximity sensor with analog output.

13.18   Body Weight Scale

The goal of this application is building a body weight scale we use in our homes. Basically, we need four load cells (also called strain gauge) to be screwed to legs of the scale. These load cells convert the applied force on them to the electrical voltage. The XADC module on the Basys3 board can be used to convert this voltage to digital form. Note that an instrumentation amplifier may be needed between the sensor and Basys3 board depending on the sensor output. The weight of the user should be displayed on the seven-segment display. The user should also be able to store the last measurement in memory of the device. To avoid false measurements, the system should wait for the sensor to get stabilized. Several measurements should be taken and their average should be displayed on the seven-segment display or saved in memory.

13.19   Intelligent Billboard

Nowadays, companies want to measure the impact of an advertisement published on a billboard. One way of doing this is extracting statistics based on who viewed or paid attention to the billboard while the advertisement is on. We can develop a prototype system for this purpose. Our system is composed of a proximity sensor and the GSM module. The proximity sensor with digital output that we used in previous applications in this book can also be used for this system as well. The proximity sensor will be faced to people passing by. When someone gets closer to the billboard to read the advertisement, a counter in the system will increase. Since the billboard is an autonomous device located in the public area, we have to integrate GSM capability to send the count results to the company. Here, the GSM Click module offered by Microelektronika can be of use since it has an interface with the UART communication [61]. Previously introduced UART modules can be used for this purpose.

13.20   Elevator Cabin Control System

We can design a prototype elevator cabin control system using the Basys3 board. Our elevator works in a building with six floors. We will use the first three switches on the board (sw[0] to sw[2]) to identify which floor we are calling the cabin from. We also need an elevator call button. Let's assign btnC on the board for this purpose. We also need six buttons inside the cabin to indicate the target floor. We will use the next six switches (sw[3] to sw[8]) for this purpose. For example, after the elevator door is closed and sw[8] goes to logic level 1, the elevator should go to the sixth floor.

The system works as follows. The cabin starts at the first floor. If someone at this floor presses the call button, the door of the cabin will open. If someone from another floor presses the call button, the cabin moves to that floor. Assume that the travel time between each floor is three seconds. When the cabin reaches the target floor, its door opens and stays in that state for ten seconds. Since this is a prototype system, we assumed one user at a time. Therefore, scheduling issues within the elevator control are avoided. However, we suggest that the reader think about the possibility as well of a more advanced elevator cabin control system.

The first six LEDs on the Basys3 board will show which floor the cabin is at. The seventh LED shows if the elevator is busy or not. The eighth LED indicates whether the door is open or closed. At this stage, these are sufficient for a prototype system. We can improve the system further by adding a proximity sensor for the cabin door. Hence, if a user is at the door, it stays open. Besides, we can add two stepper motors: one to move the cabin and the other to open/close the cabin door.

13.21   Digital Table Tennis Game

This project aims to develop a digital table tennis game to be run on the Basys3 board. Rules of the game are as follows:

- There are two players controlling sw[15] and sw[0] on the board. These switches act as rackets to send the ball to the other side.
- The ball is represented by a moving LED.
- When the first user (controlling sw[15]) sends the ball, the game starts.
- The second user should respond to the coming ball by raising the racket (turning on the switch) when the ball reaches there. Since Basys3 has 16 LEDs, the racket can be raised on last two LEDs. To avoid any confusion, both players should keep their rackets low before striking the ball.
- If the racket is raised early by a player, it is taken as a fault and the other user gets the point.
- If one player misses the ball, the other player gets the point.
- The game has four difficulty levels controlled by btnU and btnL. The difficulty level is directly related to the speed of the moving ball.
- The difficulty level is shown on the rightmost seven-segment display digit.
- The score of the players are shown on the two leftmost seven-segment display digits.

- The edge detector module (in Listing 10.33) can be used in the project to detect switch movements.

13.22   Customer Counter

We can design a system on the Basys3 board to count customers in a shopping mall with designated doors for entrance and leaving. To do so, we should place a proximity sensor to each door. Hence, we can detect whether a customer passing through the gate is entering or leaving. The customer counter is reset as the mall opens. The count is increased by one for each entering customer. It is decreased by one for each leaving customer. The total number of customers in the mall should be shown on the seven-segment display. As the shopping mall closes, the security check will be done via the count value. If no one is left in the mall, gates will be locked. This can be simulated by an LED on the board. We can further expand the system by adding a bluetooth module to each proximity sensor section such that they communicate with a main module. Hence, the two follower modules and one leader module will be needed in developing the system.

13.23   Frequency Meter

Frequency meter is a device to measure frequency (repetition rate per second) of an analog periodic signal. We can design such a system using the Basys3 board. Assume that, the average value of the analog signal is discarded. Hence, it oscillates around zero. We can use the XADC module to detect zero crossings of the periodic signal. The total number of zero crossings within one second can be used to calculate the frequency of the signal. We can display the measured frequency on the seven-segment display of the board. Let's assume that we assign three digits for the measured frequency value. Hence, the frequency values between 0 and 999 Hz can be measured by the system.

13.24   Pedometer

Pedometer is a device that counts steps when you carry it on. We can design a pedometer using a three-axis accelerometer sensor and the Basys3 board. We can pick one of the available sensors working in similar ways. They communicate over the I^2C interface providing 16 bits of data in each direction. Hence, the I^2C module introduced in Chap. 12 will be of use here. You can connect the sensor to one of the available PMOD connectors on the board. Once you get the acceleration data, you have some work on it to extract steps. There is a good article explaining how to do this [62]. Therefore, we strongly suggest applying the method described there. Once you understand the steps, you can count them and let the user know when the total number of steps reach a limit (let's say per day). We can expand this module by using a bluetooth module (such as HC-06) to send step counts to your cell phone.

CHAPTER 14

What Is Next?

Digital systems introduced up to now were fairly complex such that they can be described by an hardware description language (HDL) (either Verilog or VHDL). There may be complex digital systems needing more powerful and high-level description. Xilinx offers such a platform called Vivado High-Level Synthesis (HLS). Through it, the user can describe operational characteristics of a digital system either in C or C++ language. The result can be converted to an IP block to be used in the Verilog or VHDL description. We will explore how this can be done in this chapter. To do so, we will start with Vivado HLS. Then, we will develop a project under it to generate an IP. Finally, we will show how the generated IP can be used in an HDL in Vivado. Therefore, the reader can understand steps to be followed for such an implementation. Topics introduced in this chapter are more complex compared to the ones introduced in previous chapters. Moreover, it is not possible for us to cover them in depth here. Therefore, we titled this chapter "What Is Next?" to emphasize that topics covered in this chapter should also be included in an advanced book focusing on these issues.

14.1 Vivado High-Level Synthesis Platform

Vivado HLS is the platform that can be used to develop a complex digital system benefiting from the power of C or C++ languages. Moreover, it allows the user to represent the developed system as an IP block to be used in Vivado. As we were writing this book, Vivado HLS was coming within the free Vivado WebPACK edition. Therefore, there is no need of extra installation process for it.

Although Vivado HLS is a powerful platform, it is fairly complex to master. Xilinx offers valuable references for this purpose [63–65]. We strongly suggest the reader to review them. Xilinx also offers several example projects under Vivado HLS to be used as a starting point. These will serve as valuable sources in using the platform. We will provide a simple project (modified from one of Xilinx's examples) to explain how a fresh project can be developed in Vivado HLS next.

14.2 Developing a Project in Vivado HLS to Generate IP

After installing Vivado WebPACK, the reader should see a separate icon (titled Vivado HLS 2016.3) on his or her desktop for the Vivado HLS. As this icon is pressed twice, Vivado HLS will start with the welcome screen as in Fig. 14.1. Through this screen, the user can create a new project; open an existing project; or open an example project provided by Xilinx. Moreover, tutorials and user guides for Vivado HSL can also be reached from this screen.

FIGURE 14.1 Vivado HLS welcome screen.

Let's create a new project in the welcome screen. Assume that we want to add two eight-bit numbers. As we press "Create New Project" in the welcome screen, a new window appears asking for the "Project Name" and "Location." Let the project name be `adder_HLS`. The reader should also find a suitable location. We pick this location as `H:\Xilinx_Projects`. As we press "Next," a new pop-up window titled "Add/ Remove Files" appears asking for the "Top Function" in the project. As for now, let's leave it empty. As we press "Next," the pop-up window asks for the testbench file to be used. Let's leave this one also empty. As we press "Next," a new pop-up window titled "Solution Configuration" appears as in Fig. 14.2. Here, we should set the "Solution Name," "Clock Period," and "Uncertainty." Let's leave them as they are. We should also select the FPGA platform from "Part Selection." Let's target the Basys3 board. Hence, set the part name as `xc7a35tcpg236-1`. We can press "Finish" to create the project.

After the project is created, a new screen appears as in Fig. 14.3. The user can adjust all properties of the project through this interface. Let's first add the main file titled `adder.c` to the project. To do so, right-click on the "Source" item in the "Explorer" section; select the "New File" option and create the file. Let's copy the C source code in Listing 14.1 to the created file.

As can be seen in Listing 14.1, the `adder.c` file only has a function definition `adder`. Input to this function are two eight-bit numbers `inA` and `inB` defined by type `int8`. The output of the function is `out1` defined as a pointer to another eight-bit number. The `adder.c` file also refers to a header file which we named as `adder.h` available in Listing 14.2. We should also add this file to the "Source" directory under the project following previous steps. This structure should be kept since Vivado HSL requires the function to be defined in the header file. Hence, it can be converted to an IP block.

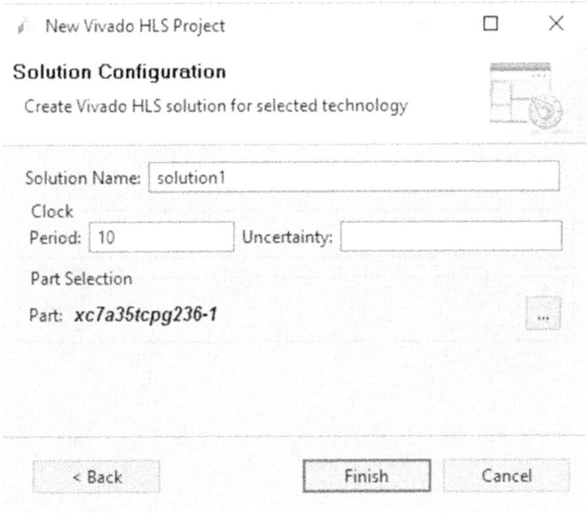

FIGURE 14.2 Solution configuration window.

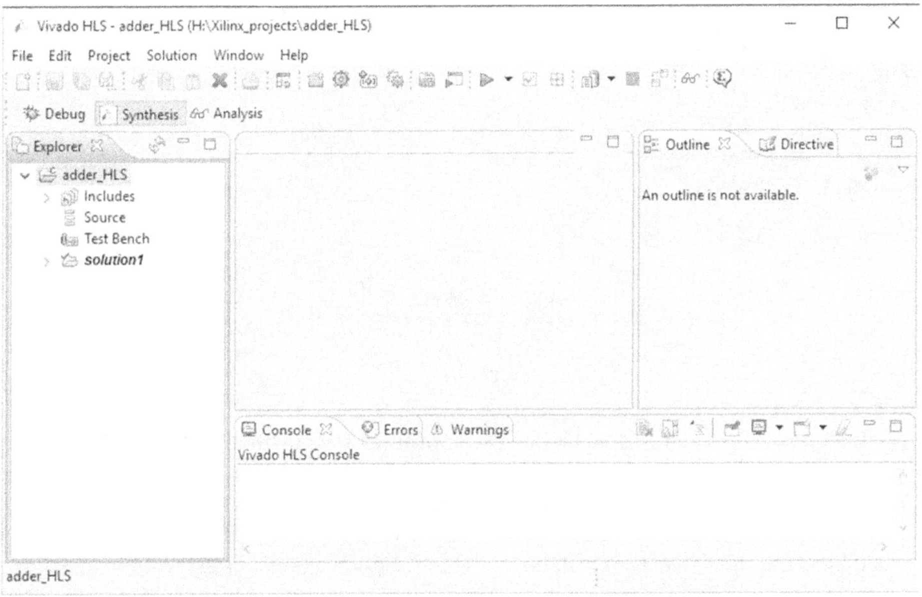

FIGURE 14.3 Project explorer window.

Vivado HLS requires a testbench file to test the C code. Let's call the testbench file adder_tb.c. This testbench file will be as presented in Listing 14.3. We should also add it to the "Source" directory as explained before.

As all three files are added to the project, we should adjust "Project Settings" by pressing the related icon in the project explorer window. There, we should declare the "Top Function" under the "Synthesis" section. After this operation, the window should

Listing 14.1 The adder.c Source Code

```c
#include "adder.h"

void adder(int8  inA, int8  inB, int8 *out1) {

        *out1 = inB + inA;
}
```

Listing 14.2 The adder.h Header File to Be Used in adder.c

```c
#ifndef _APINT_ARITH_H_
#define _APINT_ARITH_H_

#include <stdio.h>
#include "ap_cint.h"

void adder(int8 inA, int8 inB,int8 *out1);

#endif
```

Listing 14.3 Testbench File to Be Used in the Vivado HLS Project

```c
#include "adder.h"

int main () {
  int8 inA;
  int8 inB;
  int8 out1;

  inA = 1;
  inB = 9;

  // Call the function to operate on the data
  adder(inA,inB,&out1);
}
```

look like as shown in Fig. 14.4*a*. We should also add the "Testbench Files" under the "Simulation" section. After this operation, the window should look like as shown in Fig. 14.4*b*. As we press "Ok," we are ready to proceed.

We should generate an IP block corresponding to the project. Therefore, we should follow the steps Run C Simulation → Run C Synthesis → Export RTL. Vivado HLS offers extra test and validation steps at this point. The reader can check the mentioned references on how these operations can be done. The generated IP block can be found in the folder H:\Xilinx_Projects\adder_HLS\solution1\impl\ip. Next, we will use this IP block in Vivado.

14.3   Using the Generated IP in Vivado

To use the generated IP block, let's form a Vivado project following steps in Chap. 4. Let's call the project adder_Vivado. We should first add the generated IP block to the IP catalog following the steps in Sec. 4.7. Based on the Vivado HLS project settings, the specific

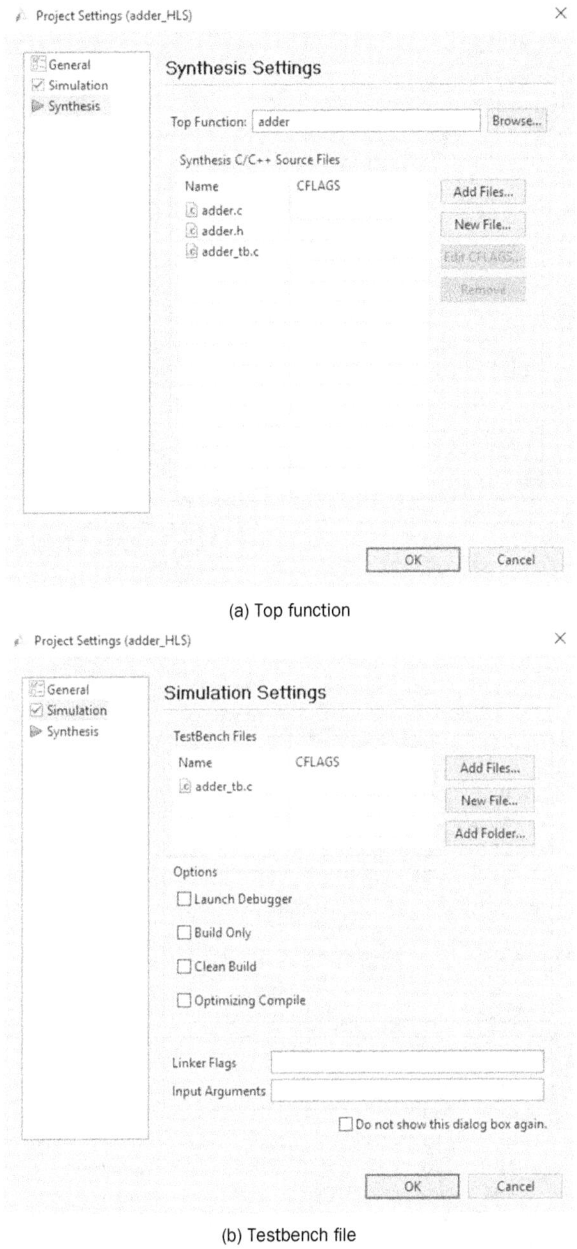

(a) Top function

(b) Testbench file

FIGURE 14.4 Project settings window after adding C source and testbench files.

IP directory to be added will be at `H:\Xilinx_Projects\adder_HLS\solution1\impl\ip`. Afterward, the generated IP should be seen in the IP catalog under "User Repository" and "Vivado HLS IP." We can add it to the project by double-clicking on it.

Let's form a top module and add the instantiation of the `adder` IP. The result will be as presented in Listing 14.4. The reader can form a testbench file to test this top module. The VHDL version of the top module will be as presented in Listing 14.5.

Listing 14.4 Verilog Top Module Using the Adder IP Generated in Vivado HLS

```verilog
module adder_top_module(y,x1,x2,vld,done,idle,ready,start);

input [7:0] x1, x2;
output [7:0] y;
input start;
output vld, done, idle, ready;

adder_0 adder(
  .out1_ap_vld(vld),    // output wire out1_ap_vld
  .ap_start(start),     // input wire ap_start
  .ap_done(done),       // output wire ap_done
  .ap_idle(idle),       // output wire ap_idle
  .ap_ready(ready),     // output wire ap_ready
  .inA(x1),             // input wire [7 : 0] inA
  .inB(x2),             // input wire [7 : 0] inB
  .out1(y)              // output wire [7 : 0] out1
);

endmodule
```

Listing 14.5 VHDL Top Module Using the Adder IP Generated in Vivado HLS

```vhdl
library ieee;
use ieee.std_logic_1164.all;

entity adder_top_module is
port(x1 : in std_logic_vector (7 downto 0);
     x2 : in std_logic_vector (7 downto 0);
      y : out std_logic_vector (7 downto 0);
  start : in std_logic;
    vld : out std_logic;
   done : out std_logic;
   idle : out std_logic;
  ready : out std_logic);
end adder_top_module;

architecture dataflow of adder_top_module is

component adder_0
port(out1_ap_vld : out std_logic;
        ap_start : in std_logic;
         ap_done : out std_logic;
         ap_idle : out std_logic;
        ap_ready : out std_logic;
             inA : in std_logic_vector(7 downto 0);
             inB : in std_logic_vector(7 downto 0);
            out1 : out std_logic_vector(7 downto 0));
end component;

begin

adder : adder_0 port map (out1_ap_vld => vld,ap_start => start,ap_done
    => done,ap_idle => idle,ap_ready => ready,inA => x1,inB => x2,out1
    => y);

end dataflow;
```

14.4 Summary

Verilog and VHDL are not the only options in describing complex digital systems. The Vivado HLS offers an advanced platform to develop complex systems in C or C++ language. The developed system can be converted to an IP block to be used in either Verilog or VHDL. We briefly introduced in this chapter methods on how this can be done. We strongly suggest that the reader master these topics using references offered by Xilinx.

14.5 Exercises

14.1 Modify the application introduced in Sec. 14.2 to
   a. subtract two eight-bit numbers.
   b. multiply two eight-bit numbers.
   c. divide two eight-bit numbers.

14.2 Modify the application introduced in Sec. 14.2 to add two 16-bit numbers.

14.3 Repeat Exercise 14.1 using two 16-bit numbers as input.

14.4 Pick an example project under the Vivado HLS offered by Xilinx. Apply steps introduced in this chapter to implement and run the project. Observe how a complex digital system can be developed this way.

14.5 Using Exercise 14.4, analyze how the C testbench file can be used for a detailed analysis in Vivado HLS.

References

1. Xilinx. (2015). *7 Series FPGAs Select IO Resources User Guide*, ug471 (v1.6) ed.

2. Xilinx. (2014). *7 Series FPGAs Configurable Logic Block User Guide*, ug474 (v1.7) ed.

3. Xilinx. (2014). *7 Series DSP48E1 Slice User Guide*, ug479 (v1.8) ed.

4. Xilinx. (2015). *7 Series FPGAs Clocking Resources User Guide*, ug472 (v1.11.2) ed.

5. Xilinx. (2015). *7 Series FPGAs and Zynq-7000 All Programmable SoC XADC Dual 12-Bit 1 MSPS Analog-to-Digital Converter User Guide*, ug480 (v1.7) ed.

6. Xilinx. (2013). *Efficient Implementation of Analog Signal Processing Functions in Xilinx All Programmable Devices*, wp442 (v1.0) ed.

7. Xilinx. (2015). *Driving the Xilinx Analog-to-Digital Converter*, xapp795 (v1.1) ed.

8. Xilinx. (2014). *7 Series FPGAs GTP Transceivers User Guide*, ug482 (v1.8) ed.

9. Xilinx. (2016). *7 Series FPGAs Integrated Block for PCI Express v3.3 LogiCORE IP Product Guide Vivado Design Suite*, pg054 ed.

10. Digilent, `https://reference.digilentinc.com/basys3:refmanual`. Accessed January 2, 2017.

11. Microchip. (2006). *PIC24FJ128GA Family Data Sheet*, ds39747c ed.

12. FTDI. (2012). *FT2232H Dual High-Speed USB to Multipurpose UART/FIFO IC Datasheet*, FT_000061 ed.

13. Spansion. (2013). *S25FL032P 32-Mbit CMOS 3.0 Volt Flash Memory with 104-MHz SPI (Serial Peripheral Interface) Multi I/O Bus*, s25fl032p_00 rev. 9 ed.

14. Digilent, `https://reference.digilentinc.com/arty:refmanual`. Accessed January 2, 2017.

15. TI. (2015). *DP83848x PHYTER Mini / LS Single Port 10/100 MB/s Ethernet Transceiver*, snls250e ed.

16. Micron. (2014). *Micron Serial NOR Flash Memory*, n25q128a ed.

17. Micron. (2016). *Micron DDR3L SDRAM*, mt41k128m16 ed.

18. Digilent, `http://www.xilinx.com/support/university/boards-portfolio/xup-boards/Basys3Board.html`. Accessed January 2, 2017.

19. Digilent, `https://reference.digilentinc.com/_media/arty/arty_sw_btn_led.zip`. Accessed January 2, 2017.

20. Xilinx. (2016). *Vivado Design Suite User Guide: Designing with IP*, ug896 (v2016.2) ed.

21. Xilinx. (2016). *Vivado Design Suite: Designing with IP Tutorial*, ug939 (v2016.2) ed.

22. Xilinx. (2016). *Vivado Design Suite User Guide: Creating and Packaging Custom IP*, ug1118 (v2016.2) ed.

23. Xilinx. (2016). *Vivado Design Suite: Creating, Packaging Custom IP Tutorial*, ug1119 (v2016.2) ed.

24. Cummins, C. E. (2000). Nonblocking assignments in Verilog synthesis, coding styles that kill!, in *SNUG 2000 San Jose*.

25. Hamid, M. (2010). *Writing Efficient Testbenches*, Xilinx, xapp199 (v1.1) ed.

26. Brown, S., and Vranesic, Z. (2009). *Fundamentals of Digital Logic with VHDL Design*, 3rd ed. McGraw-Hill, New York.

27. Pedroni, V. A. (2014). *Circuit Design and Simulation with VHDL*, 2nd ed. The MIT Press, Cambridge, MA.

28. Xilinx. (2016). *Vivado Design Suite User Guide: Synthesis*, ug901 (v2016.2) ed.

29. Mano, M. M., and Ciletti, M. D. (2006). *Digital Design*, 4th ed. Prentice Hall, Englewood Cliffs, NJ.

30. Xilinx. (2015). *Distributed Memory Generator v8.0 LogiCORE IP Product Guide*, pg063 ed.

31. Xilinx. (2016). *Block Memory Generator v8.3 LogiCORE IP Product Guide*, pg058 ed.

32. Ünsalan, C., and Gürhan, H. D. (2014). *Programmable Microcontrollers with Applications: MSP430 LaunchPad with CCS and Grace*, 1st ed. McGraw-Hill, New York.

33. Brown, S., and Vranesic, Z. (2014). *Fundamentals of Digital Logic with Verilog Design*, 3rd ed. McGraw-Hill, New York.

34. Xilinx, https://www.xilinx.com/ipcenter/processor_central/picoblaze/member/. Accessed January 2, 2017.

35. Xilinx. (2011). *PicoBlaze 8-bit Embedded Microcontroller User Guide*, ug129 ed.

36. Chapman, K. (2014). *PicoBlaze for Spartan-6, Virtex-6, 7-Series, Zynq and UltraScale Devices (KCPSM6)*. Xilinx.

37. Tracton, P., https://github.com/ptracton/Picoblaze. Accessed January 2, 2017.

38. Tracton, P., https://github.com/ptracton/Picoblaze/tree/master/PicoBlaze_GPIO_Example. Accessed January 2, 2017.

39. Xilinx. (2016). *Vivado Design Suite Tutorial Embedded Processor Hardware Design*, ug940 ed.

40. Xilinx. (2016). *MicroBlaze Processor Reference Guide*, ug984 ed.

41. Xilinx, https://www.xilinx.com/products/design-tools/microblaze.html. Accessed January 2, 2017.

42. Duckworth, R. J., http://users.wpi.edu/~rjduck/Microblaze. Accessed January 2, 2017.

43. Digilent, https://reference.digilentinc.com/learn/programmable-logic/tutorials/arty-getting-started-with-microblaze/start. Accessed January 2, 2017.

44. FPGArduino, http://www.nxlab.fer.hr/fpgarduino/. Accessed January 2, 2017.

45. Imagination, https://community.imgtec.com/university/resources/. Accessed January 2, 2017.

46. ARM, `https://www.arm.com/products/designstart/index.php`. Accessed January 2, 2017.

47. Motorola Inc. (2003). *SPI Block Guide*, s12spiv3/d ed.

48. Digilent, `https://reference.digilentinc.com/reference/pmod/pmodals/start`. Accessed January 2, 2017.

49. NXP. (2012). *UM10204 I2C-Bus Specification and User Manual*, rev. 5th ed.

50. Larson, S., `https://eewiki.net/pages/viewpage.action?pageId=10125324`. Accessed January 2, 2017.

51. Digilent, `http://store.digilentinc.com/pmod-cmps-3-axis-digital-compass/`. Accessed January 2, 2017.

52. Digilent, `https://learn.digilentinc.com/Documents/269`. Accessed January 2, 2017.

53. Digilent, `https://reference.digilentinc.com/basys3/refmanual`. Accessed January 2, 2017.

54. Digilent, `https://reference.digilentinc.com/learn/programmable-logic/tutorials/arty-getting-started-with-microblaze-servers/start`. Accessed January 2, 2017.

55. Xilinx. (2016). *Integrated Logic Analyzer v6.2 LogiCORE IP Product Guide*, pg172 ed.

56. Xilinx. (2016). *Vivado Design Suite User Guide: Programming and Debugging*, ug908 ed.

57. Xilinx. (2016). *XADC Wizard v3.3 LogiCORE IP Product Guide*, pg091 ed.

58. MIT, `http://appinventor.mit.edu/explore/`. Accessed January 2, 2017.

59. Digilent, `https://reference.digilentinc.com/reference/pmod/pmodstep/start`. Accessed January 2, 2017.

60. Sparkfun, `https://www.sparkfun.com/products/13316`. Accessed January 2, 2017.

61. Mikroelektronika, `https://shop.mikroe.com/click/wireless-connectivity/gsm`. Accessed January 2, 2017.

62. Zhao, N. (2010). Full-featured pedometer design realized with 3-axis digital accelerometer. *Analog Dialogue*, 44 (06): 1–5.

63. Xilinx. (2016). *Vivado Design Suite Tutorial High-Level Synthesis*, ug871 ed.

64. Xilinx. (2016). *Vivado Design Suite User Guide High-Level Synthesis*, ug902 ed.

65. Xilinx. (2013). *Introduction to FPGA Design with Vivado High-Level Synthesis*, ug998 ed.

Index

Note: Page numbers followed by *f* denote figures; by *t*, tables.

373

CPSIA information can be obtained
at www.ICGtesting.com
Printed in the USA
BVHW010751080819

555337BV00014B/97/P

9 781259 837906